1985

Wittgenstein on the Foundations of Mathematics

Wittgenstein on the Foundations of Mathematics

Crispin Wright

Harvard University Press
Cambridge, Massachusetts
1980

Contents

Preface

Wittgenstein, so the legend goes, was present at Brouwer's lecture, 'Mathematik, Wissenschaft und Sprache',[1] in Vienna in 1928 and was moved thereby to resume philosophy. Certainly the philosophy of mathematics intensely occupied him during the following sixteen years. It was a major area of attention in his large typescripts of 1929–30 and 1932–3 which we now have as the *Philosophical Remarks* and *Philosophical Grammar* respectively; and he continued to write extensively about topics in this part of philosophy for all but the final seven years of his life.

For whatever reason, he included hardly any of this material in his final draft of the *Philosophical Investigations*; which, however, concludes with a remark which seems to advertise it:

> An investigation is possible in connection with mathematics which is entirely analogous to our investigation of psychology. It is just as little a *mathematical* investigation as the other is a psychological one. It will *not* contain calculations, so it is not for example logistic. It might deserve the name of an investigation of the 'foundations of mathematics'.

In 1956 his literary executors published an edited collection from those of Wittgenstein's manuscripts, written in the period 1937–44, which concerned these 'foundations' of mathematics. It is the leading philosophical themes of this text which form the subject matter of my book.

The laconic, fragmentary and disorganised character of the *Remarks on the Foundations of Mathematics*, and the apparent eccentricity of the views therein expressed, has made it easy for philosophers, including many who would profess to find much of value in the *Investigations*, to dismiss and, in some cases, to abuse it. When I first read the book, I too thought Wittgenstein wrong in virtually every opinion which I could confidently identify him as expressing—and I cannot pretend to know today that I was not right. But since its thought is, as the editors originally stressed, of a piece with that of the *Investigations*, the easy stance of eclecticism with respect to Wittgenstein's later ideas is not an option; the philosophy of mind of the *Investigations* may possibly be largely right and the philosophy of mathematics of the *Remarks on the Foundation of Mathematics* largely wrong, but that is a view to which no one is entitled who has not probed their deep common sources in

[1] Reprinted in L. E. J. Brouwer, *Collected works* (see References, entry 12).

Wittgenstein's later philosophy of language. This is, first and foremost, what I have attempted to do in this book.

To understand any philosophical view involves knowing what best can be said on its behalf. Accordingly, my approach has been mainly sympathetic; an approach I have not always found it easy to maintain. I am, moreover, deeply aware that in seeking to locate in Wittgenstein's thought the system and argumentation necessary if such sympathy is to be seen to be deserved, I run the risk of representing him as engaging in philosophical theorising of the kind he distinctively forswore.

Wittgenstein's later thought about mathematics was preoccupied at bottom with the two most fundamental philosophical questions to which pure mathematics gives rise: the apparent *necessity* of mathematical truths, and the nature of our apparent knowledge of them. It is, of course, in large part a terminological issue whether these questions should be regarded as falling within the philosophy of mathematics, the philosophy of language, or the theory of knowledge. What is clear is that they are in no sense technical questions. Wittgenstein's treatment of them is conspicuously free of technicality, and those whose intellectual conscience leads them to want the philosophy of mathematics to look like mathematics will not much enjoy this book either. Many issues in this area of philosophy call for formalised exposition and investigation; but the questions with which Wittgenstein was most concerned call only for hard thought and a natural language to express it in.

My book is based on material presented to graduate classes in Oxford during the Trinity terms of 1974, 1975, and 1977. Originally, when I thought I might make a book of it all, it was my intention wholly to jettison the format of those lectures and attempt systematically to rewrite my notes from a unified point of view; but it soon became clear that this would be a massive task, quite out of proportion to any likely gain. I found, too, that it went against the grain to try to impose my later perspective on the earlier material; for example, that I wanted the reader to think through for himself the advantages and disadvantages of the ways in which Wittgenstein's ideas about following a rule are expounded at different stages. The upshot is a text which remains fairly close in content and organisation to the original lectures. The three parts are in some measure self-contained, but the whole is a document of ideas in progress and should be read accordingly.[1]

Save for a couple of absolutely elementary natural deduction examples, I presuppose familiarity with no parts of logic or mathematics which the *Remarks on the Foundations of Mathematics* does not presuppose. Something of an analytical table of contents seemed desirable, both as a guide and as an aide-mémoire to the reader. For the sake of anyone who might want to use my book as a seminar text, I have usually listed at the start of a chapter various sources taken from Witt-

[1]Prototypical versions of I–VIII were given in 1974; of IX, X, XII, XIII and XVI–XIX in 1975; and of the remainder in 1977.

genstein's later published writings and reported sayings. These are passages in which Wittgenstein addresses questions, or expresses ideas, germane to those with which the ensuing chapter is concerned. It seemed right that the scope of the selection should include all the post-Tractarian texts, though the reader unfamiliar with those books should beware of assuming that Wittgenstein's later thought about mathematics forms a unity. In general, I have cited only passages which I believe assist interpretation of the *Remarks on the Foundations of Mathematics*; and, with a couple of exceptions, only passages available, or shortly to be available, in English translation. One or two passages are, by way of reminder, cited more than once. The selection could have been much more extensive.

I am indebted to the suggestions and questions of those who attended the original classes, and to Frederick Benenson, Gareth Evans, Jane Howarth and Christopher Peacocke for their comments on sections of an earlier version; to Edna Laird who typed both the earlier version from tape-recordings and the final draft; to Michael Wrigley who compiled the Index and supplementary Bibliography; and, especially, to Michael Dummett whose ideas about Realism have enormously influenced my approach to these questions and constitute, indeed, almost as much as Wittgenstein's own ideas the subject matter of this book.

All Souls College, Oxford   C. J. G. W.
June, 1978

Abbreviations

RFM   *Remarks on the Foundations of Mathematics,* 2nd edition, eds. von Wright, Rhees, Anscombe, trans. Anscombe. Blackwell 1964.

BGM   *Bemerkungen über die Grundlagen der Mathematik,* 3rd edition, eds. von Wright, Rhees, Anscombe. Suhrkamp 1974.

PI   *Philosophical Investigations,* eds. Anscombe and Rhees, trans. Anscombe. Blackwell 1953.

PG   *Philosophical Grammar,* ed. Rhees, trans. Kenny. Blackwell 1974.

OC   *On Certainty,* eds. Anscombe and von Wright, trans. Paul and Anscombe. Blackwell 1969.

Z   *Zettel,* eds. Anscombe and von Wright, trans. Anscombe. Blackwell 1967.

PB   *Philosophical Remarks,* ed. Rhees, trans. Hargreaves and White. Blackwell 1975.

BlB/BrB   *The Blue and Brown Books.* Blackwell 1964.

LFM   *Wittgenstein's Lectures on the Foundations of Mathematics Cambridge, 1939,* ed. Diamond. Harvester Press 1976.

WWK   *Ludwig Wittgenstein und der Wiener Kreis,* shorthand notes of Waismann, ed. McGuinness. Blackwell 1967.

T   *Tractatus Logico-Philosophicus,* trans. Pears and McGuinness. Routledge 1961.

Analytical Table of Contents

XIII: REVISIONISM

XIV: NON-REVISIONISM AND 'ANTECEDENCE TO
      TRUTH'

XV: THE IDEA OF A 'THEORY OF MEANING'

PART THREE: NECESSITY

PART ONE

Mathematics as Modifying Concepts

I

Platonism

Sources

RFM: I. App. *II*, 3, 10; III. 11; IV. 5, 16; V. 34.
  PI: I. 426.
  PG: I. 42; II. 11, 42.
LFM: lectures XII, pp. 112–14; XIV, pp. 131–3; 140–1.

1. The images with which the practice of classical mathematics is surrounded are familiar. Pure mathematics is thought of as descriptive of an external, independent reality, in much the manner whereby, we are encouraged to think, natural science describes the physical world. But this 'world' described by pure mathematics is not the physical world. It is abstract, changeless and what holds there holds necessarily. Pure mathematics is nevertheless a project of discovery; its goal is first and foremost to chart the mathematical realm.

This picture of the nature of pure mathematics seems to demand a special epistemology. Thus Gödel[1] postulated a special intuitive faculty, akin to a kind of perception of mathematical objects, to explain our capacity to know mathematical truths. Such a postulation, of course, *explains* nothing of the sort. The picture, indeed, threatens to push our recognition of the truth of a mathematical statement beyond philosophical account. And it is immediately objectionable in at least two other related ways. What sort of explanation are we going to be able to give of the *necessity* of pure mathematical truths, if it is merely a reflection of a feature of the domain which pure mathematics allegedly describes? And what account is going to be possible of the *application* of pure mathematics? In particular, how is it that truths concerning an alleged special *abstract* domain carry over into the physical world also?

Even those who favour the defence of classical mathematics against, say, the mathematical intuitionists (and other 'revisionists') are bound for such reasons to feel uneasy about the platonist picture. Yet remarkably little has been done to provide an unpictorial, substantial account

[1] See for example 38, 2 pp. 271–2.

of what platonism as a philosophy of mathematics comes to in essentials. It is not obvious how such an account should go. This is in large part due to the tendency of platonists to concentrate upon felt inadequacies in positions held by their contemporary opponents rather than upon explanation of their own position. For example, the polemical parts of Frege's *Grundlagen* might give the impression that a platonist conception of arithmetic is unavoidable if one is not prepared to accept formalism, psychologism, or the empiricism of J. S. Mill.[1] But, whatever the reason, we do not have ready to hand a worked-out philosophy which (had the order of things been reversed and intuitionist mathematics been historically prior) could now be used to recommend the supplanting of intuitionist methods by classical ones. The platonist imagery is, above all, crude.

Wittgenstein thought it worse; he thought it pernicious. When mathematics turns away from its application outside mathematics and in on itself, to prove, for example, that the rationals cannot be enumerated in order of magnitude, Wittgenstein is provoked to think in terms of a comparison with *alchemy* (IV. 16):

> Is it the earmark of this mathematical alchemy that mathematical propositions are regarded as statements about mathematical *objects,* and so mathematics as the exploration of these objects?

And then

> What is typical of the phenomenon I am talking about is that a mysteriousness about some concept is not straightaway interpreted as an erroneous conception, as a mistake of ideas, but rather as something that is not at any rate to be despised, that is perhaps to be respected.

Of the same example about the rationals Wittgenstein says elsewhere (App. *II*, 10):

> This proposition seems to belong simply and solely to mathematics, seems to concern, as it were, the natural history of mathematical objects themselves. One would like to say that e.g. it introduces us to the mysteries of the mathematical world. This is the aspect against which I want to give a warning.

Again (v. 34)

> What harm is done by saying e.g. that God knows all irrational numbers, or that they already are all there even though we only know certain of them? Why are these pictures not harmless? Well, for one thing they hide certain problems.

---

[1] *Locus classicus*: 62, Book I, ch. VI.

And (III. 11)

> Arithmetic as the natural history, or mineralogy, of numbers. But who talks like this about it? Our whole thinking is permeated with this idea.

Finally (App. *II*, 3),

> The dangerous and deceptive thing about the idea, 'the real numbers cannot be arranged in a series' or 'the set of real numbers is not denumerable' resides in its assimilation to a fact of nature of what is in reality the determination or formation of a concept.

For Wittgenstein, then, it is a dangerous error to think of pure mathematics as descriptive of some objective domain. It is an error which he thinks affects our entire way of thinking about mathematics and which leads us to give to its results a skew and erroneous form of expression. He wants to get right away from the idea of pure mathematics as *descriptive*. In several places he questions whether we might not dispense with the notion that genuine *propositions* are dealt with in pure mathematics, and in the final passage just quoted he proposes an alternative picture: we are to think of the conclusion of the diagonal argument not as expressing a new discovery about the nature of the real numbers, but as marking the determination of a new concept.

Of course, this new picture is every bit as opaque as that which it is intended to supplant. The idea of proofs as instruments not of discovery but of conceptual change is a recurring theme throughout *RFM*, and is one of the factors which have prompted some commentators to regard Wittgenstein's philosophy of mathematics as conventionalist. But the interest of Wittgenstein's alternative picture depends at least in large part upon the strength of his reasons for rejecting the platonist picture. And his rejection of platonism, at any rate as illustrated in the passages just cited, might excusably be thought to be a shallow rejection. Little or no overt effort seems to be made in *RFM* to understand the motivation of the platonist view and to see whether there is not underpinning it some more substantial doctrine than the familiar images of exploration and discovery. Accordingly I want to begin by sketching, in rather less figurative terms, what the essential strands in platonist thinking about mathematics consist in. Only when that is clear will it be possible to assess whether Wittgenstein succeeded in developing any ideas on which a reasoned rejection of platonism may be based.

2.   Kreisel[1] suggests that the essential issue between platonist and constructivist philosophies of mathematics concerns not so much the existence of mathematical objects as the *objectivity* of mathematical truth. Of course, this notion of objectivity is prima facie little more clear

[1] 54.

than the general notion of an object, so that Kreisel's suggestion itself requires to be interpreted before it may be used as an instrument of liberation. But it is certainly the case both that the platonist picture suggests a conception of pure mathematics as something objective and that constructivists do not typically deny the existence of mathematical objects; at all events, they do not repudiate the existence of an abstract reference for mathematical singular terms. But do constructivists—the intuitionists, for example—reject the objectivity of mathematics either? Viewed superficially, the picture of mathematics with which they are usually associated might suggest that they do. To repudiate the objectivity of morals, for example, would be to agree with Nietzsche that 'there are no moral facts',[1] that there is no domain of reality to which moral opinion is answerable. But Brouwer's talk of mathematics as 'created by a free action'[2] suggests that it might at any stage be, with equal right, developed in different directions—a notion seemingly echoed by Wittgenstein throughout *RFM*—and is naturally taken as a comparable view about mathematical facts. On such a view there would seem to be no more objectivity in pure mathematics than in writing fiction.

This notion, however, is obviously a travesty of the *practice* of intuitionist mathematics and, indeed, of anything recognisable as pure mathematics at all. When innovation takes place in intuitionist mathematics, it is generally allowable only if it meets previously accepted criteria for the correctness of proofs and the admissibility of concepts. Nothing would count as mathematics, from whatever philosophical standpoint it emanated, in which this was not so. If Brouwer's talk of free creativity on the part of the mathematician had been meant to suggest that 'anything goes', he would have been disarmed from making the kind of *criticism* of the concepts and methods of classical mathematics which the intuitionists offer; that is, a selective criticism. So unless 'objectivity' is given a more refined sense than the general currency of certain standards of correctness and of error in proofs, the intuitionists cannot be represented as rejecting the objectivity of mathematics; indeed, nothing will count as mathematics which is not objective in *this* way.

But we ordinarily read more into the notion of objectivity than the existence of shared standards of admissibility and error. There is a general counter-idealist streak in the way we ordinarily think. In general we grant the legitimacy of the question whether our ordinary standards of acceptability for beliefs and theories actually yield statements which really do depict how things are (so that Nietzsche's conviction that they do not do so in morals at least would seem intelligible, even if moral standards were free of their distinctive cultural

[1] 65, p. 55, for example.
[2] 10, concluding summary (12, p. 97). Almost fifty years later he writes (11) of mathematics as 'an autonomic interior constructional mental activity' (12, p. 551). Cf. Heyting's (47) 'free, vital activity of thought' (2, p. 42).

variability). More than that, we generally believe that our ordinary standards of acceptability *are* adequate, that the statements which they let through will be a true reflection of the world.

The second belief is less important for our present purpose. The crucial element in the concept of objectivity is the admission of a conceptual distinction between how things seem to us when assessed by the most refined criteria which we possess, and how they may actually be. It is on our willingness to allow such a hiatus that our capacity to be troubled by traditional forms of scepticism depends.

The intuitionists sometimes write as though the proper response of a mathematician to the issue whether mathematics is objective in this way would be to ignore it.[1] There is no need for a mathematician to raise the question of the ulterior adequacy of his standards of proof, of whether the conclusions of reasoning which satisfies his normal standards are a faithful reflection of how things are in some external sense. Mathematics is rather, in Brouwer's phrase, 'inner architecture';[2] there ought to be no confrontation of the mathematician with the question of the objective truth of his results. From this point of view, the central error of classical mathematics would lie in taking sides on an issue with which the mathematician, *qua* mathematician, has no proper concern.

Whether this is an adequate account of the intuitionists' motives does not concern us immediately. What it does at least suggest is a less metaphorical account of the philosophical motivation of classical mathematics than is usual. The central question is not so much whether pure mathematics should be viewed as describing an external, changeless, abstract reality as whether we ought to admit, at least in principle, a distinction between meeting the most refined crieria of mathematical acceptability and actually being mathematically true.

3.   The objectivity, then, of our beliefs in a particular area is not merely a matter of the currency of agreed criteria of acceptability and error for them. It is a matter of allowing the question of whether there is correspondence between the results of applying these criteria and how things really are. As suggested above, it would thus appear to be a sufficient condition for the objectivity of statements of a particular species that we leave scope for the traditional sort of scepticism concerning our knowledge of them. But it is not a necessary condition. Even where we regard it as possible that we can conclusively verify a statement, a conception of its truth or falsity as objective is still a possibility. We employ such a conception if we hold that the statement may be determinate in truth-value irrespective of whether we can recognise what its truth-value is. Thus, many people would be

[1] See, for example, Heyting's 'Disputation' in 46, p. 3 sq.; (reprinted in 2).
[2] 9. (12, p. 49; 2, p. 84).

prepared to allow that, while a proof of a mathematical theorem certainly decisively shows that what makes it true obtains, its doing so is independent of our capacity to recognise the fact; that is, to devise the proof. The theorem is not *made* true by its proof; it is merely conclusively shown to be true.

Such a conception of truth for particular statements is easily intelligible only if we can give some other account of their truth-conditions than will make them coincide with conditions under which we should regard ourselves as capable of verifying those statements (or in which we should be capable of coming to believe them on optimally rational grounds). In the case of effectively decidable statements, therefore, it is unclear whether the contrast between a belief in objectivity and a repudiation of it can be fully made out, at any rate by means of the characterisation of these attitudes so far given. It is, of course, still open to someone to try to make good a sense in which, even for effectively decidable statements, conditions of truth and conditions of verification (although necessarily coinstantiated) do not necessarily *coincide*. But the nature of a belief in objectivity is obviously far less elusive when we are concerned with statements to which we would ordinarily apply the notions of truth and falsity but of which we recognise that no means of verification or falsification necessarily exists.

For one who believes in objectivity—and, perhaps, on any sane view —the making of a statement is essentially an attempted depiction of how things are. No one could be said to have understood a statement who did not understand the nature of the circumstances which it purported to depict. But if how things are is thought of as a hard, objective issue transcending determination by our criteria, or, at least, not *necessarily* determinable by our criteria, possession of this understanding cannot be thought of as essentially a recognitional ability. It is natural to protest that on such a view the making of a statement becomes a somewhat nugatory business, comparable to shooting at a target which no one can see. But this objection is questionable. For one thing, it *may* be possible to tell that one has hit the target, for example, by proof; for another, even where the possibility of verification is foreclosed from the outset, for example with some unrestrictedly general contingent hypothesis, there may still be a *point* in hitting the target, although one can never have the satisfaction of knowing for sure that one has done so.

A belief in the objectivity of a certain class of statements thus carries with it a certain conception of what it is to understand the members of that class. Such a statement can be understood only by someone in possession of a concept of the fact which it putatively describes, that is, of the circumstances under which it would be true, of a kind not essentially reducible to a capacity to recognise those circumstances should they obtain.

4.   This brings us upon the suggestion made in several places by Dummett that the hard philosophical core to the platonist conception is to be found in the notion that the meaning of a statement is fixed by determining the circumstances under which it is true or false.[1] Certainly such a notion of statement meaning was endorsed both by Frege and by the Wittgenstein of the *Tractatus*.[2] But it is doubtful whether the view that truth is the central notion in determining the meanings of statements involves of itself a commitment to their objectivity. It would, for example, have been consistent with Logical Positivism to have supposed that the meaning of a statement is fixed by determining its truth-conditions, but with the crucial proviso that truth-conditions will not be admitted as having been intelligibly specified which are not such that, if they are actualised, we can come to recognise that they are so. There would have been no harm from the positivists' point of view in supposing that the sense of a sentence is determined by fixing its truth-conditions so long as the scope of the supposition were restricted to statements whose truth guarantees their verifiability. But the point of the positivists' emphasis on verification was, presumably, in the present terms, essentially anti-objective. It was to repudiate any notion of statement-understanding dissociated from experiences which we may have and procedures which we may carry out. In the positivists' view, it was only by reference to such experiences and procedures that the idea of the correct use of language could have any content for us. However, the restriction on the idea of truth-conditions just entertained would preserve the desired connection between understanding and human practice and experience; for when truth implies verifiability, to know under what circumstances a statement is true is to possess information entailing how its truth might be recognised.

     It would appear, then, that if a belief in objectivity, in the current sense, is indeed an essential aspect of platonism, Dummett's proposal requires to be understood like this: the crucial thing is not simply the acceptance of a truth-conditions theory of meaning as such, but a particular conception of what truth is. The platonist implicitly subscribes to a strong 'correspondence' theory of truth, crediting us with the capacity to form concepts whose application to reality is something which we need not be able to determine, which characterise the world or fail to do so, autonomously, and which may then be employed in fixing the truth-conditions of statements. If the view of statement-meaning in terms of truth-conditions is to entail platonism, a conception of truth as objective has to be built in from the outset; there has to be in general no equivalence between determining the truth-conditions of a statement and determining what counts as verification of it. There has to be no presumption that only if it is thereby apparent how one

---

[1] See especially 22, 23, 25, 26, 30.
[2] For example 36, vol. I, §32; and *T* 4. 431.

might set about verifying a statement—or apparent at least how one might, with good fortune, succeed in verifying it, though not as the culmination of a procedure which one can effectively initiate—have its truth-conditions been intelligibly specified. Of course, any explanation of the truth-conditions of a statement places a condition upon what verification of it has to determine, or what recognition of its truth has to be recognition of. The point is that, for a platonist, the fixing of truth-conditions is not *aimed* at teaching someone a procedure of verification or at effectively determining for him under what circumstances he may regard himself as having recognised the statement's truth. If truth-conditions for a statement are defined in a platonist spirit, it is no objection to their intelligibility if it is not apparent how one might set about determining whether they obtain, or what would count as the experience of their obtaining.

In summary: someone who holds that to know the meaning of any statement is to know under what circumstances it is true or false is not thereby committed to a platonist philosophy of mathematics unless he also regards it as an irrelevant objection to an alleged specification of the truth-conditions of a statement that it entails no explanation of how they might be recognised to obtain. His concept of truth has to be of an intrinsically objective character: we are held to have the capacity to understand a specification of circumstances under which a statement is true irrespective of whether we know how it might be determined to be true. If we have such a capacity, it must be a comprehensible possibility that certain statements be true for which we may have no method of verification. It also follows that we may in general comprehensibly raise the question whether the methods of verification or, more loosely, of assessment which we employ for statements in a certain class do provide a faithful reflection of how the world is, do measure up to our concept of truth for those statements. A belief that we have this capacity to grasp concepts in a sense irreducible to a recognitional ability renders inaccessible any version of idealism which holds as senseless the attribution to the world of properties independently of whether, or how, we might know it to have them.

5.   If in these considerations we have hold of a general version of what is essential to mathematical platonism, it ought to be possible to trace the distinctive features of classical logic, in contrast, that is, with those of, for example, intuitionist logic, back to acceptance of the objectivity of statements to which it is applied. Where, then, in the framework of classical propositional and predicate logic can we locate consequences of crediting us with the capacity to grasp a verification-transcendent notion of truth?

If one were to judge superficially by the differences between conventional ways of formalising classical and intuitionist logic,[1] it would seem that the essential differences are all to be found at the level of propositional logic; for the introduction and elimination rules for the quantifiers are the same for both the classical logician and the intuitionist. Indeed it would seem that there is only one essential difference. Add the law of excluded middle to intuitionist logic and the whole system inflates into classical logic. The classical quartet of equivalences expressing transpositions between the quantifiers are then all readily forthcoming:

$$(x)Fx \quad \leftrightarrow \quad -(\exists x)-Fx$$
$$(\exists x)Fx \quad \leftrightarrow \quad -(x)-Fx$$
$$(x)-Fx \quad \leftrightarrow \quad -(\exists x)Fx$$
$$(\exists x)-Fx \leftrightarrow \quad -(x)Fx.$$

Thus we can also easily prove the validity of such a disjunction as:

$$(x)Fx \quad \lor \quad (\exists x)-Fx$$

irrespective of the size of the range of the variable x, or the decidability of the predicate F.

It would thus appear that the whole disagreement between the intuitionist and the classical logician crystallises in the latter's unrestricted acceptance of excluded middle. But now, does the characterisation of platonism which we are presently entertaining—the belief that we have a general capacity, provided the concepts involved are otherwise familiar, to understand a description of the circumstances under which a statement is true independently of knowing how to find out whether they obtain, or even of being in a position to recognise them should they obtain—does this characterisation involve the unrestricted validity of that law?

Suppose we have such a capacity and that it is accepted that the proper form of an explanation of the meaning of any statement is to stipulate necessary and sufficient conditions for its truth. Then if $P \lor - P$ is not valid in some case, it can only be, it appears, because the truth-conditions stipulated for P and those stipulated for its negation do not exhaust the possibilities. This, of course, is the direction which arguments against the validity of the law of excluded middle for certain statements have generally taken. And there is a natural counter to any such argument. If it turns out that the truth-conditions for P and its negation are not exhaustive, that there is some third kind of possibility, let it be incorporated into the truth-conditions of the negation of P. This is not just an arbitrary stipulation. Should such a third possibility obtain, things are at any rate not as described in P.

What this manoeuvre appeals to is a concept of the negation of a

[1] As, for example, in 50, §23 (ch. V).

statement such that not-P is true in any circumstances other than those given as truth-conditions for P. Such a concept is natural; there is something amiss with the idea that someone might understand a statement without understanding what was stated by its denial, so an account of the sense of not-P must be implicit in that for P. Suppose then that we accept the following formula for specifying the content of any statement: we specify conditions for the statement to express a truth on the understanding that its negation will express a truth in *any other* circumstances. Then of any statement which we understand, it *has* to be true to say that either it or its negation expresses a truth. There is thus no question about the unrestricted validity of excluded middle.

Of course, just what range of statements we do understand will now depend upon what conditions a satisfactory specification of truth-conditions has to meet. If the above formula were constrained to conform to the demands of the logical positivists, it would not be allowable that we understood any statements other than decidable ones. But if we are credited with the capacity sketched at the conclusion of §4, there is in advance no reason why excluded middle should not apply validly not merely to statements which are not effectively decidable but also perhaps to *absolutely* undecidable statements, statements whose truth or falsity may not be recognisable by any means whatever. Or, better, there is no reason why there should not *be* perfectly intelligible such statements, to which excluded middle will apply as to any statement whose content has been satisfactorily explained.

Someone who allowed us to have the capacity in question, and who accepted the above truth-conditions schema of meaning-explanation and the orthodox elimination and introduction rules for the quantifiers, would thus appear to be committed to accepting classical logic for any satisfactorily explained statements; it would be immaterial whether they involved quantifiers, or what size of domain of quantification they involved.

6.   It might be objected that, if the foregoing is correct, we cannot yet have secured a complete account of what is essential even in number-theoretic platonism, since nothing has so far been said about the disagreement between platonists and constructivists about *infinity*, about what is involved in the platonist conception of the infinite as *actual*. So while we may have the background thinking about truth, objectivity and understanding which underpins platonist mathematics, we do not yet have an adequate account even of platonism about elementary number-theory. Indeed, to hold the background view as so far characterised would be consistent with rejecting the notion of infinity altogether, and classical number-theory along with it.

In order to be clear about the situation, we have to notice that it is in one way a misleading perspective of the differences between classical and intuitionist logic that the respective systems may be so formalised as to differ only at the level of propositional logic. It is misleading because it masks the fact that there is disagreement about the meanings of the quantifiers. Someone might think that such disagreement is an irrelevance unless it issues in an acceptance of different logical laws for the quantifiers; whereas what comparison of standard formalisations shows is that what differences there are are *consequences* of differences in the underlying propositional logic, motivated in the case of platonism by the philosophical considerations which we have been sketching. But this, of course, is a confusion. It is true that the explanations of the meanings of the logical constants in classical propositional logic proceed in terms of truth-conditions; the truth-tables stipulate the truth-conditions of complex sentences in terms of those of their constituents. But, as we have seen, there is from an anti-platonist point of view nothing in that as such to which exception should be taken. Indeed, it is clear that if only effectively decidable atomic statements are admitted, the construction out of them of complex statements by means of a vocabulary enriched solely by the classical propositional logic connectives can never generate anything other than effectively decidable statements. Classical propositional logic becomes implicitly platonist only if fed with non- effectively decidable atomic statements. Thus the disagreement between platonists and intuitionists at the level of propositional logic is only intelligible against the background of a disagreement about how certain non- effectively decidable atomic statements, to which it is to be applied, are to be understood, and so about what forms of inference involving them may be regarded as valid. In particular, if an intuitionist accepted the platonist schema of meaning explanation described above, he would seemingly leave himself no room for modifications to classical propositional logic at all.

Consider now any standard first-order language with identity, enriched with certain individual, predicate and functional constants. Suppose each predicate constant to be effectively decidable, each functional constant to denote an effectively calculable function, and that the reference of each individual constant is known. Then, evidently, every quantifier-free statement of this language will be decidably true or false. The only manner in which other than effectively decidable statements may arise is by the introduction of bound variables; that is, of quantifiers. Even then, if the range of quantification is finite, non- effectively decidable statements may still not be constructible. But it is not a necessary condition of their constructibility that the range be infinite. It is enough if we lack any criterion for saying that the whole range has been enumerated.

The language of first-order arithmetic is of course such a language, so the intuitionists' refusal to apply classical propositional logic to it is intelligible *only* as a repudiation of the classical understanding of the quantifiers. The differences which, in standard formalisations, are apparent only in the sphere of propositional logic here have to be understood in terms of divergent understandings of the quantifiers, which do not emerge at all in the primitive quantifier rules but only in the manner in which the rival quantificational logics subsequently develop. So much is clear; so our question is, what is it specifically in the classical account of the senses of the quantifiers which is inherently platonist and to which the intuitionist therefore objects?

There is in classical number-theory essentially only one place at which an appeal is made to an alleged capacity to grasp truth-conditions for a statement other than something which devolves into a purely recognitional capacity. This is the application of the orthodox explanations of the meanings of the quantifiers, as given in, for example, Church,[1] to quantification over infinite totalities. Classically, the universal and existential quantifiers are simply thought of as generating the logical product and sum respectively of the set of all statements consisting of the attribution to an object in the range of quantification of the predicate through which the quantification is made. As is now familiar, the need to treat the quantifiers, so explained, as logical primitives, as other than abbreviational devices, does not depend upon our willingness to quantify over domains for which we have no criterion of exhaustive enumeration. Even where we have such a criterion, it may only be contingent that some particular conjunction of statements gives the truth-conditions of a universally quantified statement. Thus the classical account does not essentially involve, as Wittgenstein proposed in the *Tractatus*,[2] that the quantifiers are capable of definitional *elimination*. What it does involve is the possibility of explaining the truth-conditions of a particular quantified statement as those of a particular conjunction or disjunction in all cases where a procedure is available for effectively enumerating all objects in the range of quantification, as is the case with quantified number-theoretic statements. Obviously, however, to possess the means for so explaining the truth-conditions of a quantified statement need not amount to possession of the means for effective recognition whether or not those conditions obtain; it all depends, even where only effectively decidable conjuncts or disjuncts are involved, on whether we have some criterion for declaring the conjunction, or disjunction, *complete*, for closing it off. The kernel of number-theoretic platonism is an indifference to our lack of any such criterion in the case of quantified statements whose range is the natural numbers. The notion of truth for such statements presupposed by the general form of explanation which they classically receive is thus essentially an objective one; and the grasp of their meanings thereby attributed to us irreducible to any capacity of recognition which we actually possess.

We are now in a position to interpret within the framework of our general interpretation of platonism what it is that constitutes regarding

[1] 14, section O6.
[2] *T*, 5.52–5.525; cf. Moore's report in 63, p. 297.

the natural numbers as an 'actual' infinity. It is not so much anything within the general classical account of the senses of the quantifiers as the failure of the platonist to appreciate, or, less tendentiously, his refusal to allow, that the application of the quantifiers to infinite totalities requires some other account than that which he gives. For the intuitionist, the central error of classical number-theory is one of omission: the classical mathematician fails to make a distinction, fails to see that the account which he presupposes of the senses of the quantifiers is inadmissible once the range of these operators is such that they generate statements for which he cannot guarantee means of decision. But, for the platonist, the old account extends smoothly to the new case; it does so precisely because we are supposedly able to grasp an objective conception of the truth of statements resulting from quantification over the natural numbers, and over infinite totalities in general. To think of the infinite as actual is to think of truth as applied to quantification over infinite totalities in objective terms. This way of looking at the matter cuts right through the traditional figurative contrast between mathematical objects being created by some form of constructive specification and their existing, unspecified, all along.

7.   It is clear, however, that an objective notion of truth for number-theoretic statements cannot defensibly be *wholly* dissociated from procedures of verification. Grasp of the sense of such a statement is grasp of the objective circumstances which it supposedly depicts. Two such statements differ in meaning just in case they are associated with different such circumstances. This however is presumably to be consistent with their being true, in one sense, in exactly the same circumstances, namely, if indeed true, in all possible circumstances.[1] The platonist thus requires that our conception of sameness of truth-conditions be narrower than that of necessary coincidence. Otherwise he will have no way of explaining the distinctions in sense between number-theoretic statements with the same truth-value. Indeed he will have no way of explaining distinctions in sense between any non-contingent propositions with the same modal value. We require a notion of the respective circumstances under which arithmetical statements are true which allows those circumstances to be distinct even when either statement is true whenever the other is.

The issue arises in its simplest form when we are concerned with statements involving a single quantifier through an effectively decidable predicate. Consider a pair of such statements, '$(n)\,Fn$' and '$(n)\,Gn$'. If there is a distinction in sense here at all, it can only be because the predicate F differs in sense from the predicate G; for the other parts in each sentence are the same. But, now, it is irresistible to suppose that whatever difference there is in sense between the predicates F and G

---

[1] Cf. Dummett's discussion in 25, p. 588 sq.

has to be traced back to differences in the criteria of application associated with them. This is vague, but it is evidently a narrower notion than
that of the necessary co-extensiveness of the two predicates. One way of
getting at the point might be to reflect that, intuitively, different
computer programs might be initially needed to determine the application of a pair of number-theoretic predicates which were subsequently
proved to be co-extensive. Loosely, the way of recognising that a
number is F may differ from the way of recognising that it is G,
although subsequently, as a result of proof, either method may serve to
determine the application of either predicate.

If we were pressed to explain the sense in which such a pair of
predicates differ in content, surely we should wish to proceed along
lines such as these. Such an account would be wanted to elucidate the
distinction in sense not just between statements of the forms 'Fa' and
'Ga', for some particular number, a, but also for statements involving
restricted quantification, for example, '$(n) Fn$' and '$(n) Gn$' for n smaller
than some particular k. These statements, too, are effectively
decidable, and an account of the distinction between their decision
procedures is available just in case we have such an account of the
distinction between the procedures associated with F and with G.

When we are concerned with unrestricted quantification, then, the
obvious—indeed, plausibly, the only—strategy for a platonist, if he is to
avoid the *Tractatus* doctrine of the coincidence in sense of all necessary
statements, is to insist that the identity of the procedure associated with
the predicate through which the quantification is made is to enter into
the identity of the truth-conditions of the resulting statement, even
where the latter is no longer effectively decidable. If the procedures are
different, then so are the statements.

This suggestion, of course, will have to be extensible to arithmetical
statements involving multiple generality. But such an extension raises
no new issue. Any such statement may be seen as arrived at by means of
applying to an initial pool of quantifier-free atomic statements a process
of replacing individual constants by variables—or by themselves—and
introducing quantifiers and other logical connectives. This process may
in any particular case be regarded as determining a complex function
whose arguments are the senses of the predicates in the statements in
the original pool and whose values are truth-values. The senses, then,
of statements expressing the application of such a function to the senses
of $\langle F_1, ..., F_n \rangle$ and $\langle G_1, ..., G_n \rangle$ respectively will thus be the same only
according as the senses of F_1 and G_1, F_2 and G_2, etc., are the same; that
is, according as their decision procedures are the same. This account, of
course, would apply irrespective of the degree of logical complexity (or
simplicity) of the resulting statements.

Whether or not this particular proposal is satisfactory, the platonist
requires that 'Fa' may differ in meaning from 'Ga' purely because F and
G are associated with different decision procedures; and that this

difference correspondingly infects any pair of statements, of whatever degree of logical complexity, which differ only in that one contains F in all and only the places where the other contains G. But this is not yet to say exactly what our conception of the truth-conditions of an arbitrary arithmetical statement is supposed to be, nor how this suggestion about sameness of meaning is to be incorporated into it.

Consider an example whose main quantifier governs an effectively decidable predicate; for instance the, as far as I know, still unresolved conjecture of Goldbach that every even number is the sum of two primes. The platonist conception of its truth-conditions is actually a very natural one. The conjecture is true just in case the infinite conjunction of statements, '4 is the sum of two primes,/and 6 is the sum of two primes,/and 8 . . .,' etc., is true. We are concerned with a well-defined totality of objects, the even numbers—which, in this example, we can actually effectively enumerate—and the application to each of them of a predicate of determinate sense—which, in this example, is actually effectively decidable. Because of the effective enumerability of the totality and the decidability of the predicate, we could program a machine to check the conjecture case by case. The conjecture is pictured, platonistically, as true just in case, however far the machine goes in the implementation of its program, it will never succeed in finding a counterexample, an even number which is not the sum of any pair of its prime predecessors; the conjecture is false if the machine will eventually succeed in finding such a number.

The situation is exactly similar with Fermat's so-called 'last theorem' that $x^n + y^n = z^n$ has no solution among the positive integers for $n \neq 1$ or 2. Here we have rather to consider the series of ordered quadruples of integers and a predicate effectively decidable of each such quadruple. But the platonist conception of the truth-conditions of Fermat's 'theorem' is relevantly similar. We could again program a machine to run through the (effectively enumerable) series of ordered quadruples checking on Fermat's 'theorem', which is true if, however far the machine goes, it will never find a quadruple, $\langle x, y, z, n \rangle$, ($n \neq 1$ or 2), such that $x^n + y^n = z^n$; otherwise, false.

The platonist thesis about such examples is that we have, in some such terms, a conception of the circumstances under which they are true both dissociable from our idea of what it is for us to verify them, and allowing them to be true in virtue of different circumstances, even if each is necessarily true. This idea is appreciably less far-fetched than certain features of the traditional platonist picture. Let Γ be a program for enumerating successive even numbers and checking that they conform with Goldbach's conjecture, Φ a corresponding program for quadruples of numbers and Fermat's theorem. Then it is extremely natural to insist that we just do understand the possibility that Γ, no matter how far applied, will never yield a negative result; that we understand the corresponding supposition about Φ; and that it is not at all the same supposition, just because Γ and Φ are distinct. It is natural to express ourselves in terms such as these: that it is already *determinate* what in these examples the situation is. Talking in terms of the program of a machine is one way of emphasising this idea of determinacy. The application of Φ and Γ is purely *mechanical*. Once these programs are defined, then what their outcome will be in any particular case is in effect already decided. All we, or the machine, has to do is to find out what it is.

While we are in this frame of mind, it will make no difference if the example is one of quantification through a predicate which is not effectively decidable, for example the supposition that there are infinitely many counterexamples to Goldbach's conjecture. Here we lose the connection, present in the other examples, between falsity and necessarily recognisable falsity; but, for a platonist, this need change nothing. Whether there are infinitely many counterexamples to Goldbach's conjecture is just a question of the constitution of the programme Γ. It is implicit in Γ whether there are to be any counterexamples to Goldbach's conjecture; which numbers, if so, they are to be; and thus how many there are to be.

Such a statement is, of course, decidable if we restrict its scope to a finite set of natural numbers. It is only because it concerns an infinite totality that number theory poses us any mathematical problems. But for the platonist no new issue of principle arises when we deal with an infinite totality. Whatever number-theoretic statement we are concerned with, of whatever degree of logical complexity, it is determinate and decidable whether it holds of any particular finite subset of natural numbers. How, then, can it be indeterminate whether it holds of the set of all natural numbers? At any stage at which, if a statement is restricted to a finite subset of natural numbers, its decision procedure involves examination of all members of the set, there the application of a corresponding procedure, when the statement is applied to the set of all natural numbers, will fail to be effective. There may be infinitely many such places in a particular case. But the issue raised at each of them will be determinate in truth-value. And there is in general no reason why every consequence of the concepts which we employ should be discoverable by us.

Let us then rehearse the steps leading to a platonist outlook on the truth-value of Goldbach's conjecture. Of any particular even number we should ordinarily regard it as a determinate question whether it satisfies the conjecture. In accepting, or being prepared to accept, orthodox arithmetical postulates, we enter into a commitment about each particular case, even if we do not know what the commitment is. The step from this ordinary way of looking at things to a specifically platonist conception of the truth-conditions of the conjecture is prima facie so slight that it might almost pass unnoticed. Goldbach's conjecture, the platonist reasons, is nothing but the truth-functional conjunction of all its instances. If it is allowed that each of these has a determinate truth-value, how is there scope for indeterminacy in the truth-value of the conjunction? If of any particular even number it is determinate whether or not it is the sum of two primes, how can it fail to be determinate whether or not all are? If we are concerned with a well-defined totality of objects and a predicate well defined for each of them—par excellence, associated with an effective criterion of application—there is for the platonist no question but that we are capable of understanding the allegation that all objects in the totality do satisfy the predicate in question. We are capable of understanding it, even though its 'being true' does not entail the possibility of our recognising that it is so.

8.   This is indeed an assumption which in other areas we should not ordinarily question. The generalisations of natural science, while acknowledged to be hypothetical, are not popularly thought incapable of truth. Nevertheless it is at exactly this point that the anti-platonist or, henceforward, *anti-realist,*[1] for example the intuitionist, must take his stand. We cannot pass from the determinacy in truth-value of any statement in some range to the supposition that it is likewise determinate whether all of them are true, unless we can guarantee the existence of a way of knowing whether all of them are true.

It is at this juncture, finally, that we are able to deploy against platonism one of the primary strands in *RFM*. We encounter indeed one of the main differences between Wittgenstein and the intuitionists. It is not that Wittgenstein is not sympathetic to the suggestion that quantification over infinite totalities requires some special account. Rather, unlike the intuitionists, he would seem to reject the idea of a conceptual *commitment*, appealed to by the platonist prior to his making the move which the intuitionist rejects. Wittgenstein would reject the supposition that it is determinate whether or not *any particular* even number is the sum of two primes. It is irrelevant that we have what we regard as a purely mechanical procedure for computing the question in any particular case; for it is not, even so, determined in advance what in any particular case the outcome will be, that is, what we shall *count* as correct application of the procedure if we judge correctly.

It is usual to make free use of the idea of an application of a concept *according* with, or failing to accord with, its content. It is in accordance with the meaning of 'red', as we understand it, that it should be applied to red things rather than blue ones. We think of giving the meaning of an expression in contractual terms. Once the meaning has been fixed in a certain way, we are all obliged to make a certain kind of use of the expression; only that kind of use *conforms* with the sense of the expression that was fixed. We are, so to speak, constrained by our

---

[1] The terminology is Dummett's; but the reader had better be circumspect about assuming that my understanding of it is.

understanding. If we are to use the expression in conformity with the way we understand it, or the way the community at large has generally used it in the past, we *have* to use it in certain sorts of ways.

The idea that it is already determinate of any particular even number whether it is the sum of two primes is just a special case of this tendency. We have given the predicate in question a certain meaning by associating with it a certain criterion of application, whose appropriateness follows in turn from the meanings of the simpler expressions involved in the predicate. We have certain criteria for when the test has been properly applied. So long as we conform to these criteria, and to the understanding which we have of the predicate, there is no latitude, no tolerance at all about the question whether the predicate applies in any particular case. What comes out *has* to come out.

Wittgenstein rejects this notion utterly. His grounds for doing so are to be found in the recurrent discussion of the topic of following a rule. He seems to want to disallow that it is ever pre-determinate what counts as 'doing the same thing again' or 'applying the rule in the same way'. The considerations in question are applied to rules for expanding decimals, to rules of inference and to rules of use, or semantic rules. The cases are indeed analogous in relevant respects. In each of these cases we think of someone's having learned the rule as that which *explains* his capacity to expand $\sqrt{2}$, infer properly, apply 'red' correctly; and in each case we think of it as fixed in advance what sort of behaviour will, in certain circumstances, conform to the rule. The rule for expanding the decimal, that is, the way we understand the rule, determines in advance what its nth place is to be. The rules of inference which we accept likewise determine in advance the transitions from statement to statement that may be carried out in accordance with them, and so what may be counted as the consequences of certain suppositions. Rules of language in general determine how such-and-such circumstances may correctly be described.

If Wittgenstein's critique of these ideas is sustained, it is not clear how we can avoid rejection not merely of the objectivity of truth in mathematics, in the distinctively platonist sense, but indeed the whole picture of pure mathematics as something conceptually stable, as something in which the primary objective and substantial task is not conceptual innovation but the tracing of the liaisons and connections between concepts to which we are already committed. The next thing to do is thus to investigate Wittgenstein's idea on this issue.

II

Following a Rule

Sources

RFM:	i. 1–3, 113–18; iii. 8–9; v. 32–5, 45–6
BGM:	vi. 15–49; vii. 47–60
PI:	i. 138–326 (but especially 185–242), 692–3
PG:	i. 6, 17–18, 52, 75
Z:	87, 279–308
BlB:	pp. 73–4
BrB:	ii. 5
LFM:	lectures ii; vi; xiii, p. 124

1.   We use the idioms of a particular application of an expression 'according with its meaning', of a particular derivation being 'required' by a certain rule of inference, of a particular continuation of an infinite decimal 'according' with its governing rule. These seemingly harmless ways of talking are subjected to recurrent scrutiny throughout Wittgenstein's later writtings. Repeatedly Wittgenstein seems almost to want to say that there is in reality no substance to the idea of an expression being used in accordance with its meaning, that there is no sense in which we, as language users, can be regarded as committed to certain patterns of linguistic usage by the meanings which we attach to expressions.

Suppose, for example, that we are concerned with some formal system. When it is determined what are the axioms and rules of inference, we think of that as already a determination of everything that is to count as a theorem. In accepting the axioms and rules, we think of ourselves as undertaking a commitment to accepting certain things as theorems; the mathematical task is to uncover what in particular cases our commitment is. Wittgenstein's claim is that this is a false picture. Notwithstanding the fact that proof in such a system is a mechanically decidable notion, that is, that we may programme a machine effectively to check any putative proof, there is somehow in reality no rigid, advance determination of those sentences which are theorems. To put the point in its most general form: there is in our understanding of a concept no rigid, advance determination of what is to count as its correct application.

We have, according to Wittgenstein, a philosophically distorted perspective of locutions like:

'the steps are determined by the formula',
'the way the formula is meant determines which steps may be taken',
'*Fa*' has to follow from '$(x)Fx$', if the latter is meant in the way in which we mean it',
'if you really follow the rule in multiplying, you must all get the same result'.

We think, for example, of the meaning of '$(x)Fx$' as so determined that it simply will not cohere with an application of this sentence if '*Fa*' is not allowed to follow. We think of the meaning of '$x^2 + 2$' as so determined that it is fixed in advance what the reference of this expression will be for a particular integral value of x. But for Wittgenstein it is an error to think of our understanding of such expressions as already predetermining what sort of use of them on future occasions it will be correct to allow. It is not determined in advance what we shall accept as the value of '$x^2 + 2$' in a particular case, but only that we shall not brook more than one alternative. Similarly it is not determined in advance how we shall apply '$(x)Fx$' and '*Fa*'; it is merely, so to speak, a rule of 'grammar' that if we count an individual as 'not F', we must be prepared to conclude 'not $(x)Fx$'.

Consider 1. 113/4: Wittgenstein's objector asks, 'Am I not compelled, then, to go the way I do in a chain of inferences?' Wittgenstein says, 'Compelled? After all I can presumably go as I choose.' The objector continues, 'But if you want to remain in accord with the rules you *must* go this way.' 'Not at all,' says Wittgenstein. 'I call *this* accord.' 'Then,' according to the objector, 'you have changed the meaning of the word "accord" or the meaning of the rule'. 'No. Who says what "change" and "remaining the same" mean here? However many rules you give me, I give a rule which justifies my employment of your rules. We might also say: when we *follow* the laws of inference (inference rules), then following always involves interpretation also.'

Compare this with v. 32. Wittgenstein quotes the idiom, 'if you accept this rule, you must do this'. This may mean, he says, that the rule does not leave two paths open to you here, which is a mathematical proposition. It could also mean: the rule conducts you like a gangway with rigid walls, and against this picture he urges the objection that the rule could be interpreted in all sorts of ways.

2.   What is the correct interpretation of these passages? They might be taken to suggest that Wittgenstein thought that there were always indefinitely many, equally good, alternative ways in which a rule might in a particular case be followed. But that, we should protest, is surely

quite wrong, so long as we are concerned with a rule whose sense is at all precise. Still, it might occur to someone, there is one relatively clear way in which following a rule could be regarded always as involving an interpretation. Any rule which we set someone to follow may be applied by him at some stage in a manner both consistent with his past application of it and other than that which we intended. That is, if somebody suddenly makes what seems to be a deviant application of a rule, there may be an interpretation of the instructions which we gave him which both explains his use of the terms in question which coincided up until now with the use which we intended, and also the further deviant use. We should say that such a person had misinterpreted our instructions, that he was following a different rule from the one which we intended. But it is also, evidently, a possibility that nothing which we say or do will get such a person to, for example, expand an infinite decimal as we wish. However many rules we give him, he gives a 'rule which justifies his employment of our rules': that is, he supplies an interpretation of what he has been told to do under which it can be recognised that he is indeed doing it.

It might be suggested, then, that Wittgenstein's thought here is something like the following: where we think we have grasped the rule which people intend us to follow—where we think we understand how to apply a certain predicate, how to expand an infinite decimal—and now the occasion for a new application arises, there are always open to us indefinitely many hypotheses about how the expression should be applied in the new circumstances, how the decimal should be continued, which are consistent with the applications which we have made of it so far without being corrected, and which we have witnessed others make of it. Of course, there is no reason why these alternative hypotheses should occur to us, and typically they do not; typically, the use which we make of an expression is quite automatic and unhesitating. Nevertheless what we regard as following the rule will in this sense involve interpretation, that we shall, perhaps unconsciously, have picked on one of the available hypotheses as being the one which other language-users are applying, the one which our teachers intend us to apply.

According to this suggestion, Wittgenstein's point would not be that the same rule, be it vague or not, always permits application in indefinitely many ways. Rather it would be that there are always, on the basis of any normal training in and exposure to others' applications of the concepts involved, indefinitely many equally viable interpretations of which rule it is that we are intended to follow. And Wittgenstein is drawing our attention to the possibility of someone who fixes on an unintended interpretation of the instructions which we give him, and who, having done it once, proceeds to go on doing it, despite our best efforts to clarify to him what it is that we wish him to do. This is to be expected of course if he possesses some alternative understanding of the terms in which we attempt to clarify the original instruction; and, in any case, no matter how many examples and how much patter we give him, it will all be consistent with an indefinite variety of interpretations of the intended rule.

Of course when, as in the passage cited, it is the deviator himself who supplies the alternative interpretation which explains his behaviour, it seems that, if he is not being disingenuous anyway, the difficulty cannot be a very deep-reaching one; for if he is in a position to explain to us what he thought we meant, we are likely in a position to explain to him what we really did mean. But where we are concerned with somebody who either does not understand, or who misunderstands the terms in which we might explain to him what rule it is that we intend him to follow, so that we have recourse to examples alone, there are available to him indefinitely many hypotheses to explain whatever examples we give him.

Indeed, is not the learning of a first language just this same situation writ large? But then at any stage somebody may, like the man who continues '2004 . . . 2008 . . . 2012 . . .' in response to the rule 'Add 2', reveal that he has *all along*, as we should say, misunderstood the manner in which some expression is to be applied. And if he misinterpreted his original training, he may subsequently misinterpret any attempt to re-educate him; all this while using the expression, as it seems to him, in a perfectly consistent and regular way.

Now, how might these considerations be thought somehow to bear on the locutions which we considered: 'The way the formula is meant determines what steps are to be taken', 'If he follows the rule, he must get this', etc.? One suggestion would be that they call in question our right to speak of *the* way the formula is meant, *the* rule. What reason have we to think that there *is* a shared understanding of the rule? If variations in the temperature of a room were sufficiently localised, it would make no sense to speak of the room temperature. But is not the possibility to which Wittgenstein has drawn attention just the possibility that variations in understanding are so localised? Unless we may reasonably rule out this possibility, we have no right to speak of *the* meaning of an expression; for the meaning of an expression is just the way it is generally understood.

This is, in effect, to suggest that some sort of inductive scepticism lies at the root of Wittgenstein's comments. The suggestion might be amplified as follows. Wittgenstein apparently rejects the idea that the meaning of an expression is anything which is properly seen as constraining a certain sort of future use of it. *One* way of supporting that view would be to suppose that a correct account of the meaning of an expression is at any stage to be constituted by an account of its past uses. In that case each fresh use of the expression would be independent of the account so far given, and would, indeed, require a refinement and extension of it. Of course, a decisive objection to such a view of the meaning of an expression would be its conflict with standard criteria for

what it is to misunderstand a meaning. It is by *misusing* an expression
that someone shows that he does not understand it, however accurate
his knowledge of the history of its use. Knowing a meaning is knowing
how to *do* something; we are supposed to know how *in general* the
expression is to be used. This position is thus not really available, nor
am I suggesting that Wittgenstein held it. On the contrary. The point is
rather that it is but an inductive step from an account of the past use of
an expression to a statement of its general use. The knowledge which we
derive when we learn a first language is, plausibly, nothing other than
inductively based conclusions about how expressions ought in general
to be used, drawn from our experience of how they have been used.
Thus to possess the same understanding of an expression as someone
else will be to have formed, on the basis of suitable training, the same
inductive hypothesis about its correct use. And now, what evidence is
there that widespread, local semantic variation really is a practical
possibility? It would seem, on the contrary, that all the evidence points
to the conclusion that we have each arrived at very much the same inductive hypotheses. The suggestion that it is out of order to speak of *the*
rule, that is, our common understanding of the rule, would seem to require
that the evidence that we do indeed possess such a common understanding, viz. our continuing agreement about how the rule is to be applied, is
somehow to be discounted. Is Wittgenstein, then, doing anything other
than playing at inductive sceptic concerning general conclusions about
how an expression is to be used, based on samples of its use?

If this were a correct account of it, Wittgenstein's point would seem
prima facie nothing essentially to do with meaning. The point capital-
ises upon the supposed equi-viability of an indefinite number of incompatible hypotheses each of which accommodates one's data concerning
the past use of some expression; any number of these may actually be in
play in the linguistic community, awaiting only a suitable crucial case to
bring the misunderstanding into the open. But, as Hume and Goodman[1] have emphasised, just this is the situation with *any* inductive
inference. Thus it might appear that the proper way to combat Wittgenstein's views, on their current interpretation, would be to solve the
problem of induction: to show that it is simply untrue that there are
always available indefinitely many hypotheses which, on the basis of
certain evidence, may equally reasonably be adopted. If it could be
shown, it might be thought, that where, when attempting a simple
induction, we are confronted with indefinitely many possible
hypotheses, we are nevertheless confronted with only finitely many
probable hypotheses in some sense—hypotheses which, in some objective way, it would be rational to accept, given the available evidence—then inductive scepticism in general, and so in particular inductive scepticism about meaning, would be vanquished. In the case of an
induction about the correct use of some expression, it could be

¹ Hume *locus classicus:* 48, § IV; Goodman 40, ch. III.

expected, provided the language were used in a consistent way, that all reasonable beings would, sooner or later, arrive at the same hypothesis. There would be in practice no possibility of deep, ineradicable misunderstanding of the kind Wittgenstein imagined.

It is worth noticing, however, that it is incorrect to assimilate the problem of countering Wittgenstein's view, on this interpretation, to that of meeting the traditional epistemological difficulties with induction. If Wittgenstein's position about meaning were that of inductive sceptic, it remains the case that there would be an important difference between his position and that of inductive scepticism in general. For how are we supposed to have learned which procedures are rational, which type of hypothesis, although consistent with the data which we possess, we may nevertheless rationally eliminate? If it was right to admit at all that we are confronted, in the process of learning any concept, with at any stage indefinitely many possible hypotheses about its correct application, then the same admission must be made with respect to the notion of rationality, in particular with respect to the notion of a rational inductive inference. And now our rationality cannot be invoked, at any rate by an empiricist, to cut the number of possibilities open to us down to size, since it is rationality itself that is supposed to be being explained to us.

A solution to the problem of induction which took the form of showing that one can always advance, given adequate data, to a situation where it is rational on the basis of that data to accept only one particular hypothesis would not effectively countermand a general inductive scepticism about the identity of people's concepts in some area, and in particular about our concepts of the correct use of certain expressions. If the problem is that of giving a reason for thinking that we all possess the same understanding of some expression, that we intend to use it in accordance with the same governing hypothesis, inductively arrived at, then it is no answer to suggest that that at any rate will be the position if, on the basis of a sufficiently wide experience, we have arrived at our respective hypotheses by purely rational methods. This reply would simply push the difficulty back into that of giving a reason for thinking that we operate in accordance with the same notion of a sound, or rational, inductive inference. If we represent inductive scepticism in general as essentially questioning whether it is rational to believe one, rather than any other, of the indefinitely many hypotheses which can be used to explain any particular finite set of data, then it is a peculiarity of the application which Wittgenstein, on the present interpretation, makes of it here that it is immune to what for any other application would be a solution of the difficulty. It will be no reason for supposing that we have all achieved the same understanding of some expression that we will at any rate have done so if our conclusions are all rationally based, unless a reason can be provided for thinking that they are.

That, then, is one interpretation[1] of Wittgenstein's thought on the issue. But it is not a very promising interpretation. For it consists in drawing attention to a possibility which by ordinary criteria we have every reason to exclude. The amount of successful linguistic commerce which goes on, and the variety of situations in which it takes place, constitute by any ordinary standards overwhelmingly powerful inductive grounds for supposing that we share a common understanding of most of the expressions in our language. And there are independent practical reasons, apart, that is, from the success of our use of language, for thinking that this is likely to be so. For example, we should expect there to be a practical limit on the number of alternative hypotheses which people could competently handle or imagine; so in practice it would be quite likely that one's training in a concept would determine it to within uniqueness. After a certain stage we would simply be unable to see alternatives. In addition, it is likely that certain features of a situation in which a concept is applied will always strike a learner as dramatically more noticeable than others, so that certain possible hypotheses to account for the pattern of usage which he is experiencing will simply be ignored; and we learn a first language at sufficiently tender an age to make it plausible to suppose that our dispositions to be struck by certain features and to overlook others are largely innate, and so largely shared.

To a philosopher disposed to sympathise with the sceptic such considerations will seem enormously feeble, of course. But Wittgenstein was not so disposed. Indeed, the present interpretation of his ideas on rule-following runs counter to the whole approach of Wittgenstein's later philosophy towards traditional epistemological problems. This is the philosopher who wrote (v. 75):

> The danger here, I believe, is that of giving a justification of our procedure when there is no such thing as a justification and we ought simply to have said 'that's how we do it',

and (v. 33):

> To use a word without justification does not mean to use it without right.

Wittgenstein did not believe that we owe the sceptic an explanation of the justice in our ordinary procedures of inference;[2] on the contrary, it was a symptom of an erroneous philosophical outlook, a distorted perspective upon the nature of those procedures, that we find ourselves with the capacity to take scepticism seriously. So we need another interpretation.

3.   There is at several places in Wittgenstein's later philosophy

---

[1] Cf. Blackburn 6, pp. 141–2.
[2] Cf. *PG* I. 55, 61; *BGM* VI. 31; *OC* 150; and many other passages.

evidence of sympathy with anti-realist ideas in the sense distinguished in the previous chapter, evidence, that is, of a hostility towards any notion of truth for certain statements dissociated from criteria for their verification. For example, a desire to identify mathematical truth with provability is evident in the discussion of Gödel's theorem in *RFM* (Appendix *I*). Similarly he claims (IV. 11) that to say of an infinite series that it does *not* contain a particular pattern *makes sense* (my italics) only when it is determined by the rule that the pattern should not occur. And here it is clear that he means there to be in the rule only features which we have explicitly put there, and so are in a position to recognise, for he goes on to say:

> When I calculate the expansion further, I am deriving *new* rules which the series obeys.

It is also plausible to regard the general emphasis on the necessity for proofs to be perspicuous, or *surveyable*, as an application of an anti-realist outlook. Consider, for example, II. 2:

> I want to say: if you have a proof pattern that cannot be taken in and, by a change in notation, you turn it into one which can, then you are producing a proof where there was none before.

So if it is true that something is a proof, then, according to Wittgenstein, it must be possible for us to recognise that it is so, to 'take it in'. There is no question of something having the status of a proof which, because of its length or complexity, or for whatever reason, cannot be used as a means of persuasion.

Perhaps the most conspicuous example in Wittgenstein's late philosophy, however, of an appeal to anti-realist notions is to be found in the *Investigations* in the discussion of private languages. Consider this (chopped) quotation from I. 258:

> I want to keep a diary about the occurrence of a certain sensation. To this end I associate it with the sign 'S', and write the sign in the calendar for every day that I have the sensation . . . I remember the connection between the sign and the sensation right in the future. But in this case I have no criterion of correctness. One would like to say, whatever is going to *seem* right to me is right. And this only means that here we can't talk about 'right'.

But *why* may we not talk about 'right'? I use 'S' correctly on any particular occasion just in case the sensation which I have on that day really is the same as the sensation which it was my original intention to describe as 'S'. Wittgenstein's objection is that I have no criterion for saying, no method for determining, that the two sensations really are the same, i.e. are relevantly similar; I have no way of making a distinction between seeming to myself to be using 'S' correctly and really doing so. Wittgenstein wants it to follow that there therefore *is* no distinction here between a correct description and a merely seemingly correct description. Whereas, he wants to suggest, it is a necessary condition of someone's being said to use an expression correctly that such a distinction can be significantly made. But, whatever the merits of that claim, there is a clear repudiation of an objective notion of truth in Wittgenstein's transition from the impossibility of our effectively making a distinction of a certain sort to the conclusion that there *is* no such distinction.

To avoid misunderstanding: I am not claiming at this point that Wittgenstein's late philosophy was globally anti-realist, nor even that this is quite obviously the correct interpretation of the examples which I have cited. I am claiming merely that it is a plausible interpretation of those examples, and that it is therefore in point to consider an account of his thought about the topic of following a rule of an essentially anti-realist sort.

It is not difficult to see how such an account would go. Wittgenstein's examples suggest that we have no way of conclusively verifying that we share our understanding of some expression with someone else, that on some future occasion our respective uses of the expression will not diverge so radically that we shall be driven to regard the meanings which we attach to it as different. If this is allowed, then we are implicitly taking the supposition that two people *do* agree in the way in which they understand some expression as equivalent to some open set of statements about their behaviour in actual and hypothetical circumstances. Plainly nothing does count as recognition of the truth of such a set of statements, so, for an anti-realist, nothing counts as the *fact* of its truth. From this point of view, talk of the specific way in which someone understands an expression is admissible only if we possess some means of verification of how specifically he does understand it. But we have no such means; and thus there is no fact of his understanding the expression in some particular way.

On the first, sceptical interpretation of Wittgenstein's position, the trouble with talk of someone's application of an expression according or not according with the rule was the presumption of the rule as something public, the presumption of largely shared understanding of it. The problem of the sceptic was, what right have we to suppose that there is such a shared understanding? From the point of view of the current interpretation, the very content of this question is dubious. If there is no way of knowing that people share the same understanding of an expression, then just for that reason there is for an anti-realist nothing *to be known*; there are no intelligible truth-conditions for the supposition that their understanding is shared. The difficulty, from the point of view of the sceptic, of having the right to suppose that understanding is shared transmutes, when we shift to an anti-realist standpoint, into the problem of giving *sense* to the supposition that it is shared. But, from either standpoint, the idea of the rule for the use of some expression as something which we do share, something which we all ingest during the period when we learn the language, is inadmissible. Either—sceptically—there is no way of knowing that someone's understanding of an expression coincides with one's own; or—anti-realistically—there is no such thing as the fact that it does.

4.   Quite apart from the question of the correct interpretation of Wittgenstein's remarks on following a rule, the foregoing is obviously a problem in its own right. The natural response would be to admit the notion of truth for the supposition of shared understanding which the sceptic exploits, and then attempt somehow to play down the interest of the considerations which he proceeds to advance. But where do we get such a notion of truth *from*? How do we arrive at it? For the whole point is that we have no direct experience of what it is like for such a statement to be true. On the other hand, it seems plausible that a radically realist notion of the truth of such a statement is embedded in our customary habits of thought in this area. When somebody, expanding an infinite decimal, makes a total hash of it after some point, we say that that shows that he did *all along* not understand what we meant. Would such a description still be legitimate from an anti-realist point of view? It might seem not on the following grounds. When the person in question was earlier coping quite smoothly with the expansion of the series, in what, from an anti-realist point of view, was his misunderstanding then supposed to consist? In the truth, presumably, of the supposition that at some later stage in the series he would get it wrong. But it may be that there was then available absolutely no reason for supposing that that was so, that no one could reasonably have predicted that misapplications of the rule would subsequently take place. If the anti-realist allowed, however, that the subsequent misapplications showed that our man did all along misunderstand the rule, then surely he has to concede that, had it been asserted earlier that there was a misunderstanding, the assertion would have been *true*. And now what has become of his prized connection between truth and verifiability?

One response would be that this is to interpret the desired connection too narrowly. An anti-realist who wished to retain in these circumstances the appropriateness of the description, 'so he misunderstood all along', could indeed not insist that the truth of a statement at a time required its verifiability at that time, except at the cost of severing the equivalence between 'p was true at t', asserted at a time later than t, and 'p is true', asserted at t. But why should the connection not take the form merely that if a statement is true at t, it must *sometime* be possible to recognise its truth-at-t? On this account, the idiom 'so he misunderstood all along', remains available to an anti-realist just in virtue of its being a description for which we have a criterion of correct application.

Intuitively, however, this does not seem satisfactory. What appears unavailable to the anti-realist is the intended *point* of the idiom. The whole point of saying, 'so he misunderstood all along', as we ordinarily conceive of it, is to suggest that the foundations of the misapplication which someone makes were already laid before he made it; that something was already true of him in terms of which, had we known about it, we could have anticipated and explained the misapplication. This seems to require a more substantial notion of truth for the assertion 'He misunderstands the rule', made before the misapplication, than something imposed merely by the subsequent justifiable assertability of 'He misunderstood all along' and the standard liaisons between tenses.

One way of reinforcing this is as follows. Where we are prepared to say that someone misunderstood all along, we should also say that, had he been given the crucial case earlier, he would have revealed the same misunderstanding. What grounds are there for preserving such a connection with this counterfactual conditional if his former misunderstanding just consisted in its being the case at an earlier time than he would at a later time misapply the rule?

5.   Our present concern is not with the implications of anti-realism for the notion of understanding or for the tenses, but only with its capacity to furnish an interpretation of Wittgenstein's ideas on following a rule. And it is in fact clear that neither the sceptical nor the anti-realist interpretations so far considered will do, and for a shared reason. On both of these interpretations, at any rate as far as they have been developed, the difficulty has to do with the understanding which *another* has of some expression, of the rule which another is following in writing out the expansion of some infinite decimal. For the nature of the understanding of someone else we have only the evidence of what he has so far done. Both interpretations fasten on the point that this is never a decisive verification of the nature of his understanding. But, intuitively, the situation would seem to be quite different in one's own case; nor, in order to make the contrast out, do we need to think of our own understanding of an expression as anything other than a disposition to use it in certain ways.[1] Quite simply: each of us, in his own case, is not dependent upon inductive means to know what his linguistic dispositions are for any particular expression. Knowledge of the character of one's own understanding does not seem to be inferential at all; at any rate, it does not *have* to be based on inference. It is better assimilated to knowledge of one's own intentions.

It is consistent with both interpretations of Wittgenstein's position so far considered that each of us may be credited with the capacity to know finally how he himself understands some expression, what rule he himself is following. We speak of ourselves, for example, as grasping 'in a flash' what rule governs a series when given the initial elements; and even if it is not the intended rule, we can then at any rate be absolutely

[1] —whether or not such a view is conceived to be satisfactory.

certain of the understanding which *we* have of how to continue the series. Even if I cannot know what rule another is following, I can be certain, we would ordinarily allow, of the rule which I am following and of what it requires me to do. We are tempted to think that we can in some sense survey the whole future pattern of our use of an expression, provided we use it sincerely (cf. I. 123, 130.).[1] Specify circumstances to me and, in contrast with the situation with other people, I can be absolutely certain how in those circumstances I would, if I did so sincerely, apply some relevant expression.

Both the interpretations of Wittgenstein's position so far considered leave intact the possibility of such a first/third person asymmetry with respect to knowledge of how someone understands. But the idea of privileged access to the character of one's own understanding is, I believe, Wittgenstein's primary target. Consider the following passages. III. 8:

> If you use a rule to give a description,

> —of what someone is to do—

> you yourself do not know more than you say; that is, you yourself do not foresee the application which you will make of the rule in a particular case. If you say, '. . . and so on', you yourself do not know more than 'and so on'.

And v. 35:

> Of course we say, 'all this is involved in the concept itself'—of the rule for example. But what this means is that we incline to these determinations of the concept. For what have we in our heads which of itself contains all these determinations?

For Wittgenstein, we do not have, even in our own case, advance knowledge of the manner of employment of an expression which we shall regard as conforming to the way we understand it; the knowledge which we seem to have here, and which we cannot have for a third person, somehow comes to nothing. It remains to see why.

6.   There is a natural response to the example of the man who persistently misunderstands the rule which we are trying to get him to follow. It is to protest that while, of course, someone may always hit on an unintended meaning for some expression, or extrapolate from our examples some rule for the series other than that which we meant, it does not follow that there is no room for the idea of the correctness of

---

[1] On 'grasping in a flash', see specifically: *PI* I. 138–9, 151–2, 155, 179–81, 191, 197 and 323. Cf. *LFM* lecture II, p. 26 sq.

our and his subsequent use of the expression relative to our *respective* understandings of it, or of the correctness of our and his continuations of the decimal relative to our respective understandings of which series is involved. Somebody may, of course, misunderstand our explanations, and so may not use an expression in the way that we wanted him to. He may, for example, be struck by some, from our point of view, incidental feature of the examples which we give him, and so form a notion of what justifies the application of the expression quite other than what we intended. But that does not squeeze out the idea of the *relative* correctness of his and our uses of the expression. In such cases we can always in principle recognise that if *that* is what he means, when the meaning is specified to us, then his use of the expression is entirely understandable and correct. And even if we cannot finally be certain what he does mean, we can still recognise the correctness or incorrectness of what he does, relative to some hypothesis about what he means. Perhaps it could happen, as Wittgenstein suggests, that we might never get him to see what we mean, nor be able to form any clear impression of what he might mean; but nothing has been done to rule out the supposition that, could we do both these things, we and he could recognise of each other that we were using words intelligibly and correctly. It follows that nothing has been done to *undermine* the idea of using an expression in accordance with its meaning; we still have the phenomenon of our capacity to recognise the kind of use of an expression which a particular interpretation requires.

In what sense, though, do we really know the *kind* of use of an expression which a particular interpretation requires of it? Presumably, this is just to know the kind of use which we should be prepared to make of expressions involved in specifying the interpretation, that is to know how we understand those expressions. The sense in which we feel our own understanding of an expression to be transparent to us is that of possessing a general idea of the kind of circumstances in which we should be prepared to apply it, if using it sincerely; and the application of the expression accords with the way we understand it when it is applied in circumstances which are, as it seems to us, of this kind.

Wittgenstein points out that talk of the description which our understanding of certain expressions will *allow* us to give of particular circumstances suggests certain distortions. It makes the judgment sound essentially complex, as if we somehow 'measured' the circumstances with the meaning or compared them with a set of paradigms. But the range of circumstances in which we should confidently apply the expression may very well include cases of a sort other than those we have already encountered, or which we might imagine when asked to introspect the character of our understanding of it; to *what* are we being faithful when we apply the expression to such a case? Besides that, when we are disposed to apply the expression to a new case, this need be no more than a simple response, like our responses to colours (1. 3).

We shall not, presumably, quarrel with either of these points. Recognition of something may be absolutely simple and immediate; and we may find ourselves with definite linguistic responses to situations strictly unlike any previously experienced or anticipated to which we made, or would have made, a similar response. But these considerations only tell against certain notions which we might have been tempted to invoke to *support* the idea that we are committed to certain definite patterns of use by the way we understand expressions; it is harder to see that they tell against the idea itself. To be sure, we are not always committed by exact precedents; and it is not always appropriate to see us merely as tracing out the verdict of some complex test, when we judge whether some expression applies. But, as suggested earlier, may I not know the kind of circumstances my use of an expression will signal just as a special case of knowledge of my own intentions?

To see the character of Wittgenstein's objection, it is in point to consider the following argument. Suppose we have a semantic rule for the application of a predicate F: F is predicable of an individual just in case that individual satisfies the condition of being ϕ. Now, recognition of understanding seems first/third person asymmetric just in the sense that such a rule may be recognisably correct as an account of one's own understanding, whereas as an account of the understanding of another it is always hypothetical. Of course, it is not in general possible to specify such rules in what seems to be an informative way; it is not in general possible to specify an appropriate ϕ for an arbitrary F without *using* F. But it is immaterial to the status of the rule as a correct description of the sense of F whether that is the situation; all that is affected is the potential utility of the rule as part of an explanation of the meaning of F to someone who does not yet understand it. (If F features in the specification of ϕ, there is, to be sure, a greater temptation to say that someone's undertaking to follow the rule in his use of F is worthless than if it does not. If there is a doubt whether someone understands 'green', we should ordinarily be partly reassured if he announced that 'green' to him means the colour between blue and yellow in the spectrum of white light, and totally unreassured if he tells us that he applies 'green' 'only if things are green'. This would be a perfectly reasonable reaction if there was no reason to doubt that he understood 'blue' and 'yellow' and was familiar with the sequence of colours in the spectrum.)

What then is the argument? Consider any such semantic rule. It is irrelevant to the argument whether ϕ is informatively specified, but more natural to suppose that it is. The question is, what is it for it to be true that such a rule faithfully incorporates a particular group, or individual's, understanding of F? For they and we may sincerely agree about the correctness of such a rule, and then go on, sincerely and irresolubly, to apply F in mutually inconsistent ways. It is not open to us to protest that all that follows is that they use F in accordance with a different rule. We agreed on how the rule for F was to be characterised;

but now, it seems, we want to reserve the right to offer some other characterisation of what governs their use of the expression. (Though, of course, there is no necessity that such a characterisation will occur to us.) But from their point of view the initial characterisation remains perfectly adequate; it is of *our* use of F that a re-characterisation of the determining rule is required. But then, what content is there to the idea of a *correct* characterisation of our and their respective rules? And if there is none, how can we characterise the nature of the rule which *we* follow, if we wish to describe our use of *F*?

The knowledge which we suppose ourselves to have of the nature of our own understanding of an expression is, in this terminology, know-ledge of the rule, or set of rules, which we follow in our use of it. But what is it to know what these rules are, if it is not to be able to recognise as such a correct characterisation of them? We should conclude, it therefore appears, that we do *not* know what these rules are; for in the situation just described the same characterisation was 'recognised' to apply to what turned out to be different rules.

Specious as it may seem, this argument is decisive unless we are prepared to allow that it is only relative to his own understanding of the terms involved in it that a rule adequately characterises someone's understanding of a predicate *F*. Thus each of us knows what rule he is following, and we all give the same characterisation of the rule; each of us indeed recognises the correctness of the characterisation. It is just that the characterisation may mean different things in different mouths. But the price of allowing us this capacity is that we cannot convey to *anyone else* how our understanding is properly characterised. The dilemma is therefore this. If disagreement in use is taken to show that someone gave an incorrect characterisation of his understanding, then, since there is no way of recognising who, none of us can be said to know that his own characterisation was correct, nor therefore to know how he understands the expression; unless this knowledge is to be something which he cannot *state*. On the other hand, if it is allowed that *everyone* may have given a correct account of the rule which he follows, then since, from the point of view of person A any old manner of application of *F* by B is accounted for by B's rule, no one has actually given *any* account of his rule for *F*; no specific expectations on A's part are licensed by the supposition that B follows the rule which he, B, has stated.

The two horns of the dilemma come indeed to the same thing: if I know how I understand *F*, what I know I cannot convey to anyone else. But is there not independently a great temptation to think just that? Even where the specification of ϕ contains *F*, something is intended by someone's assent to the rule which he does not succeed in conveying. If you say that you call only green things 'green', don't you at least know which circumstances you mean? You think of something green and say to yourself: 'I call *that* sort of thing green', and this information—of the *sort*, not the particular item, which you mean—is something you cannot get across to me.

This is the notion to which the supposition of certain knowledge of one's own understanding, and hence of what conforms to it, is finally reduced, and which Wittgenstein wishes to oppose. For Wittgenstein, there is nothing in the idea of my meaning something by an assertion which someone, who by ordinary criteria understood it, would not have learnt. But this is not our ordinary outlook on the matter. My assent to the rule, we feel, is always informative *for me*, as it were; I know what kind of pattern of use I am committing myself to. When I assent to the rule: *F* is to be applied only to individuals which are ϕ, I commit myself to a quite determinate way of using *F*, even if I cannot finally make plain to others what this way of using *F* is to be.

For Wittgenstein, there is no such determinate commitment. I have no privileged knowledge of the total range of circumstances in which, if I am faithful to my understanding, I shall be prepared to say that condition ϕ is satisfied. Partly it is, as was pointed out earlier, that some of these circumstances may be unlike any that I have previously visualised or encountered. Proofs are a particularly good example of this. 'But,' it will be retorted, 'they will not be unlike in any relevant respects, or you would not presumably be prepared to describe them as *F*.' And that is right. The point is, rather, that my judgments of likeness here are *consequent upon*, rather than the basis of, my judgments about the applicability of *F*. Of course, our successive applications of an expression all seem quite familiar to us, at any rate when made sincerely. But the feeling of familiarity and the disposition to re-apply the same expression are all the same thing, so to speak. There is not, in the end, any ulterior question of *justifying* the feeling of familiarity, of vindicating the disposition.

What content do I attach to my intention to call just green things 'green', which I cannot convey to you? Suppose I am asked to concentrate on that content; what kind of thing will I do? I might look around the room at the various green things in it, visualise a few others, conjure up some memories of green things. '*That's* the kind of thing I mean', I might say to myself: '*That's* the kind of thing which I always call "green".' But where in that is the commitment to a certain determinate pattern of future use? For whenever in the future I am prepared to call something 'green', I shall count it just on that account as being *that* kind of thing. The point is essentially that of the passage quoted earlier from the *Investigations* concerning the idea of a private rule of language. Naturally, when I use the word 'green' sincerely on future occasions, it will seem to me that my use is familiar, is of the same pattern to which I earlier committed myself; just that is incorporated in the adverb 'sincerely'. But how can I give sense to the idea that it *really is* of such a pattern, that there is an objective similarity in the circumstances in which I apply 'green' which God, for example, could discern?

Wittgenstein wants to insist that actually we can give no sense to this idea. We talk of our using words in the same way and knowing of ourselves at least what this way is, and how we intend to continue. But if there were really any objective pattern which I follow in my use of the word 'green', then it is conceivable that while seeming to myself to be using the word in an essentially consistent way, my employment of it might actually be quite chaotic and irregular. In that case my private assent to whatever it is which I think I mean, and cannot convey to you, by my intention to apply 'green' only to green things fails totally as an effective commitment to a certain way of using 'green'. All that I can effectively intend to do is to apply 'green' only when it *seems* to me that things are relevantly similar; but that is not a commitment to any regularity—it is merely an undertaking to apply 'green' only when I am disposed to apply 'green'.

Asked to concentrate on what we mean by an expression F, we tend, as remarked, to concentrate upon past occasions on which we have applied it, and to visualise others in which we would apply it. We *focus*, it might be said, on our idea of what it is to be F; and if we do not employ the picture of carrying the idea with us as a means of assessment of future circumstances, we come very near to it. The truth is, however, that to describe our successive sincere uses of an expression as based on the intention to preserve a certain pattern, a pattern which for each of us is knowable only to himself, is to misdescribe what using an expression in accordance with one's understanding of it amounts to. We simply find ourselves with a sincere disposition to apply F again in this new case, and that is the whole of the matter. If sense could be given to the idea of an objective similarity between the cases, then, to adapt a simile from the *Investigations*, God could look into our minds, see what similarity it was which we intended to describe by the use of F and so predict the future occasions on which we should use F, provided we used it in accordance with our intentions. But without such an objective notion, even God cannot know what specifically, if we carry out our intentions, we shall do; if my intention to apply F in ϕ-circumstances is to license any specific expectations, there has to be some other account of when circumstances are ϕ than in terms of my disposition to apply F. If there is no other account, then just to know my intentions is to know nothing at all.

What, then, is the correct account of what seemed to be a first/third person asymmetry with respect to knowledge of how someone understands? The asymmetry came out in the temptation to say that if you tell me that you intend to apply 'green' only to green things, you tell me nothing at all about how I may expect your use of 'green' to go; but, from your point of view, there can be something which you intend to say by this, and which you know to be true. Now, however, it appears that this seeming-knowledge is spurious. It is spurious because *whatever* sincere use you make of 'green' in the future will seem to you to be doing what you tried to tell me that you would. And the point, of course, applies equally to expressions whose senses are complex and for which an informative criterion of application could actually be formulated; for in employing such a criterion simple classifications must at some level be involved, and with respect to these we have again no notion of objectivity, save that afforded by the backdrop of communal consensus, with which to underpin our classificatory dispositions.

There is thus no respectable sense in which I can be sure how I shall use some expression, if I use it sincerely, and in which I cannot be so sure about you. Of course, I *cannot* be sure about you; that is, I cannot rule out the possibility that you will later sincerely make what seem to me quite baffling uses of some expression, which force me to suppose that you understand it differently. And that much, at least, I can rule out in my own case; I shall not be baffled by whatever sincere uses I make of some expression in the future. What is wrong is the tendency to inflate this knowledge into assumed knowledge of a pattern in my use of the expression which it is my intention to preserve. I know of no such pattern for I have no way of determining whether it is continued; I merely find myself inclined to reapply the word in this new case.

The effect of all this is that it is wrong to think of our understanding of an expression as something determinate, something which beyond a certain point does not grow and which is then applied. If the verdict of my fellows can alone supply a standard whereby it can meaningfully be enquired whether I am using words in the same way on successive occasions, then I cannot, just by personally *meaning* words in a certain way, bind myself to a certain pattern of use. The requisite notion of what it is to continue the same pattern is undermined.

False pictures, then, stand in the way of a correct grasp of the idioms with which we started: 'The way the formula is meant determines what steps are to be taken', 'This use of the expression accords with the rule', etc. These pictures seduce us into thinking of adopting a certain understanding of a formula, of accepting a rule, in quasi-promissory terms, as undertaking a personal commitment to uphold certain determinate patterns of application. But this supposed 'commitment' cannot finally be made out. For when we are concerned with concepts of sufficient simplicity, to which, in the end, any case reduces, anything any of us sincerely does will seem to him at least to be a fulfilment of his commitment. Wittgenstein's objection to the pictures in question is that they lead us to impose upon our primitive linguistic responses to new phenomena a wholly bogus conception of objectivity.

These ideas are as fundamental to Wittgenstein's later philosophy of mathematics as to his philosophy of mind.[1] We shall return to them repeatedly.[2]

[1] How unfortunate, then—and puzzling—that the Editors chose to omit from the original edition of *RFM* the material on this crucial topic which has since appeared in parts VI and VII of *BGM*; and that they continue to withhold other of Wittgenstein's mss. on the subject. Cf. Hacker and Baker 44, p. 292.

[2] It has been suggested to me that chapter XI might be read at this point, before chapter III.

III

Mathematics as Modifying Concepts

Sources

RFM: ɪ. 8–9, 25–32, 74, 165–7; App. *I*, 16; App. *II*, 2;
ɪɪ. 24, 27, 30–1, 41; ɪɪɪ. 29–30, 45; ɪᴠ. 9, 40; ᴠ. 7
BGM: ᴠɪ. 6–8, 10–14
PI: ɪ. 371–3
PG: ɪɪ. 13, 22–5, 42
Z: 695–8
PB: xɪɪɪ, especially 148–53
BlB: p. 41
LFM: lectures ᴠ, p. 54; ᴠɪ, p. 63 sq.; ᴠɪɪ pp. 73–6; ɪx;
xɪ pp. 103–4; xɪᴠ, pp. 137–9
WWK: pp. 174–5

1.   Time and again throughout *RFM* Wittgenstein entertains the theme of proofs as essentially a source of new concepts, as establishing, or changing, meanings. The idea contrasts sharply with the ordinary view that proof effects an exploration within existing concepts; that the content and correctness of a proof are fixed in terms of concepts whose character is given independently of the proof.

It is this strand in Wittgenstein's thought which is largely responsible for the impression of many commentators that he held some sort of conventionalist view about necessary truths. Shwayder[1] describes such an interpretation of Wittgenstein, as given by Dummett,[2] as 'widely disbelieved'. But a correct account of Wittgenstein's thought here has to accommodate passages like the following:

What, then; does mathematics just twist and turn about within these rules? It forms ever *new* rules, is always building new roads for traffic by extending the network of the old ones. But then doesn't it need a sanction for this? Can it extend the network *arbitrarily*? Well, I could say that a mathematician is always inventing new forms of description, some stimulated by practical needs, others by aesthetic needs, and yet others in a variety of ways. Here imagine a landscape gardener designing paths for the layout of a garden. It may well be that he draws them on a drawing

[1] 71, p. 69.
[2] 23 (2, p. 495).

board merely as ornamental strips without the slightest thought of someone sometime walking on them. The mathematician is an inventor, not a
discoverer. (I. 165 sq)

I am trying to say something like this: even if the proved mathematical
proposition seems to point to a reality outside itself, still it only expresses
acceptance of a new measure [of reality] . . . Why should I not say: in the
proof we have won through to a *decision*? (II. 27)

I go through the proof and say: 'Yes, this is how it *has* to be; I must fix the
use of my language in *this* way'. I want to say that the 'must' corresponds
to a track which I lay down in language. When I said that a proof
introduces a new concept, I meant something like: the proof puts a new
paradigm among the paradigms of the language; like when someone mixes
a special reddish-blue, somehow settles the special mixture of the colours
and gives it a name. But even if we are inclined to regard a proof as such a
new paradigm—what is the exact similarity of the proof to such a
concept-model? One would like to say: the proof changes the grammar of
our language, changes our concepts. It makes new connections and it
creates the concepts of these connections. (It does not establish that they
are there; they do not exist until it makes them.) (II. 30 sq)

The idea that proof creates a new concept might also be roughly put as
follows. A proof is not its foundations plus the rules of inference, but a
new building—although it is an example of such-and-such a style. A proof
is a *new* paradigm. The concept which the proof creates may, for example,
be a new concept of inference, a new concept of correct inferring. . . .
The proof creates a new concept by creating or being a new sign. Or—by
giving the proposition which is its result a new place. (II. 41)

What is the transition that I make from 'it will be like this' to 'it *must* be like
that'? I form a different concept. One involving something that was not
there before. When I say: 'If these derivations are the same, then it *must*
be that . . .', I am making something into a criterion of identity. So I am
recasting my concept of identity. (III. 29)

Can I say: the proof induces us to make a certain decision, namely that of
accepting a particular concept formation? Do not look on a proof as a
procedure which *compels* you, but as one which *guides* you.—And what it
guides is your *conception* of a [particular] situation. (III. 30)

The proposition, 'p is unprovable', has a different sense afterwards—from
before it was proved. If it is proved, then it is the terminal pattern in the
proof of unprovability. If it is unproved, then what to count as a criterion
of its truth is not yet clear, and, we can say, its sense is veiled. (Appendix
I, 16)

The question—

whether '770' occurs in the decimal expansion of π—

I want to say, changes its status when it becomes decidable. For a
connection is then made which formerly *was not there* . . . However queer
it sounds, the further expansion of an irrational number is a further
expansion of mathematics . . . I want to say: it looks as though the ground
for a decision were already there; and it has yet to be invented. (IV. 9)

These passages are a striking illustration of how powerful a prima facie case can be made for supposing that Wittgenstein was the 'full-blooded' conventionalist which Dummett took him to be, for the conventionalist intentions of this sort of talk seem quite explicit. A mathematician does not explore; he invents. He supplies us with new measures of reality, new paradigms; he changes the 'grammar' of our language by creating new connections between concepts, which do not pre-exist the proofs in which they are established. Our acceptance of a new proof is best seen, indeed, as a *decision*—a decision to apply the concepts involved in a new kind of way.

Wittgenstein's talk of 'decision' runs totally counter to what we feel to be the phenomenology of accepting a proof. We want to say that the whole essence of seeing the force of the proof is recognition that at each stage we have no alternative course save to grant the soundness of movement to the next stage. Proof excludes loopholes. One is 'bundled along' by the proof. Of course in one sense we can decide to reject a step, as we can decide, for example, to turn over the chessboard when we find ourselves in a losing position. But that does not constitute winning at chess.

It would seem, however, that the general appropriateness of the notions of decision and invention as descriptive of what goes on when proofs are given and received just follows from Wittgenstein's idea, variously expressed in the above passages, that the effect of proof is to modify concepts. For unless one's understanding, for example, of the conclusion of a proof is left unaltered by the proof, we cannot picture that understanding as enabling us to recognise the relevance of the proof to precisely that statement, and so as requiring us to recognise it as true. Similarly, unless our understanding of the notion of proof, and of the criteria for the correctness of the steps in a proof, are left unaltered by the discovery of a new proof, it cannot be solely in virtue of that understanding that the proof is recognised as sound.

That is: to suppose that the effect of accepting a proof is to change the sense of its conclusion, or to modify the concept of proof—or any other concepts which we should ordinarily regard as involved in our recognition of the proof—seems to leave no room for the orthodox notions that there is a question of fidelity to those concepts as they were before the proof was accepted, that it is right to accept the proof just in case doing so conforms to those concepts, that it is in virtue of one's understanding of those concepts that one is able to recognise the proof. If it is in accordance with the sense of its conclusion that a proof should be accepted, then how can doing so change that sense? Behaviour in conformity with a certain concept cannot change that concept. Conversely, then, if the sense of the conclusion is changed, then nothing in the way in which we understood it before can have *required* us to accept the proof; and similarly for our criteria for the correctness of the steps. To accept the proof is a new step in no way imposed on us by our prior understanding either of the notion of correct proof or of the concepts in the conclusion. Hence the appropriateness of the picture of *decision*.

If we accept this reasoning, then, Wittgenstein's recourse to conventionalist phrases has to be understood in association with his attraction to a view of proof as essentially a means of conceptual innovation and change. If we are to understand the apparent conventionalist strand in *RFM*, we have therefore to supply an interpretation of the idea that proof is essentially an instrument of conceptual change. But first let us make absolutely sure that this idea *imposes* a conventionalist view of proof and necessity in general.

2.   As suggested, to think of a proof as changing various of the concepts which we should ordinarily regard as involved in its 'recognition' would appear simply to squeeze out the standard idea that the acceptability of a proof depends upon its conformity to such concepts, which must therefore survive unaltered our actually accepting the proof. Someone, then, who sympathises with Shwayder's suspicion of a conventionalist interpretation of Wittgenstein has somehow to be able to make out a sense in which Wittgenstein's tendency to regard proofs as changing concepts, creating new conceptual connections, laying down new tracks in language, recasting criteria, etc.—the various forms of expression which I am lumping together as the thesis of 'concept modification'—may after all be reconciled with the standard view that the acceptability of a proof is a matter of fidelity to the sense of its conclusion and the accepted criteria for the soundness of its steps.

Why, then, should accepting a proof not both conform to the sense of its conclusion *and* modify it? Or, in general, why should action in conformity with a certain concept not also change it? Part of the incongruity of this idea is no doubt due to metaphorical association, —as though, for example, a suitcase should be stretched by filling it with objects which fitted into it perfectly. But there are more substantial underlying points. The idea of conceptual change is intelligible par excellence in the sort of case where we can trace the development of a concept through alterations in the conditions under which its application is considered to be justified, where, that is, we can compare the old and the new conditions. A clear example would be a decision to call 'green' a range of shades of colour which we previously regarded as indeterminate in colour, which we previously were undecided whether to call 'green' or 'blue', for example. Such a decision changes the meaning of 'green', our concept of when something is green. But we may intelligibly ascribe such an effect to the decision only because we can compare the old and the new concepts. There will be shades of colour of which we can say that before we changed the meaning of 'green', it would not have been clearly correct so to describe them, but now it is correct. In such a case it is fair to say, as Wittgenstein himself at one point suggests (III. 30), that the old concept lingers on in the background. It lingers on precisely in the sense that the former use of the word is clear. We could, indeed, introduce a new expression to mean what 'green' formerly meant. 'Green' has changed its meaning but the meaning which it formerly had is still there to be meant, so to speak. It is only because it is so that we are able to understand the claim that the meaning of the word has changed.

Now if with Wittgenstein we attempt to maintain that accepting a proof of a statement changes its meaning, then it ought to be possible, after we have accepted the proof, satisfactorily to convey what our understanding of the statement used to be. It ought to be possible to give an account of how certain concepts have been modified. Part of one's natural resistance to Wittgenstein's suggestion is, of course, that this does not seem to be possible. It seems to us that nothing changes as a result of the proof; indeed, that if we could discern an alteration in our concept of, for example, the pattern of application of a particular rule of inference, brought about by the application of it made in the proof, then the proof would fall short of complete cogency precisely at the point where that rule is applied.

The counterintuitive character of Wittgenstein's idea is perhaps best brought out as follows. Talk of propositions, concepts, meanings—where we are concerned not with psychological notions but with information a grasp of which is purportedly constitutive of language-mastery—is naturally held reducible to talk of patterns in the use of expressions; grasp of the sense of an expression is knowing how to use it *correctly*, knowing what sort of application of it successive situations may require. 'Concept modification' thus ought to mean: a change in the pattern of use of an expression constitutive of its correctness, for grasp of which we should therefore expect fresh training to be required. Such is the situation illustrated by the example of 'green'. The reason why we want to say, however, that it is typically *in virtue of* the understanding which we already have of the expressions and rules involved that we accept new proofs is that further explanations are not in general required. Just this simple fact is what underlies our ordinary idea that correct proofs accord with the way the concepts involved are understood already. Proofs are recognised without the need for retraining; indeed, we take the capacity to recognise a good proof as a criterion for grasp of the concepts involved.

Our immediate concern, however, is not with the plausibility of Wittgenstein's idea but with the possibility of *co-adopting* it with the ordinary view that any acceptable proof of a statement S will conform to the sense of S as it is before the proof; that acceptance of the proof is mandatory for anyone who rightly understands S. In fact it is easily seen that Wittgenstein's idea, if implausible on its own, now becomes impossible. The question is: after the sense of S has been modified as the result of its proof, what account can be given of what S used to mean?

Suppose that there were a statement T, which expressed, after the proof, what S used to express before it. And now ask what answer are we to give to the question whether T should be regarded as proved by the proof of S? No answer seems possible. For if the proof is not a proof of T, then it was wrong to accept it as a proof of S; it was wrong to accept it at the stage when S expressed what T now expresses and the proof was a candidate for acceptance or rejection. But if we therefore concede that T must be regarded as proved by S's proof, how can it be regarded as now expressing anything other than what S now expresses? If the effect of accepting the proof as a proof of S was to generate a modification in concepts involved in the sense of S, so that S came to express what it now expresses, how can we now avoid allowing that accepting the proof as a proof of T effects precisely the same modification in *its* sense? It appears, therefore, that there cannot in principle be, even by outright stipulation, *any* way of expressing what was formerly meant by a statement whose meaning changes as a result of its receiving a proof which conforms with its meaning.

What this brings out is that anyone who wishes to combine Wittgenstein's view with the standard idea that the acceptability of a proof is a matter, among other things, of fidelity to the sense of its conclusion, cannot avoid setting up a crucial disanalogy between the idea of concept modification as it applies in the example of 'green' above and the kind of conceptual change which, he wants to say with Wittgenstein, is occasioned by new proofs. On such a view the proof was a proof of S only because this statement expressed what it used to express; it was in virtue of S's former sense that the proof was acceptable, and that the sense itself changed as a result of our accepting the proof. Hence, as we have seen, there cannot be a statement which after S's proof expresses what S used to express before its proof. In contrast we should not necessarily, in changing our use of 'green', affect the meanings of 'verde', 'grün', 'vert', etc. When we fix a new kind of application for 'green', we change its meaning in the same sense in which, by remarrying, I change my wife. My former wife, that is to say, 'lingers on'. This is plausibly the only intelligible model for the idea of conceptual change—at any rate if we think of the meaning of an expression as fixed by its pattern of use. But it will not serve for the conceptual changes generated by new proofs on the present interpretation of the matter. On this interpretation the modification in concepts effected by a proof cannot be seen merely as a change in the pattern of use of certain expressions, the former patterns then becoming vacant, so to speak. The model breaks down because the vacancies essentially cannot be filled.

The same point applies to the other concepts which Wittgenstein suggests undergo modification as a result of the giving of a proof: the concepts which sanction particular steps in the proof, and the concept of proof itself. Suppose that a proof contains a mathematical induction, and that as a result of the proof our concept of such a pattern of argument undergoes alteration. Can we, having accepted the proof, give any satisfactory account of what our concept of this form of reasoning used to be? It would seem that we cannot, for reasons analogous to those sketched above. For suppose that such an account were given, and ask of the concept conveyed by the account whether or not the relevant part of the proof exemplifies it. If not, then our belief that the proof contained a valid induction, as we understood that notion before we had the proof, is unfounded; but if we then say that the proof does exemplify the pattern of inference described in the account, how is it that what is described is not our current concept of mathematical induction? For what we have given is an account of a procedure which is *equivalent* to the account that would have been given before the proof; and it was inherent in the sense of the latter to alter as a result of our accepting the proof.

It thus emerges that our original feeling, that Wittgenstein's view about proof and concept modification entails the appropriateness of the images of invention and decision as applied to our acceptance of proofs, is vindicated. We cannot coherently combine our usual ways of thinking about proof and meaning—that accepting a proof must *accord* with the sense of the conclusion, that our acceptance of a proof of a statement *commits* us to regarding any synonymous statement as proved, indeed that an application of any expression commits us to a like application of any synonymous expression—with the view that the effect of proof is to modify concepts. To attempt to do so is to push the resulting notion of concept modification beyond all possibility of explanation.

3.   Now Wittgenstein, as we know from the previous chapter, will challenge these 'usual ways of thinking' in any case. For him, there is no sense in which our understanding of an expression *commits* us to a particular application of it; in particular, therefore, no sense in which our understanding of the conclusion of a proof and our criteria for the correctness of the steps can commit us to accepting the proof. But it is a reasonable question whether, even if the orthodox ways of thinking are rejected, the principle, that a proof of any statement is eo ipso a proof of any synonymous statement, will not continue to hold good. And if it does still hold good, then will not the argument of the previous section apply just as well to Wittgenstein's view *unencumbered* by orthodoxy?

Surely, the synonymy principle must survive. But its whole character is altered by the rule-following considerations. There can no longer be a legitimate sense in which by understanding expressions in the same way we oblige ourselves to be ready to apply one in any manner in which we are prepared to apply the other. It is rather the other way about. Synonymy is an equivalence relation between expressions obtaining *in virtue of* our readiness to use one in any context in which we are prepared to use the other. Thus we do not in accepting a proof of S commit ourselves to regarding as proved any statement whose meaning is the same as that of S; rather it is a necessary condition for a statement's now having the same meaning as S that we count it also as proved. The principle, that a proof of a statement is a proof of any synonymous statement, thus becomes incapable of the kind of application implicitly made of it in the argument of the preceding section. To accept the proof of S was not a matter of 'fidelity' to the sense of S at that time; nor, therefore, did it involve an open commitment to accepting as proved any statement whose meaning is that of S at that time. What is true is that no statement will count as synonymous with S now which we do not now regard as proved.

This response can only have the restricted objective of pre-empting any attempt to apply the argument of the preceding section to Wittgenstein's view unencumbered by orthodoxy. It makes it not one whit clearer *how* the changes are to be properly described which, in Wittgenstein's view, are occasioned by proofs. (That is a question which will exercise us enough in due course.) Still, it blunts both horns of the dilemma which was posed: if people say that S's proof is not a proof of T, it will now be no criticism of that verdict that, by hypothesis, T means what S used to mean; for accepting S as proved by that construction was nothing which they *owed* to its sense—S, simply, has undergone a modification of sense while T has not. On the other hand, if they allow that T also is proved by S's proof, then—enormously mysterious though it remains how (and indeed what it means to say that) T might nevertheless mean what S used to mean—it is no longer allowable to think of their verdict as, so to speak, *automatically* mutating the sense of T into the present sense of S.

The concept-modification thesis is thus not as immediately vulnerable as it would be if interpreted in the manner most grateful to Shwayder's ear. But a new and formidable difficulty with it now emerges. Consider this passage:

> Now how about this—ought I to say that the same sense can only have *one* proof? Or that when a proof is found, the sense alters?
> Of course some people would oppose this and say: 'Then the proof of a proposition cannot ever be found, for if it has been found, it is no longer a proof of this proposition'. But to say this is so far to say nothing at all. (v. 7)

One can imagine a debate in which an opponent of Wittgenstein developed this complaint into precisely the line of reasoning described in the previous section; and was then rebutted by the counter that his objection was making use of the very type of misconception which it is the point of the discussion of rule-following to explode. And it is the rule-following considerations, it is plausible to suppose, which underly Wittgenstein's off-hand treatment of the prototypical form of the objection quoted. Talk of 'this proposition', where we intend the meaning rather than the symbol, amounts, if it amounts to anything at all, to talk of the way in which the symbol is used, to talk of the kind of use which we regard as its correct use. But that is a notion which lacks objective content in Wittgenstein's view; there is no objectivity to the idea of our continuing to use an expression in the same way. Of course, there is such a thing as *seeming* to continue the same kind of use, or application, and of *seeming* to change the use of an expression. But this distinction is given sense by each of us merely in terms of his disposition to agree about the use of the expression, or to disagree about it, in a range of cases. What does it mean, then, to talk of 'the proposition' which S expressed before its proof? What, indeed, does the whole idea of *the way* in which we used S before the proof amount to?

Whether or not this correctly represents Wittgenstein's response to the v. 7 objection, it is clearly a response he was in a position to make. And it is tantamount to a *rejection* of the very notions—proposition, sense, concept—in terms of which his thesis about the rôle and effect of proof needs to be expressed. So the difficulty is simply this: if there is no such thing as the determinate, objective pattern of use constituting correct employment of a particular expression, then what is it which, if Wittgenstein is right, changes as the result of a proof?

Are we to take seriously the ordinary notion of a *correct* use of an expression, the notion of a use which conforms to the general pattern determinant of its proper use—or are we not? If we do take it seriously, then the view of proofs which Wittgenstein is commending runs up against the everyday fact, already noted, that it is not in general necessary to re-explain concepts in order to secure assent to a proof, that *ceteris paribus* it is a criterion of his understanding the concepts in play in a proof that someone is disposed to assent to it for himself. If, on the other hand, we repudiate with Wittgenstein the notion of a predeterminate pattern of use to which a particular use of an expression may or may not conform, in what, after a proof is accepted, is change supposed to have taken place?

The difficulty is acute. It would have been natural to wonder whether the thesis of concept modification and the ideas concerning following a rule are separate paths in Wittgenstein's thought towards rejection of the standard notion that the acceptability of a new proof is a matter of fidelity to the sense of its conclusion and the criteria for the soundness of its steps. But now it appears that, more than being separate paths, these strands in Wittgenstein's thought are actually prima facie inconsistent. Unless one's understanding of an expression may be thought to have a determinate character, it seems to make no sense to speak of a modification in it; but if it may be allowed to have a determinate character, it would seem that it would at least have to *make sense* that certain linguistic moves made with it should accord with that character. How, then, are we to reconcile Wittgenstein's sloganising about concept modification with his repudiation of the idea that our understanding of expressions reaches ahead of us to so far unconsidered situations in a predeterminate way?

4.   A possible suggestion would be that the idea of conceptual change is meant only as a *picture*, whose rôle is precisely to counterbalance the orthodox, and distorted, picture of mathematics as an exploration of determinate conceptual structures, and of proofs as bringing to light what our concepts commit us to. That proofs modify concepts is a *figure*, of which the substance is to be found exactly in the arguments concerning the non-objectivity of following a rule, of implementing one's understanding. The opposing image, that proof recognition is mechanical, that we are, as it were, prisoners of the way in which we understand expressions, is, on Wittgenstein's view, based upon philosophical errors about meaning. But we know that Wittgenstein thought of philosophical error as in as much need of *therapy* as of corrective argument (see, for example, Appendix II. 4). Can we not see the remarks about proof which are puzzling us as part of such therapy? If so, then, so far from being in tension one with the other, these two elements in Wittgenstein's thought go together; one is a picture for, a therapeutic preparation for adoption of, the other.

The first step in the therapy is to encourage us to think in terms of new images, to try to break the hold which the traditional pictures have on us. We shift from 'discovery' to 'invention', 'compulsion' to 'decision', from concept exploration to concept formation. What we then have, however, are still only pictures, themselves in some measure a distortion. For it is not as if we are to try to interpret our new pictures in a framework in which it still makes sense to speak of the objective implications of certain concepts, or the manner of use of an expression to which a certain understanding of it commits us; a framework in which change in concepts is still to be understood by contrast with the idea of their remaining the same. Such a framework would require us to assimilate accepting proofs to the adoption of new conventions, as with 'green' above, which is just what it so far seems impossible plausibly to do. The character of the therapy is rather more like this: do your utmost to think of proofs in terms of the model which Wittgenstein suggests, and you will be provoked to wonder what is really meant by a 'commitment' in understanding to accepting certain sorts of sequences of inference, and to making certain kinds of application of an expression; and, once so provoked, it will be easier for you to see how little substance there really is to these ideas, which dominate our whole thinking about the nature of pure mathematics.

On this view of the matter there is in the thesis of concept modification no hard philosophical substance other than the claim that errors about the nature of understanding, in particular about the sense in which one may be said to know the character of one's own understanding of an expression, underlie the usual (platonist) view of proof. But the difficulty with such an interpretation is that the thesis in consequence ceases to be of special relevance to mathematics. The same errors, if such they be, are made elsewhere. We think of ourselves as committed to accepting the relevance of certain experiments to certain scientific hypotheses; so it would be as appropriate to urge that the effect of verification of a prediction furnished by a physical theory would be to modify the concepts on which it is based. The idea of being committed by the way we understand words to certain applications of them is, indeed, absolutely general; we might as well allow that *every* agreed assertion occasions a change in certain relevant concepts.

Surely we have lost track of Wittgenstein here. *RFM* leaves an overwhelming impression that Wittgenstein was trying to focus attention upon things *specific* to the nature of proof and necessity, rather than merely bringing to bear upon our preconceptions about those notions totally general theses about the philosophical character of understanding. So it still seems desirable to try to interpret Wittgenstein as saying something like the following. Notwithstanding the facts that we talk of 'recognising' proofs, that we do indeed almost always agree about new proofs without further explanation of concepts—or about propositions whose necessity is not recognised by proof but immediately—and that we regard as a criterion for understanding the concepts involved the capacity to be brought to accept the proof; notwithstanding these facts, such cases of 'recognition' still have more in common with manoeuvres which we should agree are best described as 'decisions', which change our use of language—for example the adoption of new conventions—than with other paradigms of recognition (for example, of a building one has visited before). *Within* the framework of the ordinary distinction between recognition of when a concept applies and stipulation of a new kind of use of it, accepting a proposition as necessary, either directly or on the basis of proof, is more *like* the latter.

In proposing to look for such an interpretation we do not dismiss the 'therapy' view of the concept-modification thesis altogether. The orthodox view of proofs, that they typically uncover so-far unrecognised commitments in our concepts, surely requires a type of objectivity for our judgments of when concepts are correctly applied which the rule-following ideas proscribe. So *some* other view of proof is requisite for anyone inclined to accept those ideas. However, in order fully to understand Wittgenstein's later philosophy of mathematics, we need to explore two further issues: first, *before* we attempt to incorporate the rule-following ideas into our thinking, is there already appreciable a substantial disanalogy between recognition of new proofs and other kinds of recognition? And secondly, *after* we attempt to do so, is there (still) such a disanalogy—is it the case that, with respect to whatever, if anything, remains of the distinction between applying a concept and changing it, accepting a proof is best regarded as a move of the latter kind? But it will be a while before we come to the second question.[1] For the rest of this chapter we are concerned with the first.

5.   Let us consider the first question from an anti-realist point of view. For an anti-realist, sense is given to a statement of any kind only by reference to conditions which we can determine to obtain. The general conception of what it is to understand a statement thus has to be: to know the conditions under which its assertion is justified, rather than the conditions under which it is true. For in the case of a great many statements which we should ordinarily regard as having a clear sense, our idea of their truth-conditions is such that, were they to obtain, we would not be able to recognise that they obtained; but even for such statements there are determinable conditions in which we should consider their assertion justified. It is indeed a condition of our attaching to them a clear sense that this should be so.[2]

In the case of mathematical statements justified assertion takes place only in response to proof. There is hence an obvious disanalogy between knowing the assertability-conditions of an ordinary contingent statement and knowing those of a mathematical statement. As precise a description as can be given of conditions which would justify the assertion of a contingent statement will not in general be a fulfilment of those conditions; but a progressively more precise description of a proof becomes itself a proof. Similarly a *picture*, or simulation, of conditions which would justify the assertion of some contingent statement will not in general itself fulfil those conditions; to give a picture of a proof, on the other hand, is to give a proof.

Our question is whether an anti-realist is free to regard us as *recognising* new proofs. On the face of it he is not free *not* so to regard new proofs, for our understanding of some unresolved mathematical statement is supposed precisely to *consist* in being able, ideally, to recognise a proof or disproof of it, should one be forthcoming.[3] But the disanalogy just noted robs the notion of recognising assertability-conditions in mathematical cases of certain supports present in contingent cases. If somebody claims to know under what circumstances a contingent statement would be justifiably assertable, we can reasonably press him for an exact account of what these would be. This would not, however, be a reasonable demand in the case of a mathematical statement; we do not *have* a precise notion of what, for example, a proof of Fermat's theorem, if there is one, will be like. If we did, we should know how to prove it.

If we are thinking of proofs as given within a non-trivial formal system, the inexactness in our idea of a proof of a particular unresolved

---

[1] We shall return to it in Chapters VI, XVII and XXIII.
[2] —whether this sense is construed in anti-realist or realist terms.
[3] See, for example, 23 (2, p. 500).

statement remains; but not so as to trouble the notion of recognition. For in a reputable formal system we shall have a mechanically decidable notion of proof; and if a machine can be programmed to recognise proofs, how can we, its programmers, be thought of as incapable of doing so? We do not need to know exactly what a proof of a particular statement will be like to know that we shall recognise it if we see it; for the proof will be a sequence of formulae of which the last is the statement in question and of which every element is either an admissible assumption or derived from such by specified rules of inference, correct application of each of which is an effectively decidable matter.

Proofs in formal systems can be effectively recognised because we have a decidable notion of a valid single inference, based upon the main connective in the conclusion, and a decidable notion of statement identity—of where we may start from and where we are trying to get to—based upon purely syntactic criteria. But the interest of such systems ultimately depends upon a capacity to take an interpretation under which their proofs coincide with informal *demonstrations*—with cogent reasoning of which an informal expression could be given. Can an anti-realist argue similarly that it is proper to think of ourselves as in general in a position to recognise such a demonstration if we see it, even though we cannot exactly characterise it in advance?

The obvious strategy would be to try to mirror the account for formal systems: a proof of Fermat's theorem would be an ordered sequence of valid reasoning proceeding from accepted premises to Fermat's theorem. But that is an account which, unelaborated, could be given by someone who had not the faintest idea what Fermat's theorem meant! When do we have a valid single inference? When is our conclusion Fermat's theorem?

In order to answer the first question, an anti-realist might be tempted to avail himself of something like the intuitionists' explanations of the logical constants. We give a general explanation of when a given single inference is valid based upon recognition of the main logical constant in its conclusion and upon understanding of the constituents in the conclusion. Admittedly, to adapt the intuitionist explanations to this purpose would require that something be done about their impredicativity; for the ideas of a proof of a negative, and of a conditional statement are both explained by appeal to an unrestricted quantification over proofs.[1] Thus an appeal is made in explaining the notion of proof for particular classes of statements to a quite unrestricted notion of proof supposedly already grasped. The intuitionist explanations are no better than *characterisations* which someone who already understood the notions in question could recognise as a faithful reflection of that understanding. In the present context, however, we require explanations which would serve to *introduce* someone to the notions of valid inference to a negative or to a conditional statement; for the anti-realist

[1] See Heyting, 46 pp. 98–9; cf. Dummett 30, 'Concluding Philosophical Remarks', p. 389 sq.

has to justify the assumption that *we* already possess such general notions.

But it is the second question which seems to pose the more serious difficulty. If a proof of Fermat's theorem is a chain of valid single inferences culminating in that statement, then we need an account of what it is for a statement to *be* that statement, or at least to be equivalent to it. But now, how are we to identify an expression of Fermat's theorem or something equivalent to it? Obviously, 'by the meanings of its constituent expressions and the way in which they are strung together'. But this reply is irrelevant to our present purpose; for what we have to recognise is that this combination of expressions produces something whose proof-conditions coincide with those of Fermat's theorem; whereas it is precisely these proof-conditions which we are trying to explain.

The position is this. Asked to justify our attribution to ourselves of a capacity to recognise a proof, should one be given, of Fermat's theorem, although we have no exact idea how such a proof will go, the obvious strategy is to assimilate the situation to that within a formal system: to attempt to give an account of when a single inference is valid, depending upon the main logical constant in its conclusion, and then to explain a proof of Fermat's theorem as a chain of such inferences preceding from accepted premises to the theorem as conclusion. But the feasibility of this strategy depends upon the possibility of giving an account of statement-identity which enables us to identify Fermat's theorem as the terminus of such a chain; we have to be able to know when we have reached Fermat's theorem, or the idea of its being the culmination of a sequence of inferences means nothing. An analogy with proof-recognition in a formal system can be constructed only if we can independently identify what we are trying to prove—independently, that is, of knowing the circumstances under which it *is* proved; for it is those circumstances which we are trying to explain.

In a formal system such independent identification is achieved syntactically. Likewise the view that understanding a sentence is knowing the conditions of objective truth of what it expresses gives us some purchase on the identity of the mathematical statement independent of knowing when it would be proved. But on an anti-realist conception of understanding—on any view which makes grasp of the proof-conditions of a mathematical statement a necessary condition of grasp of its sense (see, for example, *RFM* iv. 42)—the strategy of attempting to explain the proof-conditions of a statement by appealing to the idea of a valid chain of inferences culminating in the statement in question, falls foul of the fact that only to someone who is already aware of its proof-conditions is the identity of the desired conclusion intelligible.

If we are equipped to recognise as such a proof of Fermat's theorem, it has to be possible to give some account of this equipment; that is, it must be possible to *explain* under what circumstances the theorem would be proved. But we cannot picture, or simulate, these circumstances, as we could the assertability-conditions of an empirical statement. (Of course, in the latter case, the assertability-conditions might involve reasoning on the basis of observation. But the exact nature of the required reasoning could be characterised in advance.) And not merely cannot we describe these conditions exactly; the obvious model of a satisfactory, indefinite description—that of the account we should give of a proof of some sentence in a formal system—turns out circular from the standpoint of anti-realism.

6.   'I'll know it if I see it'. If it really is proper to speak of us as recognising new proofs, we must have a concept of what the proof would have to be like in a particular case. But what reason has an anti-realist for supposing that we *have* a sufficiently definite such concept, for example, for Fermat's theorem—sufficiently definite, that is, to justify us in describing our response to a proof of it, should we ever get one, as *recognition*?

At this point an anti-realist might question the rôle in our understanding of contingent statements of the relatively precise accounts of their assertability-conditions which, in contrast with mathematical cases, can be given in advance of finding that they obtain. Understanding such a statement cannot be held to *consist* in the capacity to give such a verbal or pictorial account, for the tests of the two abilities are logically independent. Someone who could give such an account might later so use the statement that we felt obliged to conclude that he did not, after all, understand it. Conversely, someone unable to give such an account might go on to show that he understood the statement perfectly well. What does it matter, then, if the first ability is virtually missing in the case of unresolved mathematical statements?

That is right. We cannot *identify* our understanding of the statement with the capacity to explain its conditions of justified assertability. But if it is proper to speak of recognition of proofs, it must be possible to give some sort of account of the features which such-and-such a proof will have to have. (Otherwise, how have *we* acquired the ability to recognise the proof?)

It is natural to describe new proofs as recognised because we feel that they involve, in essentials, no novelties. Everything that happens seems to be an example of a pattern to whose validity we were already committed. A proof of Fermat's theorem will yield some method for recognising of an arbitrary ordered quadruple of integers that it does not satisfy the relevant equation. We cannot circumscribe in advance the techniques that might be used to devise such a method, so a final characterisation cannot be given of what may happen in the proof. Nevertheless, we want to say, even if new techniques are involved, they will be answerable to previously accepted criteria.

But is that not how it would seem if we all just spontaneously inclined to a certain analogy? A proof of Goldbach's conjecture, to switch to a more convenient example, will accomplish for the predicate, 'is the sum of two primes', and the concept of an even number what the proof of the Fundamental Theorem of arithmetic accomplishes for the predicate, 'has a unique prime factorisation', and the concept of a positive integer. But that is an analogy. Is it absolutely clear how we should interpret it? Of course, we don't consciously think in terms of alternative interpretations when a proposed proof is given; but that need show nothing other than that we find overwhelmingly natural the applications involved of certain analogical explanations.

If the last point is endorsed, it begins to look as though we are gravitating back, after all, towards the general considerations about following a rule in new cases. But Wittgenstein intends, I believe, aside from all that, to stress the rôle which analogy plays in such an account of, for example, the proof of Cantor's theorem as could have been given before it was devised, and to contrast such accounts with what might be given, for example, for an effectively decidable statement. Consider for example this passage (Appendix *II*, 2):

> That however is not to say that the question—can the set, R, be ordered in a series?—has a clear sense. For this question means e.g. can one do something with these formations corresponding to the ordering of the cardinal numbers in a series? So if it is asked, can the real numbers be ordered in a series?, the sure answer might be: for the time being I cannot form any precise idea of that. 'But you can order the roots and the algebraic numbers in a series. So you surely understand the expression.' To put it better: I have got certain analogous formations which I call by the common name single 'series'; but so far I haven't any certain bridge from these cases to that of 'all real numbers', nor have I any general method of trying whether such and such a set 'can be ordered in a series'.

Compare this with IV. 12:

> Suppose I were to ask, what is meant by saying the pattern, '777', occurs in this expansion? . . . But don't you really understand what is meant?' But may I not believe I understand it and be wrong? For *how* do I know what it means to say 'the pattern, "777", occurs in this expansion'? Surely by means of examples . . . But these examples do not show me what it is like for *this* pattern to occur in *this* expansion. [My italics]

We arrive then at the following suggestion. Let us think of Wittgenstein here in the rôle of anti-realist. An anti-realist, as we noted, cannot appeal to the sense of, for example, Fermat's theorem in the course of trying to explain the circumstances under which it would be proved; explaining these circumstances *is* explaining its sense. So the construction of an analogy with proof-recognition in a formal system is not open to him. Rather he has to give a general account of what such a proof must accomplish, of what it is to have such a proof. This will presumably proceed along intuitionist lines: the proof will be characterised as an adaptation to the constituents of its conclusion—the predicate through which the quantification is made—of the type of proof germane to its main connective—here the universal quantifier. But such an account is very vague. If our whole concept of the sense of such a statement is thereby explained, that concept is vague too; it takes on precision when concrete circumstances arise in which the statement is deemed proved.

Of course we should want to resist the suggestion that our whole concept of what a proof of Fermat's theorem would be is given by such a general description. Or better, we understand *more* by the general description, 'a general method for demonstrating of any ordered quadruple of integers, x, y, z, n, n other than 1 or 2, that: "$x^n + y^n = z^n$" is false', than it might seem on the surface! This understanding is acquired by doing number theory. Wittgenstein's counter is then that all that doing number theory does is acquaint us with a variety of constructions which are deemed 'analogous'. In any case the intended analogy would appear to be in the *effect* of these proofs rather than in their techniques. A proof of Fermat's theorem, if we get one, may not closely mimic these other constructions; it may rather appeal to a general concept which they illustrate, and then present new methods as relevant to it. Contrast, for example, proving '$(x)-Fx$' by induction, 'F' decidable, and proving it meta-mathematically by proving the first-order undecidability of '$(\exists x)\ Fx$'. Could an intuitionist provide a more exact account of what is shared by such proofs than is given by his general characterisation of proofs of universally quantified statements? In contrast, we can circumscribe the techniques relevant to the solution of some problem of effectively decidable type absolutely exactly.

There is ordinarily felt to be a distinction, of degree perhaps, between *recognising* that the explanation given of a concept requires its application to a new case, and *interpreting* it as doing so. We should ordinarily wish to apply the second description in any case where the explanation was vague. An initial interpretation of Wittgenstein's idea that proofs modify concepts is thus as follows: for all the appearance of rigour and exactness in our forms of mathematical expression, there is a vagueness about our general notions of proof and disproof which, from an anti-realist viewpoint, must be regarded as entering into the way in which we understand mathematical statements before they are proved; this vagueness we disguise from ourselves, partly by appealing to erroneous conceptions of meaning, for example, the platonist conception, and partly by falling victim to the misconception, complained of in III. 6, and IV. 25, and countered throughout the *Investigations*, that a grammatically correct combination of familiar expressions, even into something which looks obviously part of a language-game which we already play, wholly takes care of its application.[1]

[1] See also *PG* II. 43; *BlB*, p. 10.

So interpreted, Wittgenstein's point is independent of any scepticism concerning the distinction between recognising that a concept applies, and interpreting it as doing so, which might be suggested by the discussion of following a rule. The suggestion is rather to be understood in terms of that distinction: there is a larger rôle for interpretation to play in connecting the grammar of a new mathematical sentence with its application than is usually the case elsewhere. Fully to make this thesis out would require a much more detailed treatment of many more examples than Wittgenstein gives. We shall return to the question in Chapter VIII. What is clear from our discussion, however, is that someone who, like Dummett, sympathises with the intuitionists' criticisms of classical mathematics but wishes to retain the classical attitude to proof—that a correct proof is *binding*, that it is something whose validity we may regard ourselves as recognising, etc.—cannot rest content with an unexamined notion of knowledge of the assertability- (i.e. proof-) conditions of unresolved mathematical statements as central to our understanding of them.

IV

The Analogies with Measuring

Sources

RFM: I. 5, 8–9, 93–4, 116, 118, 139, 154–55, 164; II. 21, 74–5; v. 1–2,12, 20

LFM: lectures VIII, p. 83; XII, pp. 117–18; XXI, pp. 200–1; xxx, p.287

PG: I. 133

On the possibility of alternative modes of calculation and inference:

RFM: I. 136, 142–152; II. 78; v. 8, 11–12

PI: II. xii

PG: II.18

OC: 375

LFM: lecture XXI, pp. 202–5

1.   At several places in *RFM* Wittgenstein suggests analogies between the rôles of logic and mathematics and various aspects of measuring. The first example occurs as early as I. 5. The question is raised, how should we get into conflict with truth if we made inferences different to those which we ordinarily allow? Wittgenstein assimilates the question to that of how our results would conflict with truth if our foot rules were made of very soft rubber instead of wood or steel. He feeds his opponent the obvious reply that we should not then usually be able to determine how long things were. But this, according to Wittgenstein, comes to no more than that we could not be sure of getting the readings which we now get . . .

> so if you had measured the table with the elastic rulers and said it measured 5 feet by our usual way of measuring, you would be wrong; but if you said that it measured 5 feet by your way of measuring, that is correct.

His objector protests that soft-ruler measurement is not measurement at all. Wittgenstein replies that

> it is similar to our measuring and capable, in certain circumstances, of fulfilling 'practical purposes'. (A shopkeeper might use it to treat different customers differently.)

Here, then, Wittgenstein apparently refuses any content to the concept of the *correct* measurement of the table other than the measurement which we get with our rigid rulers; the idea of correct measurement is relativised to the means used. The intended analogy would thus appear to be this: that there is, correspondingly, no content to the idea of something's *really* being a consequence of some set of statements over and above its following from them by *our* procedures of inference.

Prima facie, this seems a wild suggestion. If one statement is a logical consequence of others, its truth is guaranteed by theirs. So, we should want to say, the possibility of a 'conflict with truth', were we to use quite different principles of inference, resides in our generally possessing *other* criteria for the truth of a statement than that it is a consequence of other statements about whose truth we are independently satisfied. This is not always so. In pure mathematics recognising a statement as a consequence of others may be the only means which we have of recognising it as true. But with many contingent statements the situation is different: in such cases statements which it is possible to know by recognising them as consequences of known statements will also be knowable directly, for example by observation, by someone who does not know the premisses. To infer in accordance with rules dreamed up ad hoc would be to invite in such cases the possibility of a conflict of criteria of truth; a conclusion which 'followed' from true premisses might be false by observational criteria.

One is, in any case, uneasy with Wittgenstein's claims on behalf of soft-ruler measurement. The case which he presents for describing the ritual with elastic 'rulers' as measurement' is that it is 'similar' to what we do and capable of fulfilling certain practical purposes. But in what does the similarity reside, save in that someone goes through the normal motions of measuring—only with a piece of elastic? It is a feature of the concept of measuring that an accurately measured object will yield distinct readings at distinct times only if *it* changes; so much is implicit in the notion that measuring is to ascertain a property of the object measured. Now, the soft-ruler method is going to produce a high degree of variability in its results. Are these variations to be attributed to inaccuracy in the measurements, or to changes in the object measured? If the first, then it is conceded that it is extremely difficult to measure accurately with elastic rulers; but then, contrary to the intended point of the analogy, the notion of correct measurement is implicitly dissociated from the means used. But if changes are postulated in the length of the object measured, then the question arises, what rôle is to be played in the concept of length by our rough, directly *observational* assessments of sameness and difference in length—the assessments which measuring is supposed not merely to supplant but to *refine*? For the plain fact is that it is the rulers which *seem* to fluctuate in length, and not the objects to be measured.

The point is that measuring is intelligible only as a more exact means of determining changes or differences which, if sufficiently gross, may be recognised without measuring by direct observation. Any procedure which may be coherently regarded as effecting such a refinement has therefore to corroborate a high proportion of our inexact, observational assessments of length—the applications which we are able confidently to make of relations like 'is much shorter than', 'is about the same height as', etc., just by observing objects; where we do not get such corroboration, we start talking in terms of illusion. If measuring is to be conceived as *improving* on observational assessments of relative size, we must be allowed to have the capacity to know very often the sort of comparative readings that measurement of a pair of objects will yield—the general range of difference that may be expected. Measuring has to be understood not as making possible the application of the concept of length, but as enabling us to be relatively precise where before we could make only vague assessments, although—within their limitations—*correct* assessments.

For people who conducted all their measurements with highly elastic rulers this would not be so. Objects which seemed to be more or less the same length would, as often as not, prove when measured to be substantially different; and it would not generally be possible to anticipate the *range* of difference anything like as accurately as we can. Conversely, objects which seemed to be of quite different length would, when measured, prove as often as not to be quite similar in length or to differ the other way round, so to speak. A world in which soft-ruler measurement supplied the criteria of sameness, difference and change in respect of some dimension would be a world of observational illusion; a world in which much less could be known—many fewer sound comparisons could be made—on the basis of unaided observation.

In such a world what would be the use of knowing the result of measuring an object at some time? The information would not license exact expectations about the object's physical potentialities—to fit in such-and-such a place, cover such-and-such a surface, etc.—for lengths would *fluctuate* so; nor would it be possible in general to predict by means of such information the relative sizes which objects would *seem* to have just on the basis of observation, even if, by measurement, no changes in them had taken place. 'Length' in such a world would have lost contact with its roots in simple observation and practical needs.

It is, moreover, thoroughly unclear what becomes of the notion of a *unit* of length in the soft-ruler community. What do the people say about inconsistency in the results of reciprocal application of their 'rulers'? One ruler, when measured by another, will almost always give a proportionately inconsistent reading when compared with the reading obtained by measuring the other ruler with it; for example, one yard-stick may measure half the length of another which, measured by the first, proves one-third as long again. What do the people say? Are the yardsticks fluctuating in length? If so, there is no point in using them. Or is it that, however much care is taken, it is virtually impossible to measure accurately? If so, there is no point in trying. How is a unit of length to be established if objects have to be conceived of as randomly varying in length, or if 'correct measurement' is given no operational content?

It might be replied that Wittgenstein's claim is only that there could be a concept of length whose criteria of application were incorporated in soft-ruler methods; not that it would be desirable to have such a concept. But what is the real analogy between what is done by people who employ such a concept and what we do, in virtue of which both groups may be described as employing concepts of *length*? The concepts are not grounded in observation in the same way, and they do not admit of the same practical applications. Wittgenstein's own example of the 'practical purposes' that might be served by soft-ruler measurement is wholly inept. From whose point of view is the shop-keeper treating different customers differently? Not, presumably, his own unless he realises that his soft-ruler can yield variable results in relation to a *set amount*, for example, of fabric. But then, for the reason noted, he isn't measuring; for correct measurement gives different results only when the objects measured differ in length. But the 'practical purpose' is the shop-keeper's; he wants to favour some customers and short-sell others, and for this purpose he requires some concept of sameness of length other than is determined by soft-ruler measurement. That is, either the soft-ruler method does not provide the criterion for sameness in length or, if it does, it is not available as a way of treating different customers differently.

In summary: Wittgenstein appears to want to drive us towards the conclusion that we have no more general notion of logical consequence than is determined by the rules of inference which we actually happen to employ. This conclusion would, of course, require more than an analogy to support it; but this first analogy seems in any case not to have been properly made out. For we *do* have a more general concept of what it is to determine length than is fixed by the procedures which we actually use. It is of the essence of measurement to effect a refinement of our rough and ready visual (or tactual) assessments and to supply information adapted to certain practical requirements; but to know that a pair of objects coincide in length is, when this is determined by soft-rulers, to know nothing about the relative sizes which they will *seem* to have, or whether they will, for example, fit in the same space on a shelf.

2.   Consider a second measuring analogy which Wittgenstein uses to recommend a shift in the way we think about logic and mathematics. In I. 155, we read:

. . . the reason why logical inferences are not brought into question is not that they 'certainly correspond to the truth' or something of the sort. Rather it is just that this is what is called 'thinking', 'speaking', 'inferring', 'arguing'. There is not any question at all here of some correspondence between what is said and reality. Logic is *antecedent* to any such correspondence; in the same sense, that is, as that in which the establishment of a method of measurement is antecedent to the correctness or incorrectness of a statement of length.

The thought here would seem to be the same as that expressed in a remark which Moore quotes from Wittgenstein's lectures of 1932:[1]

Rules of deduction are analogous to the fixing of a unit of length. '3 + 3 = 6' is a rule as to the way in which we are going to talk. It is a preparation for description, just as fixing a unit of length is a preparation for measuring.

Now what does this analogy come to? The commended conclusion seems to be the same as before: there is no ulterior concept of correct inference lurking behind our actual procedures of inference and to which they are answerable. But now Wittgenstein seems to be presenting the suggestion in a slightly different way to that involved in the soft-ruler example. In that example the point seemed to be that there was no kind of correct inference transcendent of what we count as correct inference, no 'ultra-physics' to which logic and mathematics have to answer. There is no sense in this idea much as, in Wittgenstein's view, there is no sense in the idea of the 'real' length of a thing, dissociated from the methods of measurement which *we* employ. The concept of length is fixed by the way in which we measure; it does not stand apart from particular techniques of measurement, so that the question of their adequacy may arise. We had reason to question this supposition about measurement. But now the point would appear to be not merely that there is no sense in the *objective* correctness or incorrectness of certain rules of inference, but that there is no sense in describing them as 'correct' at all. '3 + 3 = 6' can no more be described as 'correct' than the fixing of a unit of length can be described as 'correct'. It is simply a precondition of the activity of giving certain sorts of description of the world.

In order to assess the analogy, imagine a tribal community who possess the notions, 'just as many as', as fixed by 1–1 correspondence, and 'just as long as', as fixed by congruence when objects are laid alongside each other, but who have introduced neither a system of numerals nor a unit of length. Suppose that a member of the tribe requires to measure a rectangular wall and count the number of tiles needed to cover it. He proceeeds as follows. He first anchors the end of a piece of cord to one corner of the wall and runs it tautly along the longer side, snipping it off exactly at the point adjacent to the other corner. He

[1] 63, p. 279.

then runs the same cord along the shorter, vertical edge and marks it at the point adjacent to the third corner. Using this cord, he can now determine whether the opposite wall is the same size, where to cut a piece of skirting board to fit along the wall, etc. But he cannot explain to another member of the tribe who has not seen the wall what its dimensions are without reference to the piece of cord.

After he has completed the tiling of the wall, our man now requires to order tiles from a fellow tribesman to cover the opposite wall which, let us suppose, he has determined with the piece of cord to be of the same dimensions. What he does is to chalk-mark each tile once only while simultaneously writing a stroke on a piece of paper for each chalk-mark that he makes. Now he can order exactly as many new tiles as he needs just by referring to the piece of paper. But he won't be able to explain to the tile-maker how many tiles he wants without reference to the piece of paper.

There are obvious and parallel ways to overcome these shortcomings. First, our man cuts a new piece of cord and folds it, marking the half-way point. Then he repeats the process until the whole cord is divided into a large number of equal, marked segments. He lays the cord out and cuts others of the same length which he marks in the same places and distributes to other members of the tribe. Now they no longer have to cut pieces of cord to show to people in order to communicate propositions about length; they can instead *state* that the length of *x* is so many segments of cord—though these statements still have to involve reference to suitable marked pieces of paper in terms of which *how many* is to be understood. So, as a second refinement, the builder introduces a number of distinct symbols and an operation upon them which generates out of them a potentially infinite series of unique symbols in a determinate order. He then explains that, instead of handing round pieces of paper, a group of objects to be numbered is to be 1–1 correlated with the symbols in the series, starting at the beginning, and that the last symbol needed is to stand for the number of objects in the group.

Against the background, then, of use of the concepts of sameness of number and sameness of length, the clear parallel, if we are concerned with the preconditions of certain sorts of description, is not between the introduction of a unit of length and agreement about any particular arithmetical propositions, but with establishing the use of a system of *numerals*. There is, for the very simplest of practical purposes, no reason why a unit of length should be established at all; so long as a concept of sameness of length, given in terms of laying things alongside one another, is in currency, a community can always resort to ad hoc paradigms—cutting up bits of cord or whatever—in order to tell each other how long things are. Similarly, so long as the notion of sameness of number, as given in terms of 1–1 correspondence, is established, the community can always resort to ad hoc paradigms—marking bits of paper or whatever—in order to tell each other how many things satisfying some condition there are. What the introduction of a system of numerals and a unit of length achieves is *convenience*; they give us general paradigms which may be adapted to any simple, practical situation of the type just illustrated.

Wittgenstein's second analogy thus seems doubly wrong. There is no clear point of similarity between the rôle of a unit of length and the rôle of arithmetical equations, and it is in any case an error to suppose that nothing amounting to measurement—to the determination and communication of judgments of length—can take place in advance of fixing a unit of length. We are thus so far no nearer understanding in what sense logic and mathematics may be 'antecedent' to truth and falsity.

3.   There is, however, an obvious refinement of the analogy just considered. If fixing a unit of length is properly compared to establishing a system of numerals, the propositions of elementary arithmetic which serve to express relations between the senses of the numerals ought presumably to be compared with propositions expressing relations between the senses of expressions denoting different units of length. Just this is, of course, another of Wittgenstein's favourite measuring analogies. Consider, I. 9:

> What we call 'logical inference' is a transformation of our expression. For example, the translation of one measure into another. One edge of a ruler is marked in inches, the other in centimetres. I measure the table in inches and go over to centimetres *on the ruler*.—And of course there is such a thing as right and wrong in passing from one measure to the other; but what is the 'reality' that 'right' accords with here? Presumably a *convention*, or a *use*, and perhaps our practical requirements.

The suggestion seems to be the same as before: that there is nothing 'exterior' for a transformation of inches into centimetres to correspond to, that correct transformation is transformation in accordance with *our convention*—nothing further. And correct inference, presumably, is to be viewed correspondingly as a matter of inferring in accordance with *our* rules of inference.

But, once again, Wittgenstein seems to have given a highly implausible account of the element in measurement which is supposed to suggest an improved way of looking at the propositions of logic and mathematics. For, we want to protest, surely it is in no sense a mere *convention* that 1 inch = 2.54 centimetres. On the contrary, what the relation between the units is must have been, in the first instance, a matter for *discovery*; it must have been a question for investigation how the imperial and metric systems corresponded. How can it be a conventional issue whether our practices correspond in such-and-such a way to those of the French? What are matters of convention are the units of length which a culture adopts, the way it chooses to describe them, etc.; but it is surely no matter of convention what ratio between the readings is obtained if an object is simultaneoulsy measured accurately in inches and in centimetres. The 'reality' to which the rule of conversion is answerable is the reality of the results of accurate measurement in both systems; as, indeed, we want to say, the reality to which rules of inference in general are answerable is that of the truth-values of the propositions which they allow us to link as premisses and conclusions.

So the net result of our discussion of the three analogies is that Wittgenstein has failed to draw attention to aspects of our orthodox way of regarding measurement which might serve to illuminate his own, unorthodox attitude towards logic and mathematics. Of course, it is in any case clear that Wittgenstein's view of logic and mathematics is not properly *argued for* by analogy. The most the analogies may do is to cast some light upon the nature of the view, whose real substance can presumably be grasped without analogy—indeed, it should be apparent from considerations adduced in its support. Yet the distinctive difficulty in reading these excerpts from Wittgenstein's notes is precisely to identify confidently *what* motivates him to advance some of the seemingly more extravagant ideas. The potential value of the analogies to the interpreter is to suggest what the character of the motivation may be. Only when this is clear will we know how to criticise these ideas relevantly. It is therefore disappointing that the parallels which Wittgenstein attempts to draw with measuring seem so wayward. The point of them is clear enough: it is to attack the notion that propositions of logic and mathematics state objective truths. Partly, the attack is that these 'propositions' are not properly seen as *capable* of truth and falsity; we do not say of a ruler, or of a unit of length, that it is true or false. But, as I argued, the parallel between rulers, or units of length, and arithmetical propositions, is ill-taken; they do not occupy similar respective places in practical measurement and assessment of number. However, the attack is also upon the objectivity of whatever 'truth' the propositions of logic and mathematics might be thought to possess; and here Wittgenstein's chosen analogues—rules of conversion between distinct metric units—while indeed in some ways similar to simple arithmetical equations, would not in orthodoxy be viewed in the way which he commends.

4.   Still, let us persevere. Wittgenstein does not always attempt specific analogies with aspects of measuring. Often he is content to talk vaguely of logic and mathematics as 'measures' of reality (e.g. ii. 21, 75). So perhaps it is a mistake to press the specific analogies more closely; perhaps it is rather that he intends something like the following general parallel:

Rulers, units of length, rules of conversion, all figure in the institutional activity of measuring, and measuring provides the criterion for the correctness of ascriptions of length. The whole point of the activity is to determine the application of certain concepts in contingent, practical contexts. Measuring is a 'game' which we play, and the various aspects of it to which Wittgenstein assimilates our rules of logical and mathematical inference are just elements in the rules, concepts and equipment of the game. It is in the character of the whole game that the nature of our concept of length is determined; and it thus makes no sense to enquire of the concept whether it shapes up to reality in some way, whether it is as it ought to be, whether the procedures and rules of the game really do provide methods whereby lengths may be ascertained. The game as a whole does not stand in need of justification —indeed it makes no sense to ask for one—and neither do the particular elements in it. If we change them, that will be a change in the game, a change in our idea of what it is to 'measure', and so in our concept of length.

In like manner abaci, systems of numerals, and simple arithmetic are all part of the institutional activity of assessing number, of determining how many. Their rôle is as elements in the rules, equipment and concepts of the 'game' of ascertaining the number of things in a group in ordinary, practical contexts. The whole game provides the criteria of correctness of ascriptions of number. In the procedures which make it up our concept of what it is for an ascription of number to be correct is encapsulated; thus this concept has no ulterior standing in terms of which these procedures might be evaluated. If we changed them, that would be to change the concept. When we think of arithmetic as dealing with a special kind of object, as giving its 'natural history', we have lost sight of the fact that what we are now regarding as propositions feature originally simply as rules of description, simply as elements in the network of procedures which we call 'determining how many'—and that these procedures together go to make up our concept of the number of things in a (finite) group. So arithmetical 'propositions', rather than being themselves descriptive of anything, enter into the fixing of criteria for the application of concepts in terms of which description can then proceed. In this sense they are 'antecedent' to description, and so to truth.

It is a natural suggestion that it is in some such general terms that we should understand Wittgenstein's talk of logic and of mathematics as 'measures' of reality, as antecedent to truth and falsity, and that the message of the more specific analogies with measuring is the same. Wittgenstein wants us to see number theory, for example, as essentially just an apparatus geared to the production of criteria for applying numerical concepts in contingent, practical contexts. The repeated emphasis in *RFM* upon the application of mathematical concepts as a source of their content is, prima facie, especially easy to understand on such an interpretation. Consider, for example, IV. 2:

It is the use outside mathematics and so the *meaning* of the signs that makes the sign-game into mathematics;

and IV. 41:

Concepts which occur in 'necessary' propositions must also occur and have a meaning in non-necessary ones.

Naturally, it is not clear how to extend such a suggestion to those parts of mathematics which were not developed with a view to any particular kind of application outside mathematics, or where mathematics seems essentially concerned with its own objects, for example transfinite set-theory. But it is just towards such branches of mathematics that Wittgenstein displays suspicion—just in those areas that he is provoked to speak of 'alchemy', or 'puffed-up' proofs, etc. And there is also a tendency, especially evident in the Appendix on the Diagonal Argument, to try to interpret even these parts of mathematics as, in essentials, no more than the introduction of new techniques and concepts 'which can now be applied to all sorts of other things' (App. *II*, 15). The governing idea of the generalised interpretation of Wittgenstein's measuring analogies draws these tendencies together. The rôle of mathematical statements, and of necessary statements in general, is on Wittgenstein's view 'grammatical'; it is to establish associations between certain forms of description. Clearly, therefore, the *content* of these forms of description must initially originate elsewhere. The rule of conversion between inches and centimetres, for example, may be used to give someone a new form in which to express results of measurement; but this is possible only if he already understands the application of one of the forms of description, only if he already understands the meaning of, say, ascriptions of length in inches.

The suggestion that we regard arithmetic simply as a component in the general language-game of assessing cardinal number, as an ever-extensible compendium of rules of description, is meant to stifle our inclination to think of it as answerable to, and indeed uniquely true of, certain special aspects of reality. Arithmetic enters into the *determination* of our concept of number; naturally, then, in a vacuous sense it is uniquely correct for that concept. But Wittgenstein appears to want to suggest that this does not preclude the possibility of other concepts, in some respects 'akin' to ours—perhaps in the practical purposes for which they are applied—but associated with significantly different techniques of application. Such, it appears, is the point of the examples of measuring with soft rulers, of pricing a pile of wood in accordance with the area of floor which it covers, etc. In these examples different concepts, for instance, of length and of quantity are involved, concepts determined by criteria of application different to ours. But we may not say, according to Wittgenstein, that these concepts incorporate error, that they somehow fail to accord with reality. There is error in the procedures of people who use these techniques only if they are thought of as determining the application of *our* concepts.

Wittgenstein's conventionalism would thus appear at this point to have at least three quite different strands. There is, first, his repudiation of the idea that our understanding of an expression commits us to certain sorts of application of it, for example to recognising as necessarily true certain statements containing it. There is, secondly, the tendency, of which the measuring analogies as presently interpreted are an illustration, to tie concepts tightly down to the techniques and procedures which we use to apply them; to count a group correctly is just to count it in accordance with our criteria of correct counting, to calculate correctly is to calculate in accordance with our rules of calculation, etc. And he gravitates, thirdly, towards suggesting the possibility of procedures *alternative* to ours; he is hospitable towards the idea that people might measure, count, infer and calculate differently. What I want to suggest now is that the second and third strands are in tension and that both are objectionable for related reasons.

5.   What would it be for a soft-ruler society to employ a concept of length essentially the same as ours? It is a question, presumably, of how they respond to what we regard as the discordant results which their procedures will give them. A book, though it continues to look much the same and to fit in the same slot on the book shelf, etc., will, when measured on successive occasions with a soft-ruler, turn out to be of highly variable thickness. A pile of wood, though containing the same number of logs as before—which in turn seem to be more or less the same size—will, when assessed by the criterion of the area which it covers when spilled out of a truck, prove to be of highly variable quantity. The clearest kind of case in which it would seem right to say that something like our concepts of length and quantity are nevertheless being employed here would be if, confronted by this variability of results, the people in question simply abandon the practice of measuring with elastic rulers, or of assessing quantity of wood by area, and move instead in the direction of rigid rulers and the assessment of quantity, say, by weight. They are employing something clearly similar to our concepts when they can be brought to accept the superiority of our techniques.

But procedures of measurement, calculation, etc., alternative to ours were not to be alternative in this uninteresting sense. And it is far from clear what to say if these people *cannot* be brought to accept that our ways are better. Partly the difficulties here are masked by Wittgenstein's tendency to present hypothetical dialogues with such people; whereas their seemingly ersatz notions of length, quantity, counting, or whatever, might in practice make us doubt whether we had correctly translated their language. But let us continue to entertain the fiction that it is somehow known that we have the best translation of their language possible, and that when the variability of their results is pointed out to them, they say things like, 'Yes, indeed; on this planet length is extraordinarily variable'; or, after we rearranged a pile of wood in order to make plain to them the absurdity of their procedure, 'Yes; now it is a lot of wood and costs more' (I. 149). Then what is not clear is what justifies the description of such people as employing different concepts of *length,* or of *quantity.* Where are the analogies anchored? What are the points of similarity?

It might be thought that the trouble is that we have not been given enough detail; and it is true that Wittgenstein presents the examples in a very sketchy way. But actually, it is important to be clear what sort of detail we want. Surely, if it is proper to speak of the wood-sellers as employing a different concept of quantity, then it has to be correct to attribute to them the belief that when the wood covers a greater area, there is *more wood there.* Now, obviously, just the fact that they pay more when the covered area is greater does not substantiate the supposition that they hold such a belief. That could be explained in all sorts of other ways. (They may have introduced the practice as a kind of lottery, for example, for paying purchase tax on wood; everyone pays the tax, but how much depends on the luck of the way the logs fall out.) So nothing of that sort is to be the explanation. But then, what sort of thing would make us want to describe them as really believing that the greater the covered area, the greater the quantity of wood?

The 'thin and unconvincing' quality of these examples, complained of by Dummett,[1] is not just a matter of lack of detail. It is a matter of Wittgenstein's having paid insufficient theoretical attention, here if not elsewhere, to the question of what kind of consideration is a legitimate *ground* for the claim that certain concepts are analogous. If the attraction for us of the idea that our mode of calculation is the only correct way of calculating is to be weakened, then we need examples of procedures alternative to our own which it is still proper to describe as 'calculation'. Similarly, we need examples of alternative ways of determining 'how much wood', and of measuring. But Wittgenstein's soft-ruler people are merely *stipulated* as doing a kind of measuring; and the wood-sellers likewise are stipulated as determining quantity by area—they just 'state' that they are. What makes it the case that they are still, in some sense, measuring or assessing quantity?

It is not that Wittgenstein has not spelled out sufficient similarities between our practices and those of the people in his examples. Rather he would appear to be unclear about the kind of similarity which it is in point to suggest. The cited examples involve strong behavioural similarities with things which we do: the laying of things against things in measurement, the exchange of coins in proportion to some parameter, and so on. But it is not in this kind of thing that the most

[1] 23 (2, p. 498).

important similarities reside. The best kind of reason for interpreting a tribe as *measuring* when they go through such-and-such movements would not be an analogy between the kind of thing which they do and the kind of movements which we go through when we measure, but an analogy in the kind of *application* which they make of the information thereby obtained—the purposes to which they put it. Suppose, for example, that members of the tribe go through a routine of holding up one hand, palm outstretched, and backing away from an object with slow deliberate steps until the palms of their hands blot it out from view. Are they measuring? Not, for example, if these movements are done in the context of a dance, in which they might symbolise the magical activity of shrinking an object and then obliterating it. But if the routine is carried out in the context of construction work, and if the demand for materials in a particular kind of situation is seen to vary depending on the outcome of the procedure—how many paces it takes to blot the object out—then a social anthropologist would be happy to report the tribe's possession of a crude method of measurement.

This point, of course, is of a genre with which Wittgenstein is perfectly familiar—which, indeed, we may owe to him. But then the tension here is just the more puzzling. The overriding purpose of the analogies with measurement, as presently interpreted, is to stress the rôle of, for example, arithmetic in determining the *meaning* of numerical expressions, in determining their criteria of application. Arithmetic supplies concepts in terms of which we carry out descriptions; it does not itself describe anything. The intended effect of the analogies is to short-circuit the question of correspondence between arithmetic, or logic and mathematics generally, and some sort of ultra-general aspects of reality, to nip in the bud the conception of logic and mathematics as a kind of 'ultra-physics'. There is no question whether our techniques and principles of numerical description are correct; rather it is in terms of them that our notion of correct assessment of number is determined. 'Calculating', 'inferring', 'measuring' are, for us, determined by the methods which we use; there is no residue in these concepts in terms of which the adequacy of our methods might be questioned. But if this is a correct account of the matter, how is it that we were able to understand and, hypothetically, to answer the question whether the members of the tribe just described were measuring?

Of course, it is analytic in philosophers' jargon that the *criteria* of application of a concept enter into the determination of its content. And in this resides the temptation to say that our notion of what it is correctly to assess length just is fixed by our entire paraphernalia of measuring —instruments, units and rules of conversion. But if we wish to have a connection between criteria and meaning, we need to be circumspect about what are to be regarded as criteria; and the most important criteria for whether or not a tribe is assessing lengths by some procedure

have nothing to do with its being a procedure—or 'akin to' a procedure—which we also use. Rather, to over-simplify, we identify
measuring by seeing what is done for, by seeing what (we may tentatively identify as) its results are applied for. The same is true of
calculating.

In several places Wittgenstein seems to give expression to this. 1. 4:

> 'Is that supposed to mean that it is equally correct whichever way a person
> counts and that anyone can count as he pleases?' We should presumably
> not call it 'counting' if everyone said the numbers one after the other
> *anyhow*. But of course it is not simply a question of a name. What we call
> 'counting' is an important part of our life's activities.

And 1. 116:

> . . . thinking and inferring, like counting, is of course bounded for us not
> by an arbitrary definition but by natural limits corresponding to the body
> of what can be called the rôle of thinking and inferring in our life.

How are we to reconcile this admission[1] with the suggestion that the
procedures which we use, the instruments, techniques, units and rules
of conversion, are exhaustive of our understanding of 'measurement',
'length', etc? And if there is nothing further to our understanding of
these concepts, how can it be coherent to suggest that other people
might measure *differently*, might calculate in accordance with other
rules, etc.?

Wittgenstein wants us to regard our ways of inferring, calculating,
etc., as techniques in which criteria for the application of concepts are
embodied, so that there is no legitimate question of the *soundness* of
these techniques; the concepts with which they deal have no independent standing to which the techniques must answer. No one would
dispute this suggestion for concepts of the particularity of, say, 'inch'.
But it is wrong for the concepts which matter: 'number', 'length',
'measurement', 'calculation', 'logical consequence'. We understand
these concepts in terms exceeding our own apparatus of calculating and
counting, measuring, and inferring exactly in the respect whose importance Wittgenstein so often stresses: their *application*. If some
habitual practice of a tribe is properly described as a kind of calculation,
measurement or inference, it will be because of the kind of task to which
they put the practice: is it involved, for example, in distribution of
goods, in building work, in problem-solving? It is primarily the application of the technique, and only in a subsidiary way its behavioural
character, which determines whether it is a kind of measurement, etc.

It is this which gives us an independent lever on the concepts of
measuring, etc.—independent, that is, of our actual techniques and
rules—in terms of which we feel our procedures to be superior to those

[1] Cf. *RFM* v. 25–6 and *BGM* vii. 24; contrast *PI* ii. xi (p. 225).

of the characters in Wittgenstein's examples. Measurement with soft rulers will be useless if the results are applied for the kinds of purposes for which we measure; but if they are not, it is seriously unclear what good grounds there could be for saying that these people who, talking apparent English, solemnly lay floppy rulers alongside things and seem to record readings are doing anything that may informatively be described as 'measuring'. Similarly, we could not apply the rule of conversion, '1 inch = 4.96 centimetres', in the same practical contexts in which such rules are generally applied—for example shopping for clothes abroad—and expect to get a decent fit. But if the rule is not applied for such purposes, what makes it into a rule of conversion between distinct metric units?

These points ought to be uncontentious. It thus remains difficult to see that Wittgenstein is not seriously confused about the rôle which his examples of procedures supposedly alternative but similar to our own can play. In the final analysis, the only kind of similarity that counts—the *master* criterion, so to speak—is similarity of application, similarity in what the procedures are used *for*. The actions of divining the Tarot pack, consulting astrological charts, looking at the medical history of the family, and examining the patient's blood cholesterol level have nothing in common in virtue of which they may all be described as attempts to assess his chances of avoiding heart disease, save that they may each be applied to such a purpose; there are institutionalised techniques for forming beliefs about the future on the basis of the results of each of these types of procedure.

Wittgenstein's examples of alternative procedures thus lay him open to the following dilemma. If we may presuppose that we share objectives with the people in his examples, that the results which their techniques yield are applied for similar purposes—they want, for example, to know how many books they can fit on a shelf, or to buy enough wood to last the week—then we can give a sense to the suggestion that their notions of length, quantity, etc., are akin to ours, that they apply similar concepts by different methods. But it also now makes good sense to ask whose procedures are superior, are better adapted to those objectives. It is like comparing interpretations of Tarot cards with a prognosis based on medical research: the concepts of the meaning of the statement being assessed are the same—there are not a pair of similar but alternative concepts of the truth of a prediction of heart disease involved. (And the wood-sellers may be viewed like children who are easily persuaded to take a longer, flatter piece of cake in preference to a shorter, thicker piece.) But if we may not presuppose that Wittgenstein's people put the results of their 'measuring', 'calculating', etc., to purposes similar to our own, then there is no longer apparent any clear sense in which what they do may be described as 'measuring' or 'calculating' at all. In particular they may not untendentiously be described, as Wittgenstein describes them, as making a use of concepts of measuring, length, and equality of length 'different from but akin to' our use. For it has still to be made out that there is in what they do anything which may rightly be regarded as the use of *any* such concepts; nothing has been done to distinguish their performance from mere ceremony.

6.   In summary: Wittgenstein's hostility to any non-conventionalist conception of necessity has at least three strands. He wants to reject the idea of a contract in understanding. He wants us to see our notions of number, calculation, etc., as grounded in, determined by the techniques which *we* employ; we have no ulterior understanding of these concepts. And he wants to suggest that alternative rules and techniques are conceivable.

Thus, to calculate properly is to calculate *our* way, to follow our rules; but there is also the possibility of other ways of calculating, measuring, etc., which are, as it were, incommensurable with ours. These ideas are obviously in superficial conflict. How can other techniques be incommensurable with ours if only ours count as correct? The answer is, presumably, that if we calculated differently, that would then *be* correct calculation; the correctness is internal to the procedure, so to speak, and the incommensurability external. What I have been concerned to draw out is a more deep-rooted conflict. We are owed an account of what would *make* allegedly alternative ways of calculating, inferring, etc., alternative ways of *calculating, inferring,* etc. If such an account can be given, we must have some understanding of what it is to calculate, etc., dissociated from implementation of our rules and procedures; but then, will it still be possible to suppose that our arithmetic sets its own standards, once it is seen as one among alternative ways of doing the same thing?

It depends upon what kind of account is given. In fact, it is clear that some such account has to be possible anyway; for we do not react to the suggestion that people might calculate differently as we react to the suggestion that they might recite Eliot's *Waste Land* in a different order of verses—'in that case they are not reciting the *Waste Land* at all'. Whether or not what they are doing is a kind of calculation depends on how they apply the results, how the activity is embedded in wider contexts. On such considerations would depend our willingness to describe some tribal activity as a mode of calculation. But if this is right, at least in outline, then our understanding of the sense in which there can be alternative modes of measuring, calculation, inference—and associated with them different concepts of length, number and logical consequence—is inconsistent with the thesis of incommensurability among such alternatives. We identify them as different ways of doing these things by recognising affinities of purpose between their users and ourselves; and it is in terms of their suitability for these purposes that our and their procedures may therefore be compared. We shall count a tribe as inferring one statement from another if, for example, they go through some rigmarole involving both statements and then act as if they believed them both where antecedently they displayed belief only in one. But once a procedure is identified as an intended means of passing from true statements to true statements, the question of how well adapted it is to that purpose may significantly be raised.

At any rate, it may be raised if we may independently decide on the truth-values of the statements. But the point which Wittgenstein really wants to press home is that we may not. Calculation and logical inference themselves supply *criteria* for truth. This brings us back to the question of the possibility of a conflict of criteria—the possibility of a statement's being true by inferential and false by observational criteria, for example, if we inferred 'anyhow'—which is what we must discuss next. If we are to understand all the facets of Wittgenstein's conventionalism, it is essential to achieve an understanding of how he can attempt to answer this crucial question. (The same, indeed, would apply to any conventionalist theory of logical and mathematical necessity.)

V

Logic and Mathematics as Antecedent to Truth

Sources

RFM: I. 155; II. 90; III. 31–4; V. 48
PI: II. xi, p. 226, 'But am I trying to say . . . in their turn?'
PG: I. 68; II. 1, 17
LFM: lectures XXIII, pp. 229–30; XXVI, pp. 249–50

1.   Each of the analogies with measuring discussed in the previous chapter seemed to have the intended point that logic and mathematics create the standards for, give meaning to the concepts of valid inference, correct calculation, etc.; that there is, so to speak, no 'independent' reality' to which valid inference, correct calculation, etc., are subservient. It was objected that, on the contrary, principles of inference and calculation are answerable to the truth-values of the statements which they enable us to connect. Our concern now is with what response Wittgenstein has to this objection.

We need, however, a rather more definite formulation of the specific account of the necessity of, for example, an individual arithmetical proposition which Wittgenstein wants to suggest. So I want to begin by enquiring in a little more detail what account we should see Wittgenstein as essentially proposing.

Consider again the rule of conversion, 1 inch = 2.54 centimetres. We protested at Wittgenstein's suggestion that this is simply the expression of a convention: it is no conventional matter how practices in distinct metric systems correspond—on the contrary, it must originally have been a matter for discovery. The relation between the inch and the centimetre must originally have been settled by empirical means. But does the rule, then, state an empirical truth? Surely not; it is no *contingent* matter how many centimetres there are to the inch. So how could we have discovered a necessary truth by empirical means?

What certainly seems a necessary truth is that, whatever unit of length we have adopted, if at time t_0 an object is measured correctly in terms of that unit and does not change between t_0 and t_1, then the same reading will be obtained if at t_1 it is again measured correctly in terms of the same unit. So much is part of what we mean by 'correct measurement'. This proposition—call it α)—holds irrespective of the unit of length involved; so it follows that whenever a correctly-measured object proves to be m inches long, it will always, so long as its length has not changed, prove if measured correctly on a metric ruler to be some constant number, n, of centimetres long. It is thus implicit in the idea of correct measurement that there should be a rigid correspondence, $m:n$, between the units in the two systems.

Now, what we discover by empirical means when we attempt to determine this correspondence is that when we measure a ruler graduated in the units of one system by means of a ruler graduated in the units of the other, and when by ordinary criteria this is done correctly—we do not allow the rulers to slip, etc.—and when by ordinary criteria they do not change in the process—there is no shrinkage or distortion—then we almost always get the same result: 1 inch on the imperial ruler corresponds to just over $2\frac{1}{2}$ centimetres on the metric ruler; in fact, if the rulers are sufficiently finely calibrated, to something very close to 2.54 centimetres. This quantitative relationship is borne out on almost every occasion on which we correctly measure, by ordinary criteria, an object on one of the rulers and then use the relationship to predict the outcome of correct measurement with the other ruler. Thus we arrive at a statistical generalisation about how seemingly correct measurements in the two systems of seemingly unchanging objects tend to correspond. By α), however, there must be some *exact* correspondence; if we really measured an unchanging object absolutely accurately in both systems, we should always get exactly corresponding results. That our results cluster around 1 inch = 2.54 centimetres strongly suggests what this result would always be. And so we *dignify* '1 inch = 2.54 centimetres'; we transform its status from that of statistical generalisation to one of *rule*; we give a specific value to the rigid correspondence which our concept of correct measurement leads us to suppose that there must be.

Wittgenstein endorses this account of the matter. Thus v. 48:

> 'Have we not determined the relative length of the foot and the metre experimentally?' Yes, but the result was given the character of a rule.

The above elaborates this remark by trying to suggest *why* we should give the result this character. It is not that we discover a necessary statement by empirical means; rather we are already pressured by our concept of correct measurement into regarding some such statement as necessarily true, and what we determine by empirical means is which candidate we may most conveniently adopt.

What is the connection between such a picture of the necessity of '1 inch = 2.54 centimetres' and the idea of concept-modification? In α) a liaison is already established between the concepts of correct measurement, lack of change in the object measured and sameness of outcome.

Thus in either system, imperial or metric, before a rule of conversion was established between the two, it would have been a criterion for saying either that mismeasurement had occurred or that some change had taken place in the object if different results were obtained on different occasions. This would be so even if, by ordinary observational criteria, no change seemed to have taken place and the measurements seemed to have been carried out perfectly properly. The users of each system thus possessed two kinds of criterion for saying that the length of an object had changed: a purely observational criterion, based just on the look or feel of the thing, and the getting of divergent results by correct measurement. Correspondingly they had two kinds of criterion for mismeasurement: overt bungling of the process, and the getting of divergent results with respect to an unchanging object. (Whether divergent results would indicate change or mismeasurement would, of course, depend on context and theory.) However now, as a result of adopting a rule of conversion, a new kind of criterion is introduced for mismeasurement or change, resting on a fresh notion of when results of measurement are divergent. Before it was fixed as a rule that 1 inch $= 2.54$ centimetres, it would not have been regarded as a criterion of inaccuracy or change in length that the same object measured 1 inch on an imperial ruler and 2.58 centimetres on a metric ruler; we should have had no expectation about the metric measurement and if, on measuring again on the imperial ruler, we again got a reading of 1 inch, there would have been no reason for us to suppose that the object was fluctuating in length or to suspect any of the measurements. Now, however, we are bound to regard these readings as divergent.

So much is clear. For a Wittgensteinian, however, this is a modification and extension of the concept of divergence among results of measuring; and with it go modifications in the concepts of change in length and accurate measurement. Once a correlation between the systems is established, we have on the basis of measurement in either of them a concept of how measurement in the other *ought* to turn out. We then say, of course, that before the establishment of the correlation, it would still have had to be the case that an object x inches long would, when measured correctly by a metric ruler, have turned out to be $2.54x$ centimetres long. But this sort of subjunctive conditional, though it purports to abstract away from the present situation to one in which we may have known nothing about the metric system, has, on a Wittgensteinian viewpoint, itself undergone a modification of sense as a result of our fixing a rule of conversion between the systems.

2.   The force of this view of the example, may become clearer if we contrast it with a view analogous to the conception of logical and mathematical inference which the Wittgensteinian view opposes. On such a view, to speak of the relation between units of length is, as usual,

to speak of the relative sizes of paradigms of those units; but since we do not disallow that the paradigms might themselves shrink or expand, what we are *really* talking about is the relation between the amounts of space involved, so to speak, to which the relation between the paradigms will correspond, provided they have undergone no change. Wittgenstein's suggestion was that when we settle on a particular rule of conversion between the inch and the centimetre, we give a new rule, in terms of results in one system, for how measurement in the other system of an unchanging object will turn out if carried through correctly; thus a modification in the concept of correct measurement has been effected. But it is obviously improper to speak of such a modification on the present, highly realist view of the meaning of '1 inch = 2.54 centimetres'. If a pair of rulers, one imperial, the other metric, really are suitable to serve as paradigms, if, that is to say, the calibrations on them really do correspond to the inch and the centimetre, then irrespective of whether we have settled on any correlation between the two systems, it is already the case in virtue of the objective correspondence of the amounts of space involved that the results obtained by simultaneous correct measurement of an object on both rulers must have a certain pre-determined numerical relationship.

Suppose that there are in this objective sense *precisely* 2.54 centimetres to the inch. Then what is being insisted on is that if an object measured simultaneously on both rulers turns out to be 2 inches and 5 centimetres long respectively, then something is objectively amiss with the measurement. Of course, it might have seemed to us, before we knew how to convert inches to centimetres, that both measurements had been carried out correctly. But this would have been a factual error. Provided both rulers were properly calibrated, it would have *had* to have been the case that at least one of the measurements taken was inaccurate. There would have had to have been something detectable in principle by ordinary criteria of inaccuracy—a failure to locate the ruler properly, a slight movement of it, a miscount of the calibrations, etc. Even if observers reported no such occurrence, something of the sort would have *had* to have happened—something which a more sensitive observer could have reported to us, which God could have known about.

Because it is in this way an objective issue whether objects have changed in length, whether measurements have been carried out properly, whether rulers used are properly calibrated, we are, when we fix a rule of conversion between the imperial and metric systems, answerable to a domain of pre-existing, objective fact. If we get the exact relationship wrong, we shall be in the position of employing a *false* criterion for these concepts; we shall time and again be in the situation of postulating change, mismeasurement or miscalibration where nothing of the sort is so. We have, when we come to fix on a rule of conversion, a responsibility to these already fully-determinate concepts, a responsibility to the way in which in any particular case they already objectively apply. What we amass, when we experimentally investigate the relationship between the inch and the centimetre, is evidence not of what convention it would be most convenient to adopt but of what is the *true* numerical ratio between those measures.

Talk of the inch and the centimetre as absolutes in this way would make most people uneasy, at any rate when it is suggested that not even the paradigms of these units may exemplify them properly. Nevertheless we do ordinarily employ an, in the relevant sense, realist concept of length: we draw a distinction between making the most painstaking and refined measurement of the length of something and actually measuring it accurately. We are pushed in this direction partly by the variability of the results which we get when we have no independent reason to suppose that the object measured is changing; and partly because whatever methods of measurement we use seem to yield non-transitive discriminations of length in certain cases—*real* lengths, of course, being allowed to differ by too small an amount for an apparatus to detect. Thus, even if we should ordinarily question the platonist notions of the inch and the centimetre just sketched, we regard questions of sameness in and change of length as objective. Judgments of sameness and change are going to be made on the basis of whatever rule of conversion is adopted between the imperial and the metric systems; so the rule picked upon must thus answer to certain antecedent facts.

3.   To what extent could an anti-realist endorse such a conception of length? Anti-realism would presumably allow ascriptions of length to an object to be intelligible only in so far as we possess criteria for determining whether or not they are correct. But we have no criterion for determining the correctness of an ascription of *real* length, in the sense just sketched. A natural suggestion would be that an anti-realist should regard statements of length as concealed generalisations concerning the outcome of actual and hypothetical measurements— perhaps, specifically, as postulations of a limiting numerical value for such measurements. The notion of truth appropriate for such statements will then be that appropriate to any inductive generalisation, that is, perhaps, no notion of truth but merely one of acceptability.

Now, the question is whether the Wittgensteinian attitude to '1 inch = 2.54 centimetres', which I have tried to indicate, is anything other than the *minimum* required by an anti-realist view of length. It is arguable that it is, and it will be enlightening to see why. If, as anti-realists, we abandon the idea of the length of a thing as an objective, determinate feature of it, it is certainly no longer intelligible to think of a rule of conversion as answerable to the results of accurate measurement of the length of an object in both units; still less as answerable to the fact of how the platonic inch and centimetre correspond. But we shall continue to employ the same practical criteria for miscalibration of a ruler that we now use, along with the same practical criteria for a measurement's having been botched. One of the latter criteria will still be the getting in suitable circumstances of divergent results. Only it will not any longer be possible to think of the connection between divergence and mismeasurement just as guaranteed by the uniqueness of the object's real length, so that at least one of any divergent pair of accounts of it must be wrong. How, then, will the connection between these concepts now be interpreted?

There seem to be possible three broadly different accounts of the matter within the framework of rejection of a realist conception of length. To begin with, a position could be adopted analogous to that of the intuitionists in mathematics. Suppose A and B have each been given what is, by ordinary criteria, a perfectly properly calibrated ruler and now simultaneously measure the same side of an object, getting marginally different results. By the contrapositive of the principal, α) it cannot be that both measurements are correct. Now we should ordinarily allow the truth of something like the following. The outcome of a process of measurement is 'stacked', in much the same way that the outcome of selecting someone from a group by means of the rhyme, 'Eeny-meeny-miny-mo', is stacked. Given the number of people in the room and the place where I am to start, and given that I go through the rhyme properly, it is determined in advance who will be selected. Similarly, given that the ruler meets ordinary criteria of correct calibration, and given that the object to be measured does not by ordinary criteria change its length, and given that by ordinary criteria I measure properly—I don't wobble, slip, miscount the calibrations, etc.—it is determined of the outcome at least that it will fall within a certain *range*. The properties of the measured object and of the ruler, as determined by ordinary criteria, and the concept of accurate measurement, as fixed in operational terms without reference to the result, fix boundaries within which the result *must* come—though unlike the situation with the rhyme, there is some measure of flexibility. Wherever, therefore, results fall outside those boundaries, it must be the case that by other ordinary criteria we have a mismeasure or a change either in the measured object or the ruler. Similarly, wherever different people in a group are selected by the rhyme, then it must by other ordinary criteria be true to say that the selectors started in different places or that at least one did not carry out the procedure properly.

What, though, is the position if the results fall *within* the margin in question? There need in that case be no independently, practicably noticeable mismeasure or change in the object measured or in the ruler. Suppose that the results of A and B differ within this margin. What account does an anti-realist give of the requirements here of the principle α)—if indeed he does not reject it?

The attitude corresponding to the position of the mathematical intuitionists would be something like this. In such a case it must still be true that something recognisable occurred in virtue of which A and B reached divergent results. Change in the object is ruled out by the hypothesis of simultaneity, so there must have been a mismeasure or something amiss with one of the rulers which could *in principle* have been recognised by ordinary operational criteria of correct measurement and suitability of rulers. We do not have to postulate a God-like being with infinitely fine perceptual powers to give sense to this idea; all that is required is a human being with abnormally fine perceptual powers, and the same concepts that we have—someone able to make much finer discriminations by eye than we can make.

Such a person, if his memory were as sharp as his vision, might well have no need of measurement in most practical contexts where we measure. Nevertheless we are still talking about a *person*; his powers are not infinitely refined, but merely sufficiently refined to cope with the actual degree of divergence between the results of A and B. The intuitionistic picture of the situation is thus that our grasp of the concepts of correct measurement, as fixed by operational criteria, and of correct calibration—though for convenience the latter will be left out of account from now on—exceed our capacity to apply them in a practical way; exactly as our grasp of the concept, 'prime', as applied to natural numbers exceeds our ability in practice to recognise whether it applies in a particular case. For the intuitionists, the sense of 'prime' is fixed by its association with an effective criterion of application; it is thought of as a determinate question how the application of this criterion works out in any particular case. The outcome, as with the rhyme, is always 'rigged', even if we cannot in practice determine what it is, by the identity of the number concerned and the character of the decision procedure; 'n is prime' thus has, for the intuitionists, a determinate truth-value irrespective of the size of n, For we can imagine beings, mentally more dexterous than we, who could in fact effectively apply our decision procedure for 'prime' to numbers much greater than we can. The proper description of such people is that they are able to discover facts which we are prevented from discovering by purely practical limitations.

On this view, we may allow that a concept applies in any circumstances in which a community of beings would sincerely report it as applying who have undergone the same training as we, but whose practical limitations—eyesight, intelligence, memory, etc.—though of the same kind as ours, are less in degree. What may not be allowed is an appeal to what would be 'recognised' by beings with some sort of *infinitary* capacity; the capacity, for example, to exhaustively examine *all* the natural numbers, or to determine visually *arbitrarily small* misalignments of rulers. The view thus produces an asymmetry between the concepts of correct measurement and of mismeasurement. If a mismeasure occurs, it will be in principle possible for us to recognise that it has done so, and in practice possible for a being of sufficiently refined, though still limited, perceptual powers. The same cannot be said of correct measurement. However subtle someone's powers of perceptual discrimination may be, it is possible, provided he has *some* such limitations, that differences—in the alignment of the rulers, or whatever—may occur which are too small for him to detect. Thus recognition of correct measurement is impossible for a being with limitations akin to our own.

Wittgenstein asked,[1] to what reality does '1 inch = 2.54 centimetres' correspond? From the present anti-realist point of view the answer will not be: to the fact that an object correctly measured to be x inches long would prove, if correctly measured on a metric ruler, to be $2.54x$ centimetres long. The notion of correct, absolutely accurate measurement is a fiction, for we lack any effective criterion for recognising when an object has been so measured. But it does not follow that there is nothing, no reality, to which the rule of conversion is answerable. The principle α) cannot survive such a repudiation of the notion of accurate measurement, but its contrapositive can. In particular, if distinct simultaneous measurements of an object give different results, it can still be insisted that one or both must be regarded as inaccurate. The contrapositive thus continues to exert the same pressure on us to adopt a *rule* of conversion; for it postulates error somewhere in any two simultaneous pairs of measurements, each involving an imperial and a metric reading, which do not stand in the same numerical ratio. And the rule which we adopt will serve as a criterion of mismeasurement in circumstances like those of A and B above. The fact of mismeasurement, however, is, on this view, always an in principle recognisable, objective matter. The reality to which the rule of conversion is answerable is thus the reality of the application of the concept of mismeasurement. The rule of conversion does not provide a *new* criterion for the application of this concept, except in the trivial sense that it is a criterion which we have not used before. The concept for which it is to be a criterion is unchanged, and on the way it applies turns the question what rule of conversion should be adopted.

In summary: the intuitionists' brand of anti-realism would appear not to entail what I am presenting as Wittgenstein's view of '1 inch = 2.54 centimetres'. For someone who adopted the intuitionist point of view, the notion of 'correct' measurement would be a fiction. But he could still conceive of the outcome of any process of measurement as 'rigged' by the concept of accurate measurement, as determined by operational criteria: the more refined an application one is able to make of the 'instructions' for measuring, the smaller the *range* of results one can get in a given case while seeming to follow the instructions. Thus if A and B get divergent results when they measure an object simultaneously, their degree of divergence will always lie outside the

[1] *RFM* I. 9.

range compatible for *some* being of sufficient perceptual refinement with neither of them having mismeasured; and *he* will then be able, independently of the results which they get, to detect an error. What '1 inch = 2.54 centimetres' has to answer to is our operational criteria of mismeasurement as applied by a being of an arbitrary, though still limited, degree of perceptual refinement.

4.   The position changes if we adopt a more austere brand of anti-realism. Such an anti-realism would claim that it is insufficient, in order for the idea of the truth of a particular statement to be intelligible to us, that we conceive that, were it true, we could at least 'in principle' recognise that it was so by recourse only to criteria which we already accept. Rather, it would be insisted that a statement may be intelligibly hypothesised to be true only if its being so implies a *practical* capacity on our part to recognise the fact.

Such an anti-realism generates the *Strict Finitist* philosophy of mathematics outlined by Dummett.[1] There is some evidence in *RFM*—notably, the repeated emphasis on the surveyability of proofs and the character of some of the criticisms of Russell's conception of a logicist foundation for mathematics—to suggest that Wittgenstein may have favoured such a view. But this is not the place at which to discuss the question.[2] Our present concern is with how Wittgenstein's view of '1 inch = 2.54 centimetres' differs from other views which may be regarded as products of an anti-realist outlook. So it is in point to consider the most extreme version of that outlook.

On such a view, mismeasurement has occurred only if we can in practice recognise by ordinary criteria that it has. There is therefore no question of our needing to adopt a rule of conversion between inches and centimetres which would accord, when used as a criterion for mismeasurement, with applications of that concept which not we but only perceptually more refined creatures could directly recognise. How, indeed, could we discharge such a responsibility? But even on this view it may be accepted that there are antecedent facts to which our choice of a rule of conversion is answerable: the fact, namely, whether in any particular case something which we could in practice recognise as a mismeasure has actually taken place. The point concerns the sense in which it may be maintained, even on this view, that the outcome of a process of measurement is 'stacked'. Even if the range of cases to which a concept applies is restricted to those where we can in practice recognise its application, it is still natural to think of there being a loose but

---

[1] 23 (2, pp. 504–6); cf. Bernays 5 (2, pp. 280–2); Kreisel 53, section 7; and Wang 75 (76, pp. 39–41). For discussion of strict finitism and the semantics of vagueness, see Dummett 28. For (purported) steps towards the development of strict finitist mathematics, see Yesenin-Volpin 81 and 82. Kielkopf's 51 purports to interpret *RFM* as expressive of a strict finitist philosophy of mathematics, but mistakes what Dummett and Kreisel had in mind.

[2] We shall do so in Chapter VII.

internal connection between the concept of correct measurement, the properties of the object to be measured, and the outcome which we get. We want to say that if by ordinary operational criteria an object is measured properly both in inches and centimetres, the results must always lie within certain vague though close boundaries around a fixed numerical relationship. Therefore if they do not do so, something practicably recognisable as a mismeasure must have occurred. It may thus be allowed that, even on this version of anti-realism, a rule of conversion is answerable to the reality of pre-determined applications of certain concepts. When we adopt a rule of conversion it is not a matter, as Wittgenstein suggests, simply of giving our blessing to an empirical regularity, of dignifying it to the status of a rule. We are committed by the understanding which we already have of the concept of properly executed measurement, even where this understanding is interpreted in terms of the present extreme anti-realism, to accepting a rule of conversion falling within a certain numerical range.

This is not, of course, to say that there is anything to choose between rules which fall within this range. Thus there could in fact be associated with this version of anti-realism a sense in which accepting the rule, '1 inch = 2.54 centimetres', effects a modification of concepts. Suppose that A and B make repeated simultaneous measurements of an object, A in centimetres, B in inches, that all their measurements are by ordinary criteria properly executed and that the results cluster around the ratio, 2.54:1, which is then adopted as a rule of conversion. It will likely be true that some of the measurements which they took did not lie exactly in this ratio. If the rule of conversion is now used as a criterion of inaccuracy in such cases, then we shall be applying the concept of mismeasurement to cases where, before we had the rule, there would have been no ground whatever for applying it—cases which, without the rule, could not in practice have been recognised as involving inaccurate measurement. Now, on the 'intuitionistic' view it was true even in such cases that the concept already applied; a being of sufficiently refined perceptions could have recognised it to do so. But if concepts get their content only by reference to conditions which *we* can actually recognise, there can no longer be any sense in the idea that, before any rule of conversion is adopted, such cases can already involve mismeasurement. Rather by adopting such a rule we have fixed a new kind of application for the concept, and thereby have modified it exactly as the sense of 'green' would be modified if it were stipulated that it was to apply to a shade of colour to which previously it was not determined that it should apply.

It is, indeed, unclear whether even the contrapositive of α) survives on this standpoint—whether measurements of an unchanging object may not give marginally different readings yet be regarded on this view as equally correct. Correspondingly, it is unclear whether there is any pressure now to adopt an *exact* rule of conversion. But any exact rule that is adopted will involve a modification in the concept of mis-measurement. It will give us a new criterion for when mismeasure-ments have been carried out, determining that the concept should apply to cases where formerly it did not.

Nevertheless, though it may be beginning to sound like it, this is still not Wittgenstein's view of the matter. For, as we saw, it is open to someone who adheres to this extreme anti-realism still to maintain that we have in choosing a rule of conversion an obligation to the range of application of the notion of mismeasurement as it was before. If it is in the nature of correct measurement, as determined by ordinary, practi-cal criteria, to produce in any given case results lying within a certain range, then the adoption of any principle which requires that only results lying outside that range can be regarded as properly reached will simply import incoherence into the concepts of correct and incorrect measurement.

Suppose, for example, that we adopted the rule of conversion, '1 inch = 3.54 centimetres'. Then whenever by ordinary operational criteria A and B measure accurately, the rule will require us to say that one, or both, has mismeasured. And whenever they get something like the right results, according to the rule, it will be patent that one or both measured in an overtly slipshod way. Nothing now will count as correct measurement for A and B. However carefully they try to measure correctly, they will get results whose mutual discordance requires them to say that one or both has not succeeded. Only when they are deliber-ately careless, when they measure in such a way that they would ordinarily be quite mistrustful of the result, will they get anything like the results which, by the rule, they ought to get—and then not *dependably*.

As remarked, it is unclear whether from a 'strict finitist' standpoint the notion of the length of an object will not become essentially ap-proximate, so that each of a range of readings resulting from, by practical criteria, properly conducted measurements will be regarded as equally correct. In that case there will be no *exact* rule of conversion from inches to centimetres; the sign of equality in the rule will rather be regarded as expressing approximate equality. But the considerations just adduced about the effects of picking a wrong rule, whether the right one(s) are considered as approximations or not, will continue to apply. So long as measuring is to serve practical purposes, the idea of correct measurement must be linked as closely as possible with safeguards which we can in practice follow out. To adopt just any old rule of conversion would be to sever this link.

5.   How, then, does Wittgenstein's account differ from this? It is not that he is committed to rejecting the suggestion that for practical purposes some rules are superior to others. What is questionable for

him is the view involved of the generalisation on which this suggestion is based: the generalisation that only if A and/or B have by ordinary criteria mismeasured will they get something like the right result if, for example, '1 inch = 3.54 centimetres' is the rule. There is nothing in anti-realism as such, even the extreme version entertained, to prevent someone from regarding this generalisation, in the orthodox way, as depending upon an internal relation between the notion of mis-measurement, as determined by ordinary practical operational criteria, and the range of results achievable by measuring an unchanging object several times; that is, as comparable to the generalisation: only if they start in different places or one of them misapplies the procedure will A and B select different objects by means of the rhyme, 'Eeny-meeny-miny-mo'. It is, however, unclear whether Wittgenstein is in a position to give to the idea of an 'internal' conceptual connection the content requisite even for the 'strict finitist' view. In order to characterise our ordinary conception of such connections, we have to do two things. First, we have to be able to make sense, for example, of an overlap, or exclusion, between the ways in which certain expressions are correctly used; we have to make sense, for example, of the generalisation that the conditions under which it is right to say that A's and B's results are in something like the ratio 1: 3.5 are all conditions under which it is right, by ordinary criteria, to describe one of them as having mismeasured. Then, secondly, a distinction has to be drawn between cases where such a relation is fortuitous, where the world just happens to have worked out that way, and cases where it is in the *character* of the senses of these expressions—the concepts of their correct use—that they should have such a relation. To talk of internal relations between two concepts is to talk of relations between the patterns of application constitutive of their correct employment, relations between the ways in which it is right to use them, *irrespective* of the circumstances which crop up warranting an application of either. An internal relation between the senses of two expressions is fixed, for example, by associating one with one rule of application and the other with another in such a way that, irrespective of how the world turns out to be, they may not correctly be applied together.

This is meant only as the crudest description of our ordinary conception of the matter. But it is enough to make it highly implausible that the ordinary conception can survive Wittgenstein's treatment of the notions of following a rule, using an expression in accordance with its meaning, etc. For Wittgenstein, as we saw, there is simply no sense in the idea of an expression's being associated with an objective pattern of use which any fresh, so far unconsidered application of it must determinately either breach or continue. There is no *objective* content to the idea of continuing to use the same expression in the same way. That being so, there is a fortiori no objective content to the notion of a relation between the ways in which expressions are to be used; or better—we may not countenance the suggestion that such a relation can obtain unrecognised just in virtue of independently established patterns of use for the two expressions. We may still, as it were, *decide* to link the use of two expressions in a certain way. What may not be accepted is the conception that, by establishing for each of them a certain pattern of correct use, we may already have implicitly taken such a decision without knowing it. Such an idea makes sense only if we think of the pattern of use as something finally established, and objective; and for Wittgenstein it is no such thing. It is therefore senseless to suppose that certain objective relations may hold between such patterns which may be veiled from us, which it may, for example, take a proof to bring to light, and which are in any case appropriate material for recognition.

On this view, there is no way in which it can be objectively implicit in our concept of correct measurement, as fixed by ordinary operational criteria, that if A and B measure correctly, their results must approximate to a fixed numerical relationship. All we can be sure of is that simultaneous measurements of an object, when carried out properly by ordinary criteria, tend to produce similar results or, if different units are involved, results which cluster closely around a certain ratio. Indeed, even this does not quite get the flavour of Wittgenstein's attitude. The following is better: we have standard inductive grounds for supposing that whenever we are prepared to allow that simultaneous measurements of an object have been carrried out properly, what we are prepared to accept as properly taken readings will be nearly identical, or, if carried out in different units, will come close to a certain fixed ratio. It is not determined in advance what (if we remain 'faithful' to our understanding of the relevant concepts) we shall count as proper conduct of the measuring or proper taking of the readings; but there is evidence of an inductive sort for the kind of supposition just described. Such a supposition we then transform into a linguistic rule.

Our discussion of this example, then, suggests the following sort of account of the concept-modification thesis. Accepting the rule, '1 inch = 2.54 centimetres', gives us a *new* criterion for when measurements have been properly carried out not just in the trivial sense that we have not used this criterion before—nor even in the sense, admitted by extreme anti-realism, of allowing us to describe as 'mismeasures' cases which previous criteria did not determine to be so—but which is new in this sense: that even if '=' is read as 'is approximately equal to,' the rule is not properly seen as a description merely of something implicit in the notion of correct measurement as it was determined by operational criteria before the rule was adopted. Of course we *now* say that, even before the rule was adopted, if A and B had simultaneously measured an object and their results had substantially diverged from the approved ratio, there would have had to have been an error by criteria then acknowledged. But what we now say is expressed in terms of the concepts as they now are, after the modification occasioned by adopting the rule. In like manner, accepting *any* statement as necessarily true will modify concepts, will provide new criteria for the correct application of the expressions involved. The criteria will be new in the sense that they may not be regarded merely as explicit formulations of something which was already implicit in the character of the concepts in question. Proofs thus essentially modify concepts because they persuade us to accept statements as necessarily true, to interweave the criteria of application of the expressions which they involve.

6.   We have yet to see how, or whether, this account may cope with the difficulties attending the concept-modification thesis with which we were concerned in Chapter III. We shall approach that question in the next chapter. First, we must apply the account to the general objection to conventionalism with which we began this chapter: that principles of inference, calculation, etc., have to answer to the truth-values of the statements which they allow us to connect.

The short answer is that this commitment could be met *whatever* principles we adopted, since the principles would contribute towards determining the truth-values of those statements. There is no question, for example, of our having as a principle of inference something which enables us to infer a false statement—by observational criteria, for instance—from true ones; if we really propose to treat the principle as a rule of inference, we shall describe any situation which seems to be of this sort as one in which the relevant non-inferential criteria of truth and falsity were misapplied—the premises were not really true, or the conclusion not really false.

Dummett takes issue with this reply. His example is that of a tribe who count as we do but have not yet devised the concept of addition, so that when asked to determine the number of things of some sort, they always count up even where we should regard the result as implicit in the results which they have already obtained.[1] We now teach these people our principles of addition; and as a result they now come to regard themselves as having miscounted in situations in which previously they would have accepted the results; viz. when their results do not 'add up'. Previously they would have regarded themselves as having miscounted only if they noticed the making of some particular mistake—or if they thought that the making of a mistake was particularly likely, because the objects were milling around, or whatever. But now, irrespective of whether they notice a particular error, they will sometimes assert that they have miscounted just on the basis of the results which they get.

Suppose that in some context a member of the tribe asserts in this way that he must have miscounted just on the basis of the results which he has obtained. Dummett's question is: is this assertion true by criteria

[1] 23 (2, pp. 498–9).

which were formerly acknowledged or is it not? That is, did something occur which the tribesman would have regarded as a mistake, had he noticed it, even if we had never introduced him to the principles of addition? If so, then according to Dummett the effect of our introducing the tribe to the concept of addition is not properly described as giving them a new criterion for when counting has not been done correctly. Rather we have taught them to recognise getting additively wayward results as a symptom that circumstances obtain, perhaps unnoticed, which would justify them by the criteria which they formerly employed in asserting that a mistake had taken place. On the other hand, if we say that there need have been nothing, other than the getting of the 'wrong' result, such that if the tribesman had noticed it, he would have been justified by his old criteria for correct counting in describing as a mistake, then we appear to be committed to saying that a mistake can occur in counting without *any particular* mistake's having occurred; it can be true that someone has miscounted without it being true that he has counted John twice, missed Jane out, etc., right through the disjunction of possible counting mistakes. But this cannot be allowed. If a disjunction is true, one of the disjuncts must be true, so it cannot be correct to admit as a criterion for the truth of a disjunction something whose obtaining does not guarantee the truth of one of the disjuncts.

Generalised, the dilemma is this. Our original objection to the proposal to regard our principles of inference just as *constitutive* of the notion of valid inference, as suggested by the measuring analogies—as answerable to nothing external—was that they are on the contrary answerable to the truth-values of the statements which they enable us to connect as premises and conclusions, and that to be valid they must not permit the derivation of a false conclusion from true premises. Wittgenstein's answer was that the truth-value of premises and conclusion are not, in the way the objection suggests, *predeterminate*. If one of our rules of inference gets us from what seemed, by non-inferential criteria, *true* premises to what seems, by non-inferential criteria, a *false* conclusion, we say that those are not really the truth-values—that the criteria involved must somehow have been misapplied. Dummett's point is then that the assertion that such a misapplication has taken place is typically to be understood as the assertion of a disjunction; if criteria have been misapplied, they must have been misapplied in a *specific* way—a specific mistake must have been made among the range of possible mistakes. And now, would ordinary criteria for the occurrence of such a mistake have enabled us, had we noticed it, to recognise something as a particular mistake in any such case? If not, then Wittgenstein appears to be committed to allowing that some specific mistake can have occurred without it being possible to recognise it as such by ordinary criteria. For the principle of inference only licenses us to say that *some* mistake has occurred; it does not show us which. But if we reject the idea that a specific mistake can occur without its being in principle possible to recognise it by ordinary criteria, then our adoption of principles of inference is answerable to such criteria; if something is going to serve as a criterion for the truth of a disjunction, we have a responsibility to ensure that it coheres with established criteria for the truth of the individual disjuncts.

7.   How should a Wittgensteinian tackle this dilemma? One strategy might be as follows. Since we regard it as part of the meaning of a disjunction that if it is true, one of its disjuncts must be true, anything which we regard as a criterion of the truth of a disjunction is ipso facto to be regarded as a criterion of the truth of one of the disjuncts. In that case the getting of results, prohibited by the arithmetic of addition is indeed a criterion for saying that we counted Jane twice or missed John out or . . . etc. We take the getting of the 'wrong' result precisely as a criterion for saying that something specific must have occurred which, had we noticed it at the time, we should have regarded as a mistake even before it issued in the wrong result.

This is to accept one horn of the dilemma. But it will provoke the protest that we can't just *make* it true that some specific error has occurred, merely by adopting certain rules of inference. Either a specific mistake has occurred or none has. If none has occurred, then the assertion that one has is false; and if one has, then since it has been conceded that it must therefore have been possible to notice it in terms of ordinary criteria, how can we guarantee this possibility just by a *convention*?

Imagine an isolated sub-culture in our society who, as a result of a freakish series of unnoticed errors, magical appearances and disappearances of objects, etc., have generally found it to be the case that whenever they count twenty-three F's and fourteen G's, a seemingly correct count of the F-or-G's gives thirty-eight. Suppose these people have adopted it as a rule that $23 + 14 = 38$, and use it as we use our arithmetical rules as a criterion for what results of counting should be in certain circumstances. Now the freak happenings peter out on them, but they go on saying in such circumstances that, unless they get '38', some definite error must have occurred. These people, we should want to say, are simply wrong. There need have been no such mistake.

What we are appealing to is the idea that the concept of a mistake in counting is already fully determined by its *operational* criteria, that the criteria accepted by the tribe, for example, before we taught them to add, already determine whether a specific mistake has occurred in any particular count-up. There is thus no room for new criteria unless it can be ensured that their results of application essentially coincide with those of the criteria which we already have.

Now, we know that such a conception, of the existence of a determinate range of cases in which by the old criteria a mistake in counting occurs, will be repugnant to Wittgenstein on general grounds. But what is also striking is that it is totally unclear how we could force members of the sub-culture to change their arithmetic. How could we demonstrate to them that when they count twenty-three girls, fourteen boys and thirty-seven children, no counting mistake need have occurred?

If the difficulties here are not immediately apparent, it is because the discussion so far has been oversimplified. Logic and mathematics furnish criteria not merely for the misapplication of certain procedures but for the way in which the world is behaving in certain other respects. If we count five girls, seven boys and then thirteen children, a miscount is only one possibility. A child may have been hiding, or may have come into the room during the counting—or something more outlandish might have occurred along the lines of hallucination, or the spontaneous creation of a child, or whatever. And these events, like miscounts, are things whose occurrence may be *observed*. But, of course, in our use of statements about the external world we draw a distinction between how things seem to us when perceptually assessed and how they may actually be. Thus the criteria for miscounting used by the tribe before we taught them to add will not have amounted to *conclusive* tests; perception never gives us indefeasible verification of how things are in the world. It is merely that we are hardly ever called upon to suppose that, with respect to the sort of statement whose truth we should regard as explaining discrepant counting results, things may be other than as assessed by perceptual criteria.

Thus to the objection, how can our adoption of a convention guarantee the existence of a specific something which pre-addition, operational criteria would decree as a counting mistake, one answer would be that, even if it cannot, we *run no risk of being found out*. For if perceptual evidence most strongly suggests that no counting error occurred, that in itself was never in any case regarded as decisive that there was none. The effect, then, of adopting the rule, $23 + 14 = 38$, cannot uncontentiously be described as enjoining the use of a criterion for the occurrence of counting errors which may actually be no such thing, which may conflict with the concept of a miscount as already determined. For, quite apart from the question of the legitimacy of thinking of concepts in this manner, the concept of a miscount has not been so fixed as to exclude the possibility that one should have occurred where ordinary perceptual criteria most strongly suggest that none has. What is at stake in the adoption of such a rule is our conception of when our perceptions may be regarded as *reliable*; and this concept is indeed one in whose conditions of application we may always be prepared to make a change.

It is in any case questionable whether Dummett is right to talk as though in any process of counting there were a definite disjunction of possible counting mistakes. It is hard to see what principles one could use for deciding that a list of possible counting mistakes included every possible way of counting five girls, seven boys and thirteen children. It is also hard to see how we should close off a list of possible physical explanations, not involving miscounting, for getting the wrong result. So even if we did have decisive methods of verification of such possibilities individually, it is not clear how we could verify that *nothing* of the sort had occurred. Of course everything may seem to be in order. But then, that is exactly how the count-up of twenty-three girls, fourteen boys and thirty-seven children seems to the sub-culture! How can we persuade them that there *really is* no mistake, or physical peculiarity, at work unless we can specify all the germane possibilities and then eliminate them, case by case?

The suggestion was that we have a responsibility to ensure, when we adopt arithmetical rules, that we are not led into postulations of miscounts, changes in the size of a group, hallucination, etc., where nothing of the sort has taken place. And the reply being suggested on the basis of the considerations just advanced is that, whatever arithmetic we adopt, we shall not be liable to conviction for failing to meet this responsibility; first, because the perceptual criteria for the individual possibilities are *defeasible*—their giving a verdict is compatible with its falsity—and secondly because, owing to its indefiniteness, the claim that *something* has occurred which would explain why we did not get what we regard as a right result is not falsifiable.

I am not here reneging on what was said in the previous chapter about alternative ways of calculating, inferring, etc. It was claimed that such descriptions made sense only if the techniques and principles involved were put to substantially the same purposes for which we calculate, infer, etc.; and that alternatives will therefore be in principle commensurable with our logic and mathematics according as they serve, better or worse, those purposes. And it is not now being claimed that we might as well, counting as we do, have adopted an arithmetic in which $24 + 14 = 38$; we have not even addressed the question how such an arithmetic might be formulated. The point is simply that people who counted as we do but accepted, for whatever reason, that $23 + 14 = 38$, could not necessarily be shown that they had adopted as a criterion for error in counting something which was actually no such thing—that they had failed in their responsibility to the notion of counting-error as fixed by their prior operational criteria. For it is concord with arithmetic that sanctions our transition from its *seeming* by ordinary criteria that no counting error has occurred to the judgment that it really is the case that there has been no such error. If counting results do not add up, we too will say that there must have been an error in counting, or some other physical explanation, even if ordinary perceptual criteria suggest nothing of the sort. But if members of the sub-culture may not be convicted of failing in a responsibility, neither may we be congratulated on having met it.

8.   These considerations cast some light, perhaps, on Wittgenstein's conception of logic and mathematics as 'antecedent' to truth, as 'measures' of reality. But they do not entirely capture what is essential to Wittgenstein's position—that is, what from a Wittgensteinian point of view is most fundamentally amiss with Dummett's objection. Consider again the objection in its most general form. The principles of logic and mathematics serve as criteria for the misapplication of criteria for the truth of other—contingent—statements; for miscounting, mismeasuring, etc. But then the claim that such a misapplication has occurred is intelligible only as the claim that some *specific* misapplication has occurred; typically, there will be a variety of ways in which to botch such a procedure.

Now, the objection runs, does a Wittgensteinian allow that whenever, by appeal to a principle of inference, we are prepared to assert that such a misapplication must have occurred, both that one has and what specific misapplication could always in principle have been recognised independently? If not, the Wittgensteinian would seem, if we refine Dummett's objection slightly, to be committed to postulating specific mistakes where nothing in principle need amount to recognition of *what* specific mistake has occurred. But if it is allowed that it must in principle be possible to locate a specific mistake, or other peculiarity, whenever a principle of inference declares there to have been one, how can this have been ensured just by the adoption of a mere rule of language?

What has so far been suggested is that a deviant rule of inference could not be *demonstrated* to involve postulation of error, etc., in a situation where nothing of the sort has occurred. But while in one way a Wittgensteinian would accept that a specific, recognisable 'something' must have occurred wherever an accepted principle of inference requires us to say that *some* 'something' has occurred, Dummett's alternatives are actually *both* inadmissible from a Wittgensteinian point of view if taken in the manner required by the objection. Logic and mathematics precisely supply criteria for how procedures work out if properly applied. There is no Olympian standpoint from which we can ask: is it *really* the case that there is always a specific, recognisable error, or other peculiarity, in applying some procedure whenever a principle of inference requires us to say that there is? Nor, therefore, does it make sense to suppose that principles of inference have to face a tribunal of what is, in this sense, really the case; that we have to 'pick' our rules of inference in such a way as to ensure that we avoid postulating errors and anomalies where none have taken place. If we had such a responsibility, we should not know how to discharge it.

Thus Wittgenstein's viewpoint allows us to say the sort of thing which we ordinarily should say in circumstances of the relevant kind; for example, that there must have occurred a specific recognisable error. But this is to concede nothing to the objector. It concedes nothing because the idea is now rejected that there is a determinate range of cases in which, independently of what our arithmetic requires, it is by operational criteria correct, or incorrect, to suppose that specific counting errors, or other peculiarities, are present and of which any satisfactory arithmetic must not enjoin misclassification. The concepts of correct counting, and of the various ways in which the size of a group can alter, are always capable of further determination just because our understanding of them in no way settles in advance what we ought in future cases to count as their correct application. Dummett's objection poses a difficulty for Wittgenstein's brand of conventionalism only if we think of our understanding of notions like 'correct counting' as frozen by procedural rules, as predetermining which situations may and may not rightly be regarded as involving counting errors, etc. But Wittgenstein's conventionalism, as so far interpreted, emanates precisely from a repudiation of such a conception of understanding.

VI

'The Deep Need for the Convention'

Sources

RFM: I. 74; II. 23; III. 15–16; v. 39–40
LFM: lecture XIII, pp. 128–30

1.   In Chapter III, when we began to try to interpret Wittgenstein's suggestion that proofs are essentially instruments of conceptual change, that they fix new rules for the use of expressions, we encountered the following dilemma. If, first, we accept as legitimate a distinction between those sorts of application of an expression which objectively *accord* with the meaning which we have attached to it and those which do not, then it may well—in that sense—accord with the way in which we understand a particular statement to regard it as demonstrated by a particular proof. To accept the proof may well be a matter of fidelity to our understanding of the concepts involved in the conclusion, and of the rules of inference appealed to in the individual steps. In that case, it will not be coherent to regard these concepts as changed by our accepting the proof; it will not be coherent because, as we saw, the attempt to *combine* the notions of accord and modification, so that one application of an expression may simultaneously both conform with the way we understand it and change that understanding, results in its being impossible to give an account of the change. If, on the other hand, we disallow that there is any legitimate and substantial sense in which accepting a proof may accord with the way we understand the concepts involved—the rules of inference, etc.—then what does it mean to talk of conceptual *change* here? What is as it would not otherwise have been? If we think of the sense of an expression as determined by its pattern of correct use, a change in the way we understand it must be a change in that pattern. But if there is in reality no such objective pattern, of which, for example, accepting a proof might or might not be a continuation, then in *what* has a change taken place? If there is ultimately no objectivity in the idea of a correct application of an expression, of our continuing to use it in the same way as before, how can sense any longer be given to the claim that the use of an expression has altered?

We left open the question whether there is a solution to this dilemma which both incorporates Wittgenstein's views on rule-following and assigns a special rôle to *proof* as a source of conceptual change, whatever then remains of the latter notion. Are we now any nearer to finding such a solution? It is hard to see that we are. Suppose we take the usual apparatus for fixing a concept for granted: the definitions and examples which we should use, the criteria of application which we should attempt to explain, and so on. The crucial question is then, how would accepting a new proof of a statement involving the concept change the character of this apparatus? The answer which we should ordinarily want to give is that there will be no essential change if it is right to accept the proof. Not that the explanations which we might offer of the concept would necessarily be wholly unaffected; the proof might, for example, give us a new technique for applying the concept which it might be regarded as useful to include in an explanation, and at the very least it will provide a new example of the concept. But the sense in which we should want to insist that the *essential* character of the concept had not changed would be just this: even after we accept the proof, the former apparatus will still be perfectly *appropriate* to serve as an explanation of the concept. It is just here that the situation contrasts with that of the kind of stipulation about 'green' which we entertained; after such a stipulation, the old range of green samples would no longer be considered sufficiently representative, sufficiently broad to provide a satisfactory ostensive definition of the new sense of 'green'.

Wittgenstein's discussion of the notion of following a rule has, if we accept its general conclusion, robbed us of the distinction between our seeming to continue using an expression in the same way as before, and our genuinely, objectively doing so. We could say that there is finally only such a thing as *seeming* to oneself and others to be applying a concept in the same way as before; we share classificatory dispositions, but this is, so to speak, a primitive fact, embroidered on rather than explained by the conception of our use of language as the implementation of the requirements of a set of semantic rules which we absorb during our training in the language, and which a specific description of specific circumstances will objectively either conform to or breach. Nevertheless, there is still such a thing as the ordinary activity of explaining a concept—of giving explicit rules for its application, perhaps—and, in certain circumstances, modifying the concept for whatever reason, changing the rules, the character of the examples, etc. Even if the picture must be abandoned that in explaining a concept we lay down an objective track, along which it is our intention that correct use of the expression must run, still it is only in terms of the explanations which we should give, and the pattern of use which we seem to ourselves to follow, that the identity of a concept has any meaning for us at all. It is therefore in such terms that the ideas of conceptual change and innovation too must make sense. If there is in the end no place for a notion of fidelity to the meaning of an expression, then we can no longer picture our acceptance of a new proof as merely an explicit ratification of something implicit in a contract which we have already signed; the notions of changing a meaning and keeping it the same are no longer objective. But, in terms of their non-objective remains, the fact is that meanings do not generally *seem* to change as a result of new proofs. The previous definitions, explanations, rules, etc., which we intuitively regard as fixing the identity of the concepts involved, all remain adequate. In terms of the ordinary criteria for what it is for a meaning to change, no change has taken place. These, to be sure, are not now criteria for anything objective; but, just for that reason, how can it now make sense to claim that, although not by these criteria, a modification of concepts has nevertheless been effected?

2.   In order to see the difficulty more sharply, consider again the character of Wittgenstein's position with respect to the rule of conversion, 1 inch = 2.54 centimetres. For Wittgenstein this is originally an empirical statement, a statistical generalisation about how pairs of properly conducted measurements in terms of inches and centimetres tend to correspond, which we proceed to transform into a rule. So transformed, it comes to express a condition upon how pairs of measurements in terms of the two units *ought* to come out—how they will come out if properly conducted. It thus embodies a criterion for when measurements have been accurately taken distinct from the purely operational and other criteria previously constitutive of that notion; if a set of readings are not as the rule requires, then, irrespective of how meticulously the operational criteria were observed, they are not all accurate. That the rule embodies such a criterion is not, of course, contentious. The crucial Wittgensteinian thesis is that this criterion is not merely something which was previously unused; it is novel in a sense which makes it proper to say that our adoption of it modifies the concept of accurate measurement.

The natural objection to this view is, in essence, that the result obtainable by accurately measuring an unchanging object is determined in advance by the concept of accurate measurement and the properties of the object measured; just as the result obtainable by correctly counting a group of objects is determined in advance by the concept of correct counting and the number of objects in the group. So, for any proposed rule of conversion, it is already the case *either* that if an unchanging object is accurately measured in terms of both units, then the results will correspond—*must* correspond—as required by the rule, *or* that they will not—that only by inaccurate measurement can the required correspondence between results be secured.

Those of intuitionist, or strict finitist, inclination can be expected, as we noted, to quarrel with the use made of the idea of 'accurate measurement' in this way of formulating the objection. But they and the realist can agree, against Wittgenstein, that we have a responsibility to pick on a rule which, when used as a criterion of inaccurate measurement, yields results coincident with the way in which that concept already applies. Such a rule cannot correctly be described as incorporating an essentially new, distinct criterion for applying the concept of inaccurate measurement, whose effect is to modify that concept. For whenever, by the new criterion, a mismeasurement has occurred, that judgment will be independently verifiable—at any rate 'in principle' —in terms of the concept of inaccurate measurement as it was before the rule of conversion was adopted. The former explanations of that concept will thus continue to suffice; and it is to the concept thereby fixed that the rule of conversion has to answer—and will indeed necessarily answer correctly if we pick the right rule.

The Wittgensteinian's response to this objection, according to our interpretation, is to reject its assumptions. He rejects the idea that the concepts of accurate and inaccurate measurement, as they were before the rule of conversion was adopted, have a life of their own, so to speak—that they determinately apply, or fail to apply, to a specific range of unratified cases. And he rejects the idea that it is objectively inherent in the concept of accurate measurement whether a particular result, or range of results, can be reached by accurately measuring a particular unchanging object. Thus a rule of conversion is not answerable to antecedent facts about how the concepts of accurate and inaccurate measurement already objectively apply; nor is it already objectively necessarily true or false, irrespective of what we say, that a pair of simultaneous imperial and metric measurements of an object will, if accurate, produce results standing in a certain (approximate) ratio. The 'correct' rule of conversion is not properly seen as a description of something implicit in the notion of correct measurement as it was before the rule was adopted. Nothing is ever 'implicit' in a concept in that way. 'Essential connections' are created by our *explicit* ratification of them. The only sense in which it is 'determined' in what ratio simultaneous accurate measurements of an object in inches and centimetres will stand is that *we* have determined that nothing is to count as accurate measurement which does not produce results standing in the ratio which we have chosen.

Well and good. But now, is it really admissible on such a view to claim that the effect of adopting the rule of conversion is to *modify* the concept of accurate measurement? For the fact is that the new criterion—the rule—is treated as a criterion for whether the old criteria—the operational techniques—have themselves been properly applied. If somebody gets results conflicting with the rule after, as it seems to him, meticulous observation of the techniques of correct measurement, we say not that meticulous observation of those techniques is no longer decisive, but that he cannot *really* so have observed them. (Likewise, where the results of counting appear to conflict with those of addition, we say not that counting is no longer decisive but that there must have been a miscount.) Naturally, then, for us who play the measuring 'game', who make use of the two kinds of criteria for accurate measurement, it remains the case that the core of the concept is fixed, as before, by reference to the techniques of application which we originally learned—as the core of the concept of finite cardinal number is fixed by correct counting. From a standpoint within the language-game in which the concept is applied, there would appear to be no room for the claim that a modification in it has taken place.

The Wittgensteinian may want to say something like: by adopting the rule we have *created* facts here about how the old criterion will work out if applied properly. And the general intention of this claim may be thought to be clear. But can it be coherently formulated? Such facts were 'created', presumably, when the rule, 1 inch = 2.54 centimetres, was adopted; but that would appear to entail that before we adopted the rule it did not have to be the case that at least one of any pair of measurements of an unchanging object, carried out on appropriately calibrated instruments, was, if the results did not stand in the appropriate ratio, incorrect by criteria then accepted. The difficulty with this stands out clearly if we ask whether it is a consequence drawn from a standpoint within or without our 'language-game' of measurement. If the former, then it is of course unacceptable; by the rule that 1 inch = 2.54 centimetres, it must already have been the case in any such situation, even before the rule was adopted, that a mismeasure had occurred. Contrast the situation with the stipulation for 'green'. Here from a standpoint within the language-game we can perfectly intelligibly compare the states of affairs before and after the stipulation was adopted; we can contrast what constituted correct use of 'green' before the new shade was incorporated and after. But this is because the stipulation of a new meaning for 'green' is not a stipulation about how it *ought* to have been applied in certain cases when it had the meaning it formerly had. Adopting the rule, 1 inch = 2.54 centimetres, on the other hand, must be seen, on the Wittgensteinian view of the matter, not as changing merely the *course* of our use of 'correct measurement' and 'incorrect measurement' but, so to speak, as *creating facts for all times and possibilities*. There is therefore no combination of tense and mood which can yield a description of how the conditions of correct application of those concepts have changed.

Suppose, however, that the consequence is drawn from a standpoint outside the language-game. It is asserted that, whatever *we* now say about the matter, it was perfectly possible, when we had only the operational criteria of correct measurement, for measurements of an unchanging object in centimetres and inches to give readings not in the ratio 2.54:1, yet both be by operational standards perfectly correct. But now, what could be the ground for this assertion? The picture on which it seems to depend is that of an *Olympian view* of the potentialities of the operational criteria; it was, so to speak, inherent in those criteria that they could be satisfied by simultaneous measurements whose results did not correspond as we should now require them to. But such a conception is precisely what Wittgenstein rejects. There is no such Olympian knowledge to be had.

Thus neither within nor without the language-game can it be truly asserted that, as a result of adopting the rule, 1 inch = 2.54 centimetres, moves are now admissible that would not have been before. Outside the language-game there is no distinction to be drawn between moves which would have been admissible and moves which would not; there is only the contrast between those which have been allowed and those which have been refused. But inside the language-game no move is permitted by the new rule which may not with perfect propriety be said to have been admissible already.

3.   Someone might wish to contest these conclusions along the following lines. Structurally the situation is this: we have a concept, ϕ, viz. accurate measurement, and an initial criterion of application for it, C_1, viz. observation of certain operational safeguards, which we proceed to supplement by a new criterion, C_2, viz. conformity of results with the rule of conversion. The difficulty for the Wittgensteinian is that C_2 is treated as a criterion for the satisfaction of C_1, that both before and after the new criterion, C_2, is adopted, the content of ϕ may correctly be explained by reference merely to C_1. Nevertheless, there *is* a change: what changes is our conception of the conditions under which C_1 has been properly observed. Before, we should have been prepared to judge that C_1 was satisfied just on the basis of observation. Now, that is no longer sufficient; it is in addition necessary, in order for C_1 to have been properly observed, that the results obtained conform to certain requirements. It remains true that ϕ applies if and only if C_1 is satisfied, and in that sense the concept, ϕ, has undergone no change. But this is only a formal sense; for circumstances which would previously have been regarded as sufficient for supposing that C_1 was satisfied are no longer regarded as such. Thus there is, after all, inside the language-game a recognisable modification in the concept ϕ. It is not that its criterion of application has changed. Rather the change is second-order; we have changed our conception of the conditions under which the criterion has been duly observed.

In one way this is unexceptionable. But, taken as intended, it involves a misrepresentation. In all cases of the relevant sort we have, before the proof or rule is accepted, no *indefeasible* method of verification that the old criteria have been properly observed. Just this is what enables us to adopt the new rule as a criterion for how the old criteria work out. Otherwise, in situations in which the new and old criteria pointed in different directions, we should simply have a conflict; and so would be beggared for a description in terms of ϕ. Thus all along it was allowed that one could seem to measure correctly and yet not really have done so, that one could seem to count correctly and yet have made an unnoticed mistake, etc.

The force of this reflection can be brought out clearly by reference to Dummett's example.[1] Suppose we find a method for determining the primality of a number other than Eratosthenes' Sieve—i.e. the technique of running through the natural numbers, trying to divide each one by its established prime predecessors—though, of course, demonstrably equivalent to it in its results. And let us suppose that the new method is more powerful in this sense: that we are able to apply it convincingly to numbers of such a size that we should doubt our capacity to compute their primality by means of the Sieve without making mistakes. We thus have as before a concept, ϕ—prime—and a criterion for applying it, C_1, Eratosthenes' Sieve. Before the proof gave us a new technique for determining prime numbers, we should have regarded the Sieve as correctly applied in any particular case just on the basis of a check on the computation; but now such a check is no longer regarded as sufficient. Even if we can find no mistake when we check the computation, we shall still say that there must be one if a number turns out composite by the new technique, C_2, which appeared prime when computed by the Sieve. Thus although it is true, both before and after the new technique is established, that the meaning of 'prime' could be explained by reference to Eratosthenes' Sieve, accepting the new technique effects a modification in our concept of when the Sieve has been properly applied. Before, it sufficed to check the computation; now, in addition, the results of the computation must coincide with those of the new, simpler technique.

Now, what I am saying is amiss with this attempt to meet the difficulty is simply that a check on the computation was never regarded as decisively sufficient for its having been carried out correctly. Of course, we regarded such a check as *corroborating* the result; we should have regarded ourselves as justified in saying that the computation contained no mistake if a careful check revealed none. But a check was never considered to establish conclusively that no error had been made; after all, the check could itself contain an unnoticed error. All along we drew a distinction between such a computation's really being correct and its being corroborated by a check. The check, we should have said, is merely a practical safeguard; we sometimes find previously unnoticed errors that way. But what the proof gives us is exactly a better way of checking such a computation.

The situation is thus not correctly described as one in which a formerly sufficient condition for supposing a criterion to have been properly applied now ceases to be so, nor, therefore, as one in which our concept of what it is properly to apply the criterion has been changed—save possibly in an irrelevant, psychological sense. Rather we should want to describe ourselves as better placed, after we have the

[1] 23 (2, pp. 504–5).

new technique, to recognise when, in cases where the application of the old criterion becomes complex, we have or have not correctly applied it. So there is still no concept of which we can recognise what in ordinary terms would count as a change in the conditions of its correct application.

4.   Let us relate these ideas to II. 30–1. Suppose we apply the Sieve to some large number—the computation may be imagined to take some three or four hours to complete—and get the result that it is prime. We check the computation over and get the same result again. Now we apply the new technique—which, let us suppose, takes a couple of minutes and less than a side of paper—and get the result that the number has factors in a certain range. We conclude that there must be an error in the original computation, which we missed when we checked it over. This 'must', according to Wittgenstein, corresponds to a track which we lay down in language. We have established a new concept of how such a computation ought to turn out. As a result of the proof's having given us the new technique, the rules for the judgment, that an application of the Sieve contained errors, have changed. This judgment has a new 'grammar'; the idea of error has been connected with a new criterion, and this connection did not exist before we made it. So it is not that the new criterion enables us to discover errors that were already there and could in principle have been recognised in terms of our ordinary standards of correct computation.

Well, it would seem at this point that the author of the *Tractatus* has here once again placed himself in a situation of trying to state what it is doubtful may coherently be stated. Only from a standpoint inside the language-game is a 'track' laid down by the proof; only from such a standpoint do we determine a pattern of use by accepting the proof. From a viewpoint outside the language-game there is, for a Wittgensteinian, no track, no pattern; no sense in which a 'decision' to 'fix the use of my language *this* way' may be seen to change the kind of use of an expression which I would otherwise have made. Outside the language-game, every move within it must be seen as primitive; there are no determinate truths about how expressions would have been used if different 'decisions' had been taken within it, or about how they *will* be used if *this* decision is kept to. But it is only from a viewpoint outside the language-game, if at all, that the 'track' can be regarded as a *new* one; as establishing a connection between concepts which did not pre-exist it, which is independent of how those concepts were before. Inside the language-game the tracks are correctly said to coincide: the new criterion only tells us things which the old criterion would have told us anyway, if properly applied. Inside the language-game we may regard the proof as establishing a pattern of application for certain concepts which we should otherwise not have followed; but we say it would have accorded with those concepts, as they were before the proof, so to apply them. For a Wittgensteinian, outside the language-game, there is no substance to this proviso, no sense in which it would have accorded with reality, even before we had the proof, to link the application of the concepts in question in the manner in which we now proceed after the proof. But from no standpoint can it be said *both* that our manner of application of certain concepts changes as a result of the proof *and* that these changes are *radical* changes, that we are not merely applying concepts in a way in which it would have been right to apply them all along.

It is hard to be clear whether this is to expose an incoherence in the character of Wittgenstein's thesis of concept-modification, or merely in the form of expression which he, and I by way of interpretation, have typically given to it. The burden of Wittgenstein's argument, as I have tried to explain it, is directed against pictures conjured by the expressions, 'determine', 'accord', etc. We think, in the present example, that the concept of correct application of Eratosthenes' Sieve is imbued by standard explanations with a certain automony: it is predeterminate what result it gives in any particular case, if correctly applied. When we appreciate the force of the proof, we see that the *essence* of the new technique is such that it cannot ever clash with the Sieve if both are properly applied. The new technique thus merely gives us a way of ratifying the application of 'prime' to cases to which it was already determined that it should apply, although we might in practice have been incapable of recognising that it did. For Wittgenstein, in contrast, all this is a muddle. The concept 'prime', has no determinate range of application, unratified by us, which any adequate new criterion must assess properly; and our acceptance of the proof is not rightly described as 'recognition' that the new technique meets this condition.

Now in one way this is all clear enough. The trouble is that the 'pictures' which Wittgenstein is opposing look to be more than pictures; they are forms of description enshrined in the language. It is by ordinary criteria—the mechanical character of the Sieve—true to say that it is determined of any particular natural number whether or not it is prime; and it is by other ordinary criteria—the proof—true to say that any lengthy application of the Sieve which classified as prime a number proved by the new technique to be composite must have contained an error appreciable by means of criteria used before we had the new technique. If 'concept-modification' is to make sense in ordinary terms, then what we have here are forms of expression which serve precisely to capture what it is for concepts to remain the same, to undergo no modification. Thus the thesis that proofs modify concepts—give us *essentially* new rules for their application—begins to sound like a radical thesis, a suggestion of change in our actual linguistic habits. And we know from Wittgenstein's general philosophical outlook how unattractive such a revisionary position would have seemed to him.

It is tempting to conclude that Wittgenstein does himself no service by advancing the thesis in the forms which he typically favours—that proofs stimulate us to recast criteria, lay down new linguistic tracks, establish connections which formerly were not there, and in general change the use of our language. On the contrary, unless we abandon certain well-entrenched idioms, there simply seems to be no standpoint from which these things can be correctly said. What, we might suggest, Wittgenstein's arguments tend to establish is better expressed along the following lines. There is capable of coherent explanation no 'ultra-physical', objective sense in which, after a new proof is accepted, the concepts involved *remain the same*; there is, that is to say, no fact in such a case of which our habitual use of certain forms of expression—the language of determinacy and accord—serves to express our grasp, and which is somehow ulterior to the fact that it is, by consensus, correct to apply the relevant forms of expression in this kind of circumstance.

This would be to endorse, after all, the 'therapeutic' view of the concept-modification thesis, sketched in Chapter III. But we shall return much later[1] to the question whether a more literal interpretation of it may not yet be defensible.

5.   There is a difficulty in Wittgenstein's view of proof and necessity on which we have not so far touched. Consider I. 74:

> 'If the form of the group was the same, then it must have had the same aspects, the same possibilities of division. If it has different ones, then it isn't the same form. Perhaps it somehow made the same impression upon you; but it is the *same form* only if you can divide it up in the same way.' It is as if this expressed the essence of the form. But I say: if you talk about *essence*, you are merely noting a convention. Here one would like to retort that there is no greater difference than that between a proposition about the depth of the essence and one about a mere convention. But what if I reply: to the *depth* that we see in the essence there corresponds the *deep* need for the convention.

Contrast this with v. 40:

> It is natural for us to regard it as a geometrical fact, not as a fact of physics, that a square piece of paper can be folded into a boat or a hat. But is not geometry, so understood, part of physics? No; we split geometry off from physics. The geometrical possibility from the physical one. But what if we left them together?—if we simply said, 'If you do this and this and this with the piece of paper, then *this* will be the result'? What has to be done might be told in a rhyme. For might it not be that someone did not distinguish at all between the two possibilities? As e.g. a child who learns this technique does not. It does not know and does not consider whether these results of folding are possible only because the paper stretches, is pulled out of shape, when it is folded in such and such a way, or because it is *not* pulled out of shape. And now isn't it like this in arithmetic too? Why

[1] In Chapter XXIII.

shouldn't it be possible for people to learn to calculate without having the concepts of a mathematical and a physical fact? They merely know that this is always the result when they take care and do what they have learned.

For practical purposes, for example, making hats and boats out of bits of paper, geometrical considerations need not obtrude at all. But *if* we have a geometry in terms of which we can describe the sequence of folds made, then that will determine what account to give of a case where someone seems to carry through the sequence perfectly properly and yet a paper hat does not result. That is, depending upon the geometry, we can blame the paper for changing shape—or being otherwise unsuitable—or for *not* changing shape. (If we are independently confident of the suitability of the paper, we shall look for a mistake in the folding.)

Now what, practically speaking, do we gain by being in a position to offer such an explanation? A natural answer would be that if we incline to the first explanation and it is practically urgent for us to be able to make paper hats, the proper strategy would appear to be to ensure that the paper is not too flexible; and this is something which is answerable to independent (physical) criteria. On the other hand if our geometry gives an account of the second sort, it might be prudent to supply all hat-folders with rubberised paper. Which geometry we adopt matters in just this sense: it is out of our hands which strategy would prove the more successful, will heighten the rate of paper hat production.

This sort of reply, however, does not meet the issue. There is no doubt that it may be of practical consequence what account of the situation is given by one's geometry. But the question is rather, why have a *geometrical* account of the situation at all? Why should we not approach the question of meeting our practical needs—the manufacture of paper hats or whatever—in a wholly *inductive* spirit, classifying paper into types just in accordance with whether assiduous attention to the folding technique seems to produce the desired results? We want to know which kind of paper we can expect to get the right results with if we try hard, and with which we cannot. Geometry can tell us; but so can induction. For the purpose of folding hats, an inductive correlation between, for example, lack of flexibility and success is all we need.

According to Wittgenstein's view, a geometry will here establish a conventional association between the concepts of folding the paper properly, the qualities of the paper, and the getting of a certain outcome. We shall establish as 'grammatically' correct such a conditional as: if such and such a sequence of folds is followed, and the paper is not too flexible, then a paper hat results. But while, if we are to have a convention, there may be practical reasons for preferring one to another, what is completely unclear is why, as far as purely practical reasons go, a conventional association should be established at all. What, then, is the source of the 'deep need' for the convention?

It is, as Wittgenstein suggests, the same with arithmetical examples. If I have to pave a yard with slabs of concrete one metre square, and on measuring it find it to be three metres by four metres, I can be very sure indeed that twelve slabs will suffice. But, for practical purposes, why should this certainty be any less use if it is merely inductive? The empiricist conception of arithmetic, proposed by Mill, is generally thought inadequate to explain the way in which we actually use arithmetical statements. But *if* we regarded such statements, as Mill suggested we actually do, simply as very well supported inductive generalisations about what tends to happen when by ordinary criteria we count and calculate correctly, what practical purposes now served by arithmetic could not then be served? What is the pressure on us—the need—to fix it as a *rule* of language that $3 \times 4 = 12$, to lay down a new track in language, establish new criteria for miscounting, etc.? Why for practical purposes would it not do to explore inductive relations between the concepts involved as they stand *before* the rule is adopted?

For Wittgenstein, a necessary statement is created when we, for whatever reason, effect a connection between the concepts which it involves. We involve them in one another's conditions of correct application. Such an account at least liberates us from the responsibility of explaining how we could recognise that such an involvement was already independently established; there is no need on such an account to explain, reverting to the previous example, how we are able to recognise that the new and old criteria of 'prime' necessarily coincide in result if properly applied. There is no problem about explaining how we recognise such a thing because we do not *recognise* such a thing; rather we come not to *count* any application of the old method as correct which diverges from a surveyable, seemingly correct application of the new method. But now it appears that we have lost one mystery only to gain another; if there is no longer any problem about how necessity can be recognised, what is now unclear is why there should be any necessary statements at all. This is a question to which the answer, according to the recognitional conception of necessity, would be trivial: the presence of necessary statements in our language is a product of the requirement that it be sufficiently rich to express every kind of fact which we are capable of apprehending. But what is Wittgenstein's answer?

6.   It might be thought that we began to move towards a possible solution to this difficulty at the beginning of the previous chapter. The problem was: if '1 inch = 2.54 centimetres' is a necessary truth, how could this necessary truth be discovered by a purely *empirical* investigation, by seeing how seemingly correct measurements in inches and centimetres tended to correspond? Wittgenstein's answer was that what was yielded by such an investigation *is*, of course, an empirical statement, which we then promote to the rank of rule, thus removing it from

the area of possible conflict with experience. Otherwise seemingly accurate measurements which conflict with it are deemed, just for that reason, not to be accurate. So our question is, what is the point of making this remarkable move? For there is no practical purpose served by the rule which could not be served equally by the empirical generalisation.

What was suggested in Chapter V was, in effect, that one source of pressure on us to accept a rule in this instance derived from our conception of measurement as determining a property of the object. If we allowed that a pair of simultaneous measurements of an object in inches and centimetres could both be accurate, yet give results standing in a variable ratio, that would entail that variations in the readings given by accurate measurements need not be associated with changes in the object measured. But it is implicit in the idea of a procedure which tells us something about an object that, if it is properly applied, different results correspond to differences in the object and vice versa. Suppose for example, that, on the first pair of measurements, the object proves to be 2 inches and 5 centimetres long and, on the second pair, 2 inches and 5.08 centimetres long, and that all the measurements are considered to be perfectly accurate. Has the object undergone a change in length between the two pairs of readings or has it not? If it has, then an object may change in length and yet accurate measurements reveal no difference. ('2 inches' was obtained twice.) If it has not, then accurate measurement of an object may give distinct readings at distinct times, although the length of the object remains the same. (Different readings were obtained in centimetres.) Thus the need for a *rule* of conversion appears to be imposed on us by the principle that accurate measurement of an object gives different readings on different occasions if and only if the object has changed; and this principle in turn is imposed on us by our conception of length as a property of the *object*, of which accurate measurement supplies the means of discovery.[1]

We might be encouraged to extend this account to any case involving a concept whose application is conceived of in this way as wholly a matter of the condition of something external; that is, a concept of which we say, when its associated criterion of application gives different results when applied to the same object, that either the criterion has been misapplied or the object has changed. Just this, of course, is a feature of our concept of the cardinal number of a group of objects as determined by counting. If a group of objects is counted correctly on different occasions and different results are achieved, then it is the number of the group which has changed.

[1] So formulated, the principle is actually stronger than α) of the last chapter; and, as there noted, α) would likely be unacceptable as it stood to a philosopher of generalised intuitionist, or strict finitist, opinion. I leave it to the reader to ponder what, if any, reconstruction such a philosopher could give of the present suggestion why acknowledgment of rules of metric conversion might seem imposed. But the problem will not, of course, arise where we are concerned with procedures, like counting but unlike measuring, where the idea of an error too *subtle* to be apprehended by us has no application.

The status of arithmetical equations as rules can thus arguably be explained along similar lines. Suppose we were to allow it to be a possibility that simultaneously conducted correct counts of the children in a classroom should on one occasion yield five girls, seven boys and thirteen children and on another occasion five girls, seven boys and twelve children. Has the number of children in the classroom changed in the meantime or has it not? If it has, then a group may change in number although an accurate count of two exhaustive sub-groups reveals no difference; and if it has not, then correct counting may give variable results with respect to an unchanging group. To think of counting a group, either completely or by counting a number of exhaustive sub-groups, as determining a property of the group, commits us to regarding the results of correctly applying the two techniques of counting as standing in fixed relations. The need to treat arithmetical statements as rules is thus imposed on us by the conception that for an arithmetical expression, standardly understood, to denote the number of some group is a matter purely of how it is with the group.

7.   The distinction between properties of objects which are in this sense thought of as wholly *of* the object, and properties which are not—properties such that, if associated with a procedure of application at all, we do not insist, when it gives different results, that either the object has changed or the procedure has been misapplied—is, if it can be made out, a fundamental distinction. It would perhaps be one way of approaching the question of exegesis of Locke's distinction between primary and secondary qualities of material bodies.[1] But, on reflection, can it really be used to explain a need to endow generalisations with the status of rules of language in certain areas? To be sure, if we think of length as a property of the object and of correct measurement as determining length, then descriptions of length in inches and centimetres must bear a certain stable relation one to another; otherwise they were not all arrived at by correct measuring. Likewise, if correct counting either of a whole group or of exhaustive sub-groups is thought of as a method for determining a property of the group, then the results of these two techniques of counting must stand in stable relations. But, of course, these considerations simply assume that we are concerned with determining a *single property*—length, number—in different ways. It is just assumed that measurement either in inches or in centimetres determines length, although different forms of description of length result; and that counting a whole group or exhaustive sub-groups both ascertain the number of the group, although again yielding different forms of description of it. So if the proposal of the previous section is to lead to an explanation of the 'deep need' for arithmetical and metric rules, then we need to unearth a deep need for thinking of

[1] *Loci classici*: 56, II. viii, 9–26; xxiii, 9–11; IV. iii, 11–13, 28.

the procedures involved—measuring in inches and in centimetres, counting a whole group and exhaustive sub-groups—as essentially determinations of the same thing.

Again, however, it appears that such a need cannot be made out to be purely *practical*; that is, it does not derive from the kind of application to which we want to put the concepts of length and of number. Suppose we were to fragment the notion of length into two notions: *length-in-inches* and *length-in-centimetres*. Then what we now regard as the discordant readings above—2 inches, 5 centimetres, 2 inches, 5.08 centimetres—could each be regarded as correct determinations of properties of the object. All that has happened is that while the length-in-inches of the object did not change between the two readings, the length-in-centimetres has marginally increased. Suppose it never occurred to us to think of length-in-inches and length-in-centimetres as analytically associated. There would nevertheless be very good inductive evidence for supposing that when length-in-inches remains constant, only minimal variations in length-in-centimetres tend to occur, and vice versa. So an assessment of either property would be in practice a very good guide as to how an assessment of the other would work out. And, for purely practical purposes, all we require is such a guide.

Similarly with arithmetical equations. Suppose we were to fragment our concept of the cardinal number of a finite group into two notions: the concept of the *total*, as determined by a complete correct count of the group, and the concept of the *n-residue*, as determined by a correct count of all but n elements of the group, n smaller than the total. Then we could think both of the total and of each n-residue of a group as properties of the group consistently with accepting as correct all the results obtained above in counting the children. The total of the group changed from twelve to thirteen but the 7- and 5- residues both remained the same. Such an occurrence would still be extraordinary; in practice we could expect it hardly ever to be the case that a change in total was not accompanied by a change in each n-residue. Indeed, we could expect law-like generalisations to hold between total and n-residue as codified in the usual recursive definition of addition. This general account of the matter would be so strongly corroborated by experience that where we had an apparent counter-example, we should be strongly inclined to suspect that error had entered into the determination either of total or of n-residue. We might say indeed, 'I must have miscounted', as a physics master says of a wayward schoolroom experiment, 'There must have been a pressure leak.'

As far as their practical application goes, there would be nothing lost by treating equations of elementary addition in this way. Our use of them would indeed strongly resemble what it is now. The difference would be that in the event of a repeatable breakdown in the correlations, associated with no independently noticed error in counting or partition, we should look for a restriction on the scope of the law; whereas now we

danger is (II. 19) that of looking at the shortened procedure as a 'pale shadow of the unshortened one'—of a proof, for example, that a number is prime involving more powerful means as a pale shadow of what can be done with the Sieve. But (II. 42):

It is not something behind the proof, but the proof that proves.

II. 27:

Even if the proved proposition

—for example, that an arbitrary true arithmetical equality can be derived in higher-order logic—

seems to point to a reality outside itself, still it is only the expression of acceptance of a new measure of reality.

On this interpretation, Wittgenstein's emphasis upon the surveyability of proofs has to be taken in close association with the thesis that proofs essentially modify concepts. It is not so much that we are to *exorcise* the whole concept of a correct but unsurveyably long proof or calculation; rather, we are to think of the content of statements which seem to have essential recourse to the idea of such proofs as determined not by some antecedent conception of a realm of fact accessible only to computers and super-prodigies, but by the *surveyable* demonstrations which we accept of them.

VIII

The Law of Excluded Middle

Sources

RFM: IV. 9–20
  PI: I. 352–6; 516–17
  PG: II. 31 (p. 400); 39 (p. 458)
  OC: 199–200
   Z: 682–3
  PB: 173

Mathematical statements as rules or commands:

RFM: I. 164; App. *I*, 1–4, 20; II. 25–6, 28; IV. 13,
    17, 18
LFM: lectures III. p. 33; IV; V. p. 55; VII, p. 70; X.
    pp. 98–9; XIV, pp. 134–5; XXV, p. 246; XXVIII,
    p. 268

1.   In IV. 9–23 Wittgenstein considers the simplest kind of case
where the law of excluded middle[1] is called into question by the
intuitionists: an alternative of the form

$$(\exists x)Fx \quad V \quad -(\exists x)Fx,$$

'x' ranging over an effectively enumerable, infinite set and 'F' effectively decidable of each of its elements. The actual example is the
question whether a certain pattern, for example '550', occurs in the
decimal expansion of π; 'x' is thus to be considered as ranging over the
set of finite, initial segments of the decimal expansion of π, and 'F' as
expressing the property of containing the sequence '550'.
   A hasty reader might form the impression that Wittgenstein wanted
to reject the principle for reasons very similar to those of the intuitionists. In fact, though, no such explicit rejection is to be found. What
there is, in common with the intuitionists, is some of the background
from which their rejection stems. Wittgenstein agrees that the acceptability of the principle for statements involving quantification over

---

[1] Wittgenstein follows the intuitionists in calling 'excluded middle' the principle that the
disjunction of a statement with its negation is valid, rather than—more intuitively—the weaker
principle that there is no third truth-value; (which, since they accept the double negation of
P V $- P$, the intuitionists can be seen not to contest). We will follow this terminology.

infinite totalities cannot be apprehended simply by an appeal to the model of finite quantification; more generally, he insists that its validity cannot be something of which we have an abstract apprehension irrespective of questions to do with the content of statements to which it is applied. IV. 12:

> In the Law of Excluded Middle, we think we have already got hold of something solid, something that at any rate cannot be doubted. But in fact the tautology has just as shaky a sense, if I may so put it, as the question whether *p*, or not-*p*, is the case.

This seems to anticipate the reservations concerning the principle which an intuitionist, or, more generally, an anti-realist, feels: the problem is not that there may be some third possibility (IV. 33) which a proponent of the law has overlooked; what he has overlooked are rather questions to do with the *senses* of the disjuncts alleged to be exhaustive of all possibilities. When people are inclined to 'hammer away' (IV. 10) with the principle as though its application, for example to the alternatives about π, could not be called into question, the probability is that they are feeding on pictures of the truth of *P* and its negation which are drawn from cases where the range of quantification is finite. There is thus a substantial question whether these pictures remain appropriate in the infinite case and, if they do not, whether the validity of excluded middle will remain unquestionable after a better model is supplied for the sense of negating an existential quantification over an infinite range.

As is familiar, the intuitionists propose that it does not. For an intuitionist, a disjunction of this type is acceptable only if we can guarantee either to be able to find an individual with the property *F*, or to be able to prove that such a search cannot succeed. There is no sense in the idea of its *just being true* that π, for example, does not contain the sequence '550', although we have no way of demonstrating the fact. In the first part of IV. 11, Wittgenstein seems to have in mind exactly these intuitionistic reservations:

> To say of an unending series that it does not contain a particular pattern *makes sense* [my italics] only under quite special conditions. That is to say, this proposition has been given a sense for certain cases; roughly, for those where it is in the *rule* for the series not to contain the pattern.

For the intuitionists, comparably, the statement, '$-(\exists x)Fx$,' '*x*' ranging over an effectively enumerable infinite series, is acceptable just in case we possess a general method for defeating the supposition of an arbitrary member of the series that it is *F*; and the status of a construction as such a method will have to be recognised by appeal to the character of the rule which generates the series in question.

Wittgenstein proceeds to anticipate the natural objection to the intuitionists' proposal—an objection which is simply an expression of the platonist conception of *truth* for such statements, sketched at the

end of Chapter I. We are concerned with a series which is wholly
determined by its rule of expansion; that is, for which it is fixed what
counts as its correct continuation at any particular stage. Thus, for any
integer n, it is determined in advance whether or not the initial segment
of π consisting of its first n digits contains the sequence '550'. Now, the
statement ' "550" does not occur in π' is just the truth-functional
conjunction of all statements of the form ' "550" does not occur in the
first n places of π'; so how can it be correct to question the validity of
'Either "550" occurs in π or it does not'? The truth-value of each of the
conjuncts, ' "550" does not occur in the first n places of π', is *predeterminate*. If each such conjunct is true, then π does not contain the sequence
'550'; and if one, and therefore infinitely many, are false, π does contain
'550'; and if it is determined of each conjunct whether it is true or false,
then it must be determinate either that all are true or that some are false.
Just this is the thought of Wittgenstein's objector in iv. 11:

> It has to reside in the rule for the series that the pattern occurs or that it
> doesn't . . . doesn't the rule of expansion *determine* the series completely
> and if so . . . then it must implicitly determine *all* questions about the
> structure of the series.

Wittgenstein retorts that the objector is thinking about finite series.
The point, it seems, is that in finite cases to have a method of expansion
which determines the correct continuation of the series at every stage *is*
to have a method of settling any question of this general type about the
series. With the example concerning π, in contrast, we have no effective
way of *transforming* our knowledge of how to continue the series at any
stage either into a proof that '550' occurs in some region—for example,
by actually exhibiting it—or into a proof that it does not occur at all. If a
statement about the entire expansion of π is to be true, then, for an
intuitionist, it must at least in principle be possible for us to recognise
its truth; but, in contrast with the situation in the case of finite series,
the determinacy of π does not guarantee this possibility. The determinacy of π is an irrelevance to the question of the validity of excluded
middle if an anti-realist view is taken of what it is for a statement to be
true of the whole expansion. And Wittgenstein appears to be endorsing
such a view.

If this were all that he had said about the matter, it seems we should
have to regard Wittgenstein's attitude as broadly coincident with that
of the intuitionists. In fact, though, it is necessary literally to read
between the lines in iv. 11, to develop this parallel. There are three
remarks in the paragraph which suggest the divergence to which I drew
attention in Chapter I:

> Moreover when I calculate the expansion further, I am deriving new rules
> which the series obeys.

> That is correct

that is, that all the places of π are determined—

> if it is supposed to mean that it is not the case that the nth, e.g. is *not* determined.

> If you want to know more about the series, you have, so to speak, to get into another dimension (as it were, from the line into a surrounding plane.) 'But then isn't the plane *there*, just like the line, and merely something to be *explored*, if one wants to know what the facts are?' No, the mathematics of this further dimension has to be invented just as much as any mathematics.

These three remarks are most naturally interpreted as follows. It is not *incorrect* to describe the rule for expanding π as determining every place in the series. (A Wittgensteinian might add, 'For that is just what we do say'.) It is not the case that at some stage we have, as it were, no instruction about what to write down next, and so can there write down just what we please consistently with the intention to write out the decimal expansion of π. Nevertheless, in the sense I attempted to explain in Chapter II, the identity of each place in π is not *objectively* determined in advance. The considerations to do with rule-following are directly applied to the decimal expansion of π. It is in this sense that when the expansion is calculated further, new rules are derived for the series, and (IV. 9):

> However queer it sounds, the further expansion of an irrational number is a further expansion of mathematics.

And the conception of the decimal expansion of π as, so to speak, objectively fluid liberates the mathematics of the 'further dimension' from the constraint of fidelity to a predeterminate conceptual structure.

Wittgenstein thus finds the platonist attitude to this example to be doubly at fault. The platonist overlooks the importance of the point that there is not in cases like that of π the connection, present in finite examples, between the rule-determined character of the series and the existence of a method for deciding any statement involving only effectively decidable basic predicates and quantification over the elements of the series. This is essentially an intuitionistic reproach. But Wittgenstein goes further: the platonist, and perhaps the intuitionist too, misunderstand the sense in which a series like π is *determined* by rule; for the conception that each of the infinitely many in principle decidable questions of the form, do the first n places of π contain the sequence '550'?, has a predeterminately correct answer is intelligible only if we avail ourselves of the kind of objectivity which the rule-following considerations prohibit. Accordingly, the sort of mathematics needed to put us in a position to vouchsafe that all those questions were to be answered negatively would not be required, whatever other constraints it would have to meet, to effect a *discovery* about the objective constitution of π.

We noted at the beginning of the last chapter that, in addition to the ideas about surveyability, there are four main strands in *RFM* which it is natural to suspect might have implications for the question of which mathematical practices are sound. Two of these were the points of sympathy with the intuitionists and the rule-following considerations, which we have just seen brought to bear on the law of excluded middle. But the other two also figure in Wittgenstein's discussion of the law. They were, first, the considerations to do with the sense of undecided mathematical statements, and the rôle which example and analogy can play in fixing that sense, which we touched on at the end of Chapter III; and, second, Wittgenstein's recurrent play with the idea of mathematical sentences as expressive of *rules* rather than statements in a strict, descriptive sense.[1]

It is on these latter two themes that we shall concentrate for the rest of this chapter. We shall test their revisionary potential by considering whether they can be made to provide any reason for accepting or rejecting the validity of excluded middle. The seemingly gross tension of the intuitionistic and rule-following ideas with Wittgenstein's non-revisionary conception of philosophy will occupy us thereafter.

2. $PV -P$ has no clearer a sense, 'just as shaky a sense', Wittgenstein suggests, as its disjuncts. But why should we suppose that there is anything 'shaky' about the sense of the example about π or that of its negation? Suppose, however, that we had to *support* the contrary claim that we understand these examples perfectly well, and that a proof, one way or the other, would in no way enlarge or sharpen that understanding. We should presumably consider it enough to show that we understood both which object—π—was involved and what was being said about it. And we should regard it as sufficient for the former understanding that someone knows how to expand π; and for the latter that he can recognise as such other examples to which the predicate 'contains "550" ', or its contrary, applies. That is, we should attempt to characterise the understanding which we think we have here by reference to other examples of comparable semantic structure in which the same object is presented as the subject of a predication or in which the same predication is made of different, comparable objects. But, Wittgenstein wants us to ask, how do such examples show what is meant *here*? How do the examples show what it is like for *this* infinite decimal to contain '550'? (IV. 12). And, presumably, our understanding in such cases goes no further than what the explanation which we should give of it—the examples—can be used to make clear.

Why, though, should it be thought that the transition from the

---

[1] The question of the sense of the disjunction, 'Either "550" occurs in π or it does not', before it is resolved one way or the other, is prominent in IV. 12, 17, and 19. And the conception of mathematics as supplying a network of rules issues—IV. 13, 17–19,—in a proposed analogy between mathematical statements and commands.

examples to the meaning here involves *special* difficulties, unparalleled by the situation with other and, especially, contingent cases? It is unclear how this example differs from any case where we explain the meaning of '*Fa*' by explaining the reference of '*a*' and illustrating by examples a criterion for the application of '*F*'. After all, mastery of any concept resides in a capacity to apply it to *new* cases. So do we have here anything other than an application of one of the essential points made by Wittgenstein in the discussion of rule-following: the training which we receive does not fix meanings beyond all possibility of misunderstanding? Or is he trying to call attention to some special difficulty with statements like the alternatives about π to do with the move from other examples to an understanding of the new case?

Wittgenstein claims that if examples of other decimals containing '550' really showed what it is like for the pattern to occur in π, they ought also to fix the sense of the negation of this statement. In one way this seems quite unreasonable; if we are giving a child an ostensive definition of 'brown', it would make his task of learning needlessly difficult if we presented him only with brown samples and supplied him with no illustration of what it was for something not to be brown. It is true, of course, that were we—harshly—to restrict his experience to positive samples, he would still be held not to have grasped the intended concept unless he both applies *and* withholds the predicate in appropriate circumstances. But that does not provide a sense in which our examples 'showed what it is like' for something not to be brown.[1]

There is a difficulty, to be sure, about this rejoinder to Wittgenstein, if we try to carry it over directly to the example about π. For the possibility of teaching the sense of a predicate by example arguably requires that a trainee who is totally ignorant of the sense of the predicate can nevertheless understand which objects are being drawn to his attention as samples of its application; whereas it is not clear that we ought not to incorporate the ability, for example, to answer correctly questions of the form, 'does "550" occur in the first n places of x?', into the criteria for understanding what series, x, is being presented. But this slight awkwardness need not detain us. Suppose that we fix 'ϕ' as a secret code name for the sequence '550', and then by examples teach children, ignorant of the code, to answer questions of the form: does ϕ occur in the first n places of x?, x some series of integers. It will be taken as a criterion for grasp of the point of the examples that the children should be able to cope with new cases. It could be asked similarly: do examples of series with ϕ in the first n places fix an idea of what it is like for the sequence *not* to occur in the first n places of π? But now there seems no interesting difference between the example and that of 'brown'. And, besides, Wittgenstein's reliance upon the quite undeveloped notion of 'showing what it is like for' seems illegitimate; for it is a contingent matter how detailed, or elaborate, successful explanations

[1] Cf. *PI*, footnote p. 14.

by example have to be. Examples make clear as much as people tend to get out of them; the children *might* pass smoothly to an understanding of 'ϕ does not occur in x' from a single, positive illustration.

I have restricted the example to n places deliberately. Wittgenstein suggests, obscurely (IV. 18), that a peculiarity of the example about π is that the two possibilities have to be *imagined* separately, that distinct ideas are needed for P and its negation here and that one idea does not as elsewhere suffice. We might wonder what the *ideas* which we form have to do with the meaning. The intended point, however, is presumably the following. Suppose one of the children has guessed the reference of 'ϕ'; then he has a single idea of the truth-conditions of P, $=$ 'ϕ occurs in the first n places of π', and its negation, in this sense: he can form a picture of circumstances under which P would be true:

$$\text{e.g.} \quad \underset{1}{3.14.} \ldots \ldots \ldots .550. \ldots \ldots \underset{n}{,}$$

and then straightforwardly transform it into a picture of circumstances under which not-P would be true:

$$\underset{1}{3.14.} \ldots \ldots \ldots \ldots \ldots \ldots \ldots \underset{n}{,}$$

just by omitting every occurrence of ϕ. But there is no such straightforward transformation *without* the restriction to n places. When x is an infinite sequence, and P is 'ϕ occurs in x', the former type of pictures continue to serve as representations of circumstances under which P is true; for if ϕ occurs in x at all, it occurs in its first n places, for some n. But the transformation just described of any such picture results only in a picture of circumstances under which P may still be true.

If this interpretation is correct, then Wittgenstein's concern about the rôle of examples in fixing the sense of the unrestricted alternatives about π depends essentially upon the infinitistic character of these statements. Simply, we have no effective technique for resolving such alternatives in infinite cases. And this does, I suggest, make good a sense in which examples of infinite decimals where ϕ is established to occur, either by simply expanding the decimal or by more sophisticated means, do not make the *same contribution* to the sense of the conjecture, 'ϕ does not occur in π', as the examples given to the children of other decimals containing ϕ in the first n places make to the sense of 'ϕ does not occur in the first n places of π'. Not that our reservations above about Wittgenstein's notion of 'showing what it is like for' were misplaced. Rather the point concerns a disanalogy in the degree of inter-relation between the criteria for understanding P and its negation in the two kinds of case. If one of the children does not know how to settle the truth-value of 'ϕ does not occur in the first n places of π', despite knowing how to construct the decimal expansion, then he cannot have understood the point of the examples which we gave him—even if these were unreasonably meagre. But in order to be credited with the knowledge that a simple verification that ϕ occurs in the first n places of π simultaneously settles the truth of 'ϕ occurs in π'—or even with an understanding of cases where such an existential statement is proved in a more sophisticated way—it is not necessary that one also knows what to count as a criterion for the truth of 'ϕ does not occur anywhere in π'. *Any* grounds for attributing to someone an understanding of a schematic picture of the truth conditions of 'ϕ occurs in the first n places of π' will be grounds for attributing to him an understanding of that statement's negation; but they may be quite inadequate to justify attributing to him a concept of what it is for *no* such picture to apply, irrespective of the choice of n.

3.   On this interpretation, Wittgenstein has a perfectly good point about the rôle of examples in fixing the sense of the alternatives about π; and it is a point which has particular application to examples involving unrestricted quantification. But the explanation of its particularity is just that it is tantamount to a re-statement of the intuitionists' complaint about the platonist treatment of infinite quantification. If 'F' is decidable and the range of 'x' finite, the assertability-conditions of both '$(\exists x)Fx$' and its negation can be explained simultaneously by reference to the idea of an exhaustive, case-by-case check and its two possible outcomes. From a platonist point of view, we are supposed to have a concept of the truth-conditions of such alternatives which enables us to grasp their import even when the range of 'x' is broadened so as to include infinite totalities—a concept, indeed, which enables us to recognise that their disjunction remains logically valid. The intuitionist denies that we have any such concept. He will allow that the truth-conditions of '$(\exists x)Fx$' can still be explained by reference to the idea of a check of cases (though the idea of an exhaustive such check is no longer intelligible); for '$(\exists x)Fx$' is true if and only if there is some finite restriction on the range of quantification such that some individual in the restricted domain has F. But when the range of 'x' becomes infinite, a concept of truth explained by means of the idea of exhaustive examination ceases to be able to serve as an explanation of the assertability-conditions—the *use*—of '$-(\exists x)Fx$'. The ground for saying so is not an empirical assumption about what human beings can or cannot get out of particular explanations. Rather, what is essential for mastery of the use of mathematical statements in general is a concept of the circumstances under which they would be proved; and to know that infinitistic '$(\exists x)Fx$' can be proved by exemplification falls short of knowledge of what to count as a proof of its negation.

This is familiar territory. But these ideas stop short of the notion that there has to be a *haziness* in the sense of 'ϕ does not occur in π'. They are consistent with the view that there is a firm bridge from other examples where proof is available to a concept of what we should count as a proof in this case. (In fact, of course, both alternatives about π would have to be established by a substantial proof; for a 'direct' verification that '550' occurred would probably be too involved to be trustworthy.) Wittgenstein, however, as we have seen, is explicitly doubtful of such an idea (IV. 13):

> If someone says: but you surely know what 'this pattern occurs in the expansion' means, namely *this*—, and points to a case of occurring, then I can only reply that what he shows me is capable of illustrating a *variety* of facts; for that reason I can't be said to know what the proposition means just from knowing that he will certainly use it in this case.

So even after the establishment of paradigms for the assertion of infinite decimals that they do not contain ϕ, Wittgenstein would hold, it seems, that there is a special indeterminacy about the application of such paradigms to new cases.

One argument which he used in the 1939 classes was this:[1] the construction for the case of π would have to be something *analogous* to the paradigms. But we are reluctant to circumscribe the respects in which the analogy will have to hold beyond saying that the *effect* of the construction will be to show of π what the paradigmatic constructions show of other decimals; viz. that it does not contain the sequence '550'. That, however, the idea of doing something analogous provides no certain bridge can be seen by the simple reflection that if someone presents some wild construction as 'analogous' for the case of π, there is no finally defeating his claim; we will only be able to protest that *that* sort of analogy is not germane to the sense of 'ϕ does not occur in π'. But if the notion of a construction analogous to *these* but adapted to *this* case is what is supposed to fix the sense of 'ϕ does not occur in π', then it is illegitimate to appeal to that sense to rule out certain sorts of alleged analogy.

It is difficult to be absolutely certain whether there really is a special point here, peculiar to recognition of new proofs, or only an application of the points about sameness of new usage which issue from the discussion of rule-following. But two considerations strongly suggest the latter view. First, an analogue of the argument just sketched could be developed for any undecided statement, mathematical or otherwise. And, secondly, these thoughts leave unchallenged the ordinary idea that a proof of 'ϕ does not occur in π' would be acceptable only if the authority of paradigms could be cited for *every one* of its ingredient steps, provided these are taken small enough; this contrasts with the scope which we should expect occasionally to have for *discretion* when

[1] *LFM* lectures VI, and VIII, pp. 84–6.

considering whether a vague statement might justifiably be asserted in particular circumstances. Quite what account the ordinary view should give of the capacity of mathematics to develop with respect of *methods* of proof is a different matter; but, trivially, not all new proofs involve such development, and there is no reason to think that a resolution of the conjecture about π would have to do so.

4.   A quite separate question concerns not the application but the *establishment* of paradigms for the assertion of mathematical statements involving quantification over infinite domains. Wittgenstein and the intuitionists agree in holding that the platonist explanations of the quantifiers, even if appropriate for quantifiers of finite range, cease to be so when the range of quantification becomes infinite. Now, ordinarily we should want to describe ourselves as *recognising* the appropriateness of, say, proof by mathematical induction as a result of an antecedent understanding of what it ought to mean to talk to *all* natural numbers. You can imagine a class of schoolchildren being set the problem: devise a way in which we might know that all members of an infinite series had a certain property, as a kind of intelligence test. But can Wittgenstein and the intuitionists coherently say this? If an epistemic 'jump' is involved in the transition from finite to infinite quantification, it must be a possible situation to understand only the former. One's concept of generality would presumably be tied to the idea of an exhaustive search and its two possible outcomes. And how could such a concept cover the transition to infinite quantification at all? How could one even begin to recognise the bearing of a technique like induction on one's understanding of 'all'?

In fact it is straightforward to see that, for someone in this situation, the appropriateness of *any* infinitary technique to his antecedent concept of generality could not possibly be a matter of recognisable fact. For one thing which would have to be recognised would be that the new technique, when applied to an infinite totality, must give results *coincident* with those of an exhaustive examination of any of its finite subtotalities. And *this* 'any' cannot be understood by appeal to the idea of exhaustive examination; to know on the basis of exhaustive examination what would be the result of an exhaustive examination of any finite sub-totality would be possible only if the original totality were itself finite. So if someone's concept of generality was free of any residue not explicable in terms of the criterion of exhaustive examination, he would be in no position to comprehend, let alone recognise as satisfied, all the conditions which the new technique would have to meet.

I have presented the point as relating to the transition from quantification over finite totalities to quantification over infinite ones. It could, of course, be argued that it is the unsurveyability of a totality, rather than its infinity, which is the essential obstruction to an explanation of quantification by reference to the idea of case-by-case examination. But whether it is infinite or merely unsurveyable quantification whose sense is in need of fresh explanation, what the foregoing establishes, if sound, is that we must think of this fresh explanation as *giving* sense, for example, to the supposition that '550' does not occur in π, rather than as establishing new conditions of assertion for such statements in *conformity* with a concept of their sense which we somehow already have. To accept the argument is to side with Wittgenstein against any intuitionistic philosopher who, while maintaining that a statement's truth had always to be explained by reference to conditions which we can in principle determine to obtain, held that in establishing truth-criteria for the negation of an infinite existential quantification, we have a responsibility to ensure that they conform to the sense of such statements already established for cases where their range of quantification is finite. This point is quite independent of Wittgenstein's general scepticism concerning the idea of the objectivity of such conformity.

For our present purpose, however, we need not pursue the matter. For it is evident that these thoughts provide no reason, additional to those of an intuitionist who took the line just sketched, for rejecting the validity of excluded middle as applied to examples like that about π. Indeed, the whole conception that there is a haziness in sense about such statements, whatever motivates it, provides no ground for rejecting the law unless it is associated with the repudiation of a platonist concept of their meaning. Otherwise it is open to us to maintain merely that, while perhaps the truth-conditions of these statements have not yet been fixed absolutely sharply, still whatever these conditions are eventually determined to be, they will be mutually exclusive and conjointly exhaustive. On such a view, excluded middle would be seen as an imposition on a fully satisfactory explanation of the truth-conditions of a statement. But the nature of Wittgenstein's inclination to think that there is an indeterminacy of sense in the examples about π—his doubts about the transition from other paradigms, his implicit emphasis on the need for a fresh explanation for infinite quantification, and the argument last adumbrated to do with the establishment of paradigms (which he nowhere, as far as I know, explicitly develops)[1]—all presuppose repudiation of a platonist concept of such statements' meaning. All suggest that there is, or has been at a stage later than we might be inclined to acknowledge, some sort of indeterminacy in our concept of circumstances under which such statements would be *proved*; and that is no ground for thinking of their senses as 'hazy' unless these are thought of as fixed by proof-conditions.

Consider, for example, the remark in IV. 18 about one's supply of pictures giving out in the infinite case. This is no worry if we think of our understanding of 'ϕ does not occur in π' as given by the platonist notion of an effectively enumerable, infinite conjunction of effectively

[1] See, however, *PG* II, 32 (p. 406).

decidable statements. What we can form no picture of is a *method of verification* corresponding to that appropriate to the finite case. And, of course, it is not just a matter of not being able to form pictures; the claim is that the procedures for the finite case, taken alone, afford no sufficient basis for an understanding of the use of statements involving infinite general quantification. In particular it is irrelevant, as far as the validity of excluded middle is concerned, whether the fresh explanation of quantification needed for infinite, or unsurveyable, cases is seen as somehow answerable to what is already understood for the finite case, or whether it is properly regarded as essentially novel. Once the idea of conditions of proof is admitted as central to our understanding of mathematical statements, the way is open for doubts about the validity of excluded middle; for $PV -P$ is then, arguably, rightly regarded as valid only if a proof *must* be available either of P or of its negation; and this there is, in general, for examples like that about π no reason to suppose.

We can conclude that this strand in Wittgenstein's thought threatens to produce no grounds additional to those of the intuitionists for revising classical logic. Only if conditions of proof are taken to be central to our understanding of mathematical statements has Wittgenstein given any grounds for thinking that there is haziness in the senses of certain undecided statements; and if proof-conditions have that place, the intuitionistic arguments against excluded middle are available anyway and receive no additional strength from Wittgenstein's idea.

5.   The suggestion that mathematical statements might be interpreted as imperatives is an expression of Wittgenstein's desire to get away from the idea of the mathematical *statement* and the associated notion that pure mathematics is descriptive. The idea occurs in several other places in *RFM*, notably at the beginning of the Appendix on Gödel's Theorem. Wittgenstein remarks that we might typically express questions, and commands themselves, in the form of declarative sentences: 'I should like to know whether . . .', 'My wish is that . . .', for example. But this would only establish a stylistic sense in which a command or question could be described as true or false—'It is English to say so of such a sentence'. There would be no substance in such a description, because, to put it crudely, these distinctions reside not in the overt grammar of the sentence but in the response required of a recipient if he is to show that he has understood it. Can we now, then, imagine mathematical sentences as indicative expressions of commands? And what should we lose if they were explicitly recast in the imperative mood?

In the present context, however, Wittgenstein seems undecided about what precise suggestion he wants to commend. IV. 13 begins:

The general proposition that the pattern does not occur in the expansion can only be a *commandment*. Suppose we look at mathematical propositions as commandments, and even utter them as such. 'Let 25² be 625.' Now—a commandment has an internal and an external negation.

Similarly IV. 17:

> Can we imagine all mathematical propositions expressed in the imperative, e.g. 'Let 10×10 be 100'? And if you now say: 'Let it be like this, or let it not be like this', you are not pronouncing the Law of Excluded Middle, but a *rule*.

But at the end of IV. 13 we find the suggestion that mathematical sentences be interpreted not as imperatives but as genuine statements about the existence of rules:

> The opposite of: 'There exists a law that P' is not 'There exists a law that not-P'; but if one expresses the first by means of P and the second by means of not-P, one will get into difficulties.

This second suggestion is taken up again in IV. 18:

> How is the Law of Excluded Middle applied? 'Either there is a rule that prescribes it or one that forbids it.' Assuming that there is no rule forbidding the occurrence—

of '550' in π—

> why is there then supposed to be one that prescribes it?

Finally, we find the further suggestion that mathematical sentences should be regarded as direct expressions of rules, sometimes (for example IV. 17) in explicitly deontic terms:

> If 'you do it' means: you must do it, and 'you do not do it' means: you must not do it, then 'Either you do it or you do not' is not the Law of Excluded Middle.

Prima facie it looks to be of some consequence which of these ideas we pursue. If the whole point of this aspect of Wittgenstein's thought is rightly interpreted as a challenge to the received idea of *statements* in mathematics, of mathematics as being concerned with a special genus of descriptive truth, then the second suggestion, that mathematical sentences be thought of as genuine statements concerning the existence of rules, would seem to be just an aberration. Moreover, the final suggestion, that mathematical sentences be interpreted as direct expressions of rules, seems to have one overt advantage over the first suggestion, that we could reconstrue mathematics imperatively. It is a constraint upon any interpretation of mathematical sentences that they be capable of functioning as hypotheses, and so as the antecedents of conditional statements. We have to account for the practice of significant conjecture in mathematics, and of the drawing of consequences from hypotheses; and the effect of the latter is to establish conditionals. On the face of it, however, sentences in the imperative mood cannot meaningfully feature as the antecedents of conditionals. A rule, on the other hand, may properly receive direct expression in the indicative mood; for example, 'The Bishops move any number of unobstructed squares along their diagonals.'

Such a statement of a rule can plausibly be understood as what Dummett[1] has called a *quasi-assertion*. Quasi-assertions are declarative sentences which are not associated with determinate conditions of truth and falsity but share with assertions properly so-called the feature that there is such a thing as *assenting* to them; where such assent is communally understood as a commitment to some definite type of linguistic or non-linguistic conduct, and receives explicit expression precisely by the making of the quasi-assertion. Now, because of its connection with undertaking a certain sort of future conduct, the making of a quasi-assertion is something which we may well have an interest in eliciting from a speaker. The occurrence of significant conditional statements with mere quasi-assertions as antecedent clauses does not require, therefore, that we construe simple 'assertions' of the latter as candidates for truth and falsity; for the antecedent of such a conditional would be understood as the hypothesis of the speaker's, or a group of speakers', assent to the particular quasi-assertion—and as a hypothesis of the truth of something only in the sense immediately consequent on such an interpretation—and one use of such conditionals would then be precisely to articulate the implications of such assent.

Such an account is exactly what we should intuitively propose for sentences expressing the making of a promise. No one would ordinarily suppose that the use of sentences of the form, 'I promise to . . .' is best understood as the making of a statement, true or false; though their being prefixed by 'it is true that . . .' is grammatical sense. Rather such sentences express explicit assent to an undertaking, and their occurrence as antecedent in a conditional is to be understood as the hypothesis of such explicit assent.

Dummett invokes the notion of a quasi-assertion to counter Geach's[2] argument that an 'emotivist' about ethics, for example, can give no explanation of the capacity of ethical statements to feature as the antecedents of conditionals. But, it seems to me, it would have to be pursued by someone who wished to take up Wittgenstein's third suggestion that pure mathematical sentences be understood as direct expressions of rules. Very sketchily, the kind of thing that might be suggested is this: the mathematical conditional is to be seen as a rule

[1] 25, ch. 10, pp. 352–4.
[2] In 37.

articulating—or stipulating—the implications of assent to the rule
hypothesised in its antecedent; the implications, that is, of the antecedent's receiving the status of rule. The making of the mathematical
hypothesis is to be understood as the concession, for the purposes of a
certain context, of the status of a rule to the sentence in question; and
the making of a mathematical conjecture a tentative expression of such
a concession in a context-independent way.

It is by no means obvious that such a re-orientation of our thinking
about mathematical sentences would be easy to implement. But what
does seem likely, even from these very limited considerations, is that
our initial reaction to the prospects of an imperatival reconstruction of
mathematics may have been mistaken. For something akin to the
proposed interpretation of mathematical conditionals would presumably continue to apply: their antecedents could be interpreted, for
example, as hypotheses that the relevant imperative was in force, while
they themselves would be interpreted as conditional commands in the
ordinary way; as commands, that is to say, whose requirements pertain
just to the situation when a particular hypothesis is true. And the kind
of account proposed for hypothesis-making and conjecturing in
mathematics is also prima facie easily remoulded for the purpose of an
imperatival reconstruction.

Wittgenstein's idea may still seem to be of a quite different character,
depending upon whether we take as primary the rule-expressive
interpretation of mathematical statements or the suggestion of an
imperatival reconstruction of them. The former is a proposal about how
mathematics should be understood as it is, the latter a programme for
recasting mathematical discourse. On reflection, though, it appears
that for our present purposes the difference is unimportant. Wittgenstein is not seriously commending a programme of reconstruction; the
suggestion is rather that the nature of mathematics would permit such a
reconstruction of its sentences without loss, as it were—the character of
mathematics would not be betrayed if the reconstruction were carried
through. Obviously, however, this is not something which could be
learned by executing the programme. The reasons for saying so are
exactly analogous to those Wittgenstein himself gave for discounting
any decisive significance in the fact that mathematical sentences, as
things are, are framed in the indicative—that it is 'English' to describe
them as true or false. Suppose we recast pure mathematics in what is
conventionally the imperative mood—perhaps this would involve no
more than embedding it within an underlying logic of imperatives—and
suppose, whatever it means to say so, that nothing is lost in the change;
that the function of mathematics, its practice and application, remain
essentially the same. Obviously nothing would thereby be shown about
the nature of mathematics, unless it could be established that its recast
theorems were functioning as *genuine* imperatives; that is, I suppose,
that the responses appropriate to them were generically different to

those appropriate to statements. Otherwise it would be the imperative rather than the indicative form which was misleading. We can sidestep the complexities of this issue. For if the recast theorems did indeed have a genuinely imperatival rôle, it appears that, granted the hypothesis of conservation of 'function', it would have had to be possible to appreciate independently of the reconstruction that the original indicative sentences did not have a genuine statement-making rôle; otherwise the reconstruction would be a falsification. So it could not be a matter of trying such a reconstruction and learning something by its success; we could not recognise it to *be* a success unless we already knew what we might thereby learn.

It is thus of little consequence whether we concentrate on the passages which express Wittgenstein's idea in terms of rules rather than imperatives, or vice versa; for the idea comes out at bottom the same. Mathematical discourse is not fact-stating; its rôle is rather to regulate forms of linguistic practice. But only if we can independently recognise this could it be recognised that something about the nature of mathematics would be faithfully reflected in an imperatival reconstruction. Wittgenstein is here concerned with what he believes to be an example of a tendency so often responsible for philosophical error: our disposition to be misled by surface similarities between forms of expression from different regions. In this case it is a matter of being led to regard a certain genre of quasi-assertion as genuine assertions.

The revisionary potential of the idea is prima facie evident. It may be that the old fact-stating picture sanctions certain assumptions and methods of inference which no longer seem cogent when we adopt instead the idea of mathematics as a network of procedural stipulations. In particular, it is unclear that a platonist attitude towards the disjunction about π could survive a wholesale repudiation of the ideas of mathematical truth and statement. Let us see.

6.   For the sake of argument, we shall waive any technical, or more general doubt and concede mathematics to be of such a character that its theorems could be reconstrued as genuine imperatives. Now, a command, says Wittgenstein (IV. 13), has an internal and an external negation. This might be interpreted as intending the contrast between the commands, 'Let it not be the case that Q' and 'Let it be the case that not-Q', But Wittgenstein elucidates his intention by drawing a contrast in terms of one of the variant suggestions: the contrast between 'There is no law that Q' and 'There is a law that not-Q'. The point would appear to be that if we take the latter as the negation of 'There is a law that Q', then their disjunction is not an application of excluded middle. As noted, the point is made explicitly—though in terms of deontic expressions of rules—at the end of IV. 17.

The general suggestion is apparently that mathematical negation is properly understood on the model of the *internal* negation of a command. Thus what seemed to be applications of excluded middle in mathematics are not actually so; hence Wittgenstein's aim is not so much to criticise the law as to point out misapplications of it. What would be an application of it would be the disjunction, 'Either there is a law that Q or there is not'; but to mirror this in a system of imperatives we require a notion of the *external* negation of a command: something which negates its imperatival force but of which the result is not another command. Even if such a notion can be made out, it is extremely unclear whether there could be such a thing as a significant disjunction of a command with its external negation. The trouble would be to give an account of the *force* of the disjunction, for it seems that it could be neither an assertion nor a command—nor, presumably, anything else. If this is right, the only significant disjunction of a command with its negation would be the disjunctive command whose disjuncts are the original command and its internal negation. But if this is the real nature of what had appeared to be applications of excluded middle in mathematics, then that principle actually provides no authority whatever for such sentences.

If it would indeed be no distortion of the nature of mathematics to cast its sentences in the imperative mood, the foregoing at least suggests that there is something spurious about our impulsion to accept as valid the disjunction about π. The point would be not that excluded middle is itself questionable, but that it is misinvoked here. No one would regard the principle as stipulating as valid the disjunction of a statement with its internal negation. But if mathematical statements are to be understood imperatively, mathematical negation is always an internal negation. 'Either ϕ occurs in π or it does not' is a disjunctive command of a command with its internal negation (there is no significant disjunction with the external negation). Thus, again:

> . . . if you now say: 'let it be like this or let it not be like this', you are not pronouncing the Law of Excluded Middle, but a *rule*.

And, in contrast to the platonist attitude to the disjunction, why should it be mandatory for us to accept this rule?

The weakness of these considerations is that they are not advanced against the background of a notion for commands corresponding to that of *validity* for statements; for of course there is no question of citing the authority of *propositional* logic for a command. Clearly, however, it will be straightforward to introduce a working notion of command validity provided we have a coherent notion of logical consequence among commands. A valid command—an *irrefragable* command—will then be a command which is a logical consequence of an arbitrary set of commands. And there seems no reason why we should not have at least as clear a notion of logical consequence among commands as we have for statements, since we can introduce the former notion parasitically on the latter: $!C_2$ is a logical consequence of $!C_1$ just in case the supposition that someone has complied with $!C_2$ is a logical consequence of the supposition that he has complied with $!C_1$—irrespective, of course, of whether these commands have actually been issued. An irrefragable command is thus a command compliance with which is a necessary condition of compliance with any command at all—a command which it is logically impossible to disobey.

It would be natural to expect that what we have taken to be true mathematical statements would go over into irrefragable commands if mathematics were reconstrued imperatively. But the difficulty is that it is unclear how there can ever be any point in *issuing* an irrefragable command—for compliance with it is guaranteed. So what, on this account, are we to interpret mathematicians as *doing*? In particular, why should anyone bother to pronounce a disjunctive command of the sort schematised by Wittgenstein, unless simultaneous disobedience of *both* disjuncts was a possibility?—which seems unlikely with any plausible imperatival reconstruction of the alternatives about π. The imperatival view must thus provide some other account, it seems, of the activity of mathematical 'assertion' than the actual pronouncement of irrefragable commands.

What should this account be? We get a possible clue if we consider the way Wittgenstein commends the imperatival view in iv. 17. The statement 'ϕ does not occur in π', is described as 'over-reaching itself'; we are prompted to ask, 'how could we ever know anything like that?' The short answer would be, 'by proof'; but that is not the conception of the truth of that statement which we invoke when we insist that either it or the statement of which it is the negation *must* be true. Wittgenstein wants to suggest that if we think of both the statement and the disjunction as commands, we eliminate a source of philosophical perplexity without distorting the use of the sentences. The statement appears to over-reach itself because our seeming-recognition of the validity of the disjunction, 'ϕ occurs in π or it does not', constrains us to attribute to ourselves grasp of possibly verification-transcendent conditions of truth for its second disjunct; we are thereby forced to attribute to ourselves a concept of our possession of which—how acquired and what consisting in—it is enormously difficult to give a satisfactory account. No analogous difficulty arises if the statement is thought of as a command because we are no longer supposed to be *recognising* anything; rather we are stipulating as correct, for example, the procedure of leaving out '550' whenever expanding π. We should be led by a proof not to recognise the non-occurrence of ϕ in π, but to stipulate that it should not occur. And on this picture, Wittgenstein evidently feels, the mathematical world is in some desirable way restored to our control.

Now, whatever we think of this idea, it is clearly essential to it that the import of the stipulation should be that 'ϕ does not occur in π' is an *irrefragable* command. It is not that when writing out π one is not to include the sequence ϕ in the same way that one is not, for example, to smoke before the Loyal Toast when at a regimental dinner. Rather what we should have stipulated, on Wittgenstein's view, is that the command, 'If writing out an initial segment of π, never include the sequence ϕ', is one which it is impossible to disobey. We should have further determined what it is *correctly* to write out π; the command would be irrefragable because if anyone attempts to break it by including '550', he will be held just on that account not to have written out a segment of π. Similarly, '$25^2 = 625$' is to be read, if interpreted in accordance with Wittgenstein's needs, not as the command: 'if calculating the square of 25, get the result 625', but as the command that that command is to be counted as irrefragable—that anyone who gets a different result is to be held just on that count not to have calculated the square of 25. In general, it appears that the form of commands in which mathematicians deal will always be to the effect that some other command is to be counted as irrefragable.

How in terms of this conception should the disjunction about π be interpreted? Not, presumably, as the command that one or other of the disjoined commands must always be complied with; but rather as commanding that one or the other of them is to be counted as irrefragable in its own right. But this needs further refinement; for the requisite content of the command is not such that someone will have complied with it provided he accepts one or the other disjunct as irrefragable, the choice of which being up to him. Rather his understanding must be that further instructions may be issued, so to speak, about which disjunct is to be counted as irrefragable.

An analogy would be this. Suppose we are devising a game, and are giving someone step-by-step instructions in its rules as we do so. We have reached a stage at which we have decided that in a particular situation one of two exhaustive courses of action is to be mandatory and the other therefore prohibited; but we have not yet decided which. So we have ruled, in effect, that one of the commands, 'When playing the game, do such-and-such in situation so-and-so', and 'When playing the game, do not do such-and-such in situation so-and-so', is irrefragable; for if our man now proceeds sometimes to do 'such-and-such' in the relevant situation and sometimes not, it is already determinate that he has on occasion broken the rules and thereby ceased to play the game. What is not yet determinate is precisely when he did so. A certain freedom of policy has been precluded; though the exact restriction has yet to be issued.

If the application of 'excluded middle' to the 'statement' that ϕ occurs in the decimal expansion of π were construed on this model, then its effect would be comparably to restrict a certain freedom. For a mathematician to endorse the disjunction would be for him to command that one or the other disjunct was to be taken as irrefragable, though it remained to be determined which. And to accept the command would thus be to accept a constraint on the activity of expanding π, unparalleled for example, by any feature of the activity of expanding a series of arbitrarily chosen objects; though exactly what constraint, it would be understood, was not yet settled.

7.   A Wittgensteinian would hold that we tend to misconceive the sense in which correct continuation of a rule-determined series, in contrast with one generated by free choice, is *constrained*. But he would not contest that there is a perfectly good such sense. So if the rôle of 'excluded middle' under an imperatival reconstruction would be, inter alia, to attest to a genuine difference between π and a free choice series, it rather looks as though, so far from being non-revisionary, a wholesale such reconstruction would actually side with certain aspects of classical, as opposed to intuitionist, mathematics. But this impression is not, I think, borne out by closer scrutiny. Consider the infinite series of pairs of conditional commands of which each nth pair is 'If expanding the first n digits of π, include ϕ' and 'If expanding the first n digits of π, do not include ϕ'. Both platonist and intuitionist will conceive that the predeterminate character of the first n places of π, for any choice of n, settles that one of each such pair of commands is predeterminately irrefragable and the other predeterminately unobeyable. But if 'π' were a designation of a series of digits to be successively chosen at random, then of course neither member of any such pair of commands would be irrefragable; though 'If expanding the first n digits of π, either include ϕ or do not' is irrefragable in either case.

   In order, then, to give expression to the contrast between π and any infinite free choice series, it would not be necessary for the imperatival mathematician to endorse a disjunctive command about the entire expansion of π; it would suffice to indicate his willingness to endorse each disjunctive command whose disjuncts were both elements of any nth pair in the above series.

   The platonist, however, can be expected to go further. It follows from what the intuitionist grants him that the command 'Continue indefinitely expanding π', entails either 'Include ϕ in the first n places' or 'Do not include ϕ in the first n places', for each choice of n. Reflecting, then, that if a command of the first kind is entailed for a particular value of n, the same will hold for each succeeding value, the platonist now moves to the conclusion:

(D1)   *Either* a command of the second type is entailed for every value of n, *or* for some value of n, and every succeeding value, a command of the first type is entailed.

So granted imperatival analogues of the principles of sentential logic:

$$(n)(S \to Fn) \quad \to \quad (S \to (n)Fn), \text{ and}$$
$$(\exists n)(S \to Fn) \quad \to \quad (S \to (\exists n)Fn),$$

which are both classically and intuitionistically sound, it now follows
(*D2*) that: 'Continue indefinitely expanding π' entails either 'Do not
include ϕ in the first n places, for any value of n', or 'for some value of n,
and every succeeding value, let ϕ occur in the first n places'.

That conclusion is still a disjunctive statement. But it is now straightforward to elicit a disjunctive command endorsement of which would
be the imperatival dual, so to speak, of the platonist's willingness to
accept excluded middle as validly applied to 'ϕ occurs in π'. So far we
have implicitly taken it that a conditional command, $S \to !C$, is irrefragable if to contravene $!C$ is necessarily to bring it about that the condition
depicted by S is not fulfilled. But this account has nothing to say about
conditional commands whose consequents it is not in our power decisively to contravene; nothing, for example, constitutes *failing* to include
a particular pattern in an infinite series, so we have as yet no notion of
irrefragability for 'If expanding π, let ϕ occur in the first n places for
some n.' But we want there to be a sense in which that command may be
irrefragable, while its analogue for an infinite free choice series would
not be. I therefore propose:

> $S \to !C$ is irrefragable if and only if *either* it is possible to contravene
> $!C$ but not without bringing it about that the condition depicted by S is
> not fulfilled, *or* it is possible to implement $!C$ but not to neglect to do so
> indefinitely without bringing it about that the condition depicted by S
> is not fulfilled.

Now consider:

> 'Either if expanding π, let ϕ occur somewhere in the expansion or if
> expanding π, let ϕ nowhere occur.'

It follows from the disjunctive statement (*D2*), above, that one (and only
one) of the disjuncts of this command is irrefragable. So the platonist
would consider himself within his rights in commanding that one or the
other disjoined command—though it had not yet been settled which
—was to be counted as irrefragable; and to be in a position to issue that
command is the same thing, under the interpretation of disjunction in
mathematics which we are entertaining, as being in a position to issue
their disjunction above. The intuitionist, in contrast, would contest the
platonist's right to issue that command; for he would contest the
transition to the disjunctive statement (*D1*), in the course of the reasoning of the previous paragraph. He could be expected to insist in general
that we are entitled to issue a disjunctive command in mathematics if
and only if we can guarantee access to a situation in which we are
entitled to command the irrefragability of one of its disjuncts in par-

ticular. And a Wittgensteinian could be expected to contest the concept of the predeterminacy of any n-fold initial segment of π on which the platonist's reasoning depends.

This is only one example, of course. But I venture to conjecture that its moral could be generalised: that to construe in imperatival terms any classically valid principle, if it were done along the most natural lines consonant with Wittgenstein's general idea that the mathematician essentially provides regulation of what correct implementation of certain procedures consists in, would of itself give absolutely no guidance whether the principle should be allowed. The old platonist/intuitionist controversy can be expected to re-assert itself throughout the framework of an imperatival mathematics. And the idea that pure mathematics is, when rightly understood, of such a character as to permit systematic imperatival reconstruction is therefore, of itself, arguably non-revisionary.

8.   Would our conclusions have been any different if we had concentrated instead on the suggestion that mathematics, as it is, is rule-expressive rather than descriptive?

With rules, as with commands, a distinction is to be drawn between an external, force-negating denial and an internal sense-negating force-preserving denial. The contrast is roughly that between the results of prefixing the operators 'it is not mandatory that . . .' and 'it is mandatory that not . . .'. Wittgenstein wants in these terms to draw a contrast between the finite and infinite cases (IV. 18):

> The opposite of 'it must not occur' is 'it can occur'; for a finite segment of the series, however, the opposite of 'it must not occur in it' seems to be 'it must occur in it'.

That is, for a finite segment, the negation of 'mandatory: P' seems to be 'mandatory: not-P'; but for the full series it is 'not-mandatory: P'.

Why does Wittgenstein think this? Suppose we are concerned with a finite segment of a series of integers generated by free choice. Then the negation of 'mandatory: ϕ occurs' is precisely an expression of permissibility—the external negation above. It is thus not a matter of the contrast between finite and infinite series. Rather, consider such a finite sequence which is rule-determined, so that, as we ordinarily conceive, there is no freedom of choice. Then the situation is not that the negation of 'mandatory: ϕ occurs' is not properly seen as the negation of its force, the external negation, but rather that here this in turn implies the internal negation. Where every place of such a sequence has been determined, then if it has not been determined that ϕ occurs, that is for it to have been determined that it does not. There is room for a contrast between its being permissible and its being mandatory that ϕ does not occur only if the character of the sequence is itself permissibly variable.

But just this, as we saw in the case of commands, is essentially the thought of someone who regards one of the rules about π as predeterminately mandatory. Wittgenstein suggests that we regard the disjunction as the proposal of a disjunctive rule: 'mandatory: ϕ occurs in π or it does not'. Clearly the operator is not in general distributive over disjunction. Such a rule, however, even if restricted to finite initial segments of π, effects no constraint on us; it would be mandatory for a finite initial segment of a free choice series. So Wittgenstein is implicitly ignoring the contribution made to our inclination to accept the disjunction about π by the rule-determined character of the series. What a platonist wants to insist on is precisely that because of the predeterminate nature of the series, one of the alternatives in the disjunctive rule *already* has itself the status of rule.

The argument is in essentials as above for commands. If it is conceded in the finite, rule-determined case that the negation of 'mandatory: ϕ occurs' comes to 'mandatory: ϕ does not occur', then—unless it is denied that one of these rules must be predeterminately in force in such cases—it has to be conceded that one of 'ϕ occurs in the first n places of π' and 'ϕ does not occur in the first n places of π' already has the status of a rule, irrespective of the value of n. For each n, one of these is entailed *as* a rule by the rule for the correct expansion of π. The dialectic is then analogous to that in the case of commands. Does the rule for expanding π entail each of the rules, 'ϕ does not occur in the first n places of π'? Or is there an n such that for it, and every greater n, the rule is entailed, 'ϕ occurs in the first n places of π'? Only a repudiation of a platonist notion of logical consequence can now save us the admission that one or the other alternative must obtain. Once it is admitted as predeterminate what the situation is with respect to any particular n, it is irresistible that the whole character of the totality of entailed rules is settled—*unless* we insist that the totality is determinate only in respects of which we can in principle assure ourselves.

9.   In summary: platonism can reassert itself, Hydra-fashion, even under an imperatival reconstruction or rule-expressive interpretation of mathematics, unless its true philosophical roots are attacked directly. The crucial basic platonist notion in the example with which we have been concerned is that, since π is a rule-generated expansion, it is firstly predetermined what is the identity of its every nth place and, secondly, *thereby* determined whether '550' occurs in it or not. If mathematical sentences are thought of as expressive of rules or commands, the platonist can adapt his basic notion accordingly: commands, or rules, are capable of being in force without being explicitly issued—when they are consequences of ones which have been explicitly issued—nor, in order for one command, or rule, to be a consequence of others, it is necessary that we be capable of verifying that it is so. Thus in establish-

ing an effective procedure for the expansion of π we have (i) *implicitly* issued infinitely many rules/commands about what to do at the nth place; and (ii) *thereby* issued instructions, for example never to write down '550', whether or not we are capable of recognising as much. An anti-realist attacks the 'thereby' of (ii). Wittgenstein distinctively attacks (i) as well. But none of the three protagonists seems either to be given any edge over his rivals by Wittgenstein's ideas about the rôle which rules and imperatives can play in the philosophy of mathematics, or to be seriously incommoded by them.

One powerful objection to this conclusion would be that it fails to do justice to the intended *scope* of this aspect of Wittgenstein's thought. For Wittgenstein, it is the whole notion of a necessary *statement* which is in error; *all* such 'statements' are more akin to imperatives. But to suppose, as do both the platonist and the intuitionist, that by issuing a command of the form 'Write out the first n places of π', we thereby *implicitly* issue one or other of the commands, 'Include ϕ' and 'Do not include ϕ' is to presuppose that there is a species of *fact* constituted by the obtaining of so far unratified logical relations among commands; a species of fact, therefore, which it will be the province of certain necessary statements to record.

It might be countered that the existence of any particular such fact may always itself be thought of as consisting in our having implicitly issued a *further* command, pertaining to what are to be counted as the logical consequences of certain general commands governing correct inferential practice.[1] But this would be to dodge the challenge to explain what it *is* for a command, *!C*, to be implicitly issued by the issuing of others; an 'explanation' which consists in any particular case merely in positing a further command, itself only implicitly issued, to the effect that correct inferential practice requires a readiness to endorse *!C* if those other commands are issued, casts no light on the matter.

We have to acknowledge, I think, that if the very widest scope is given to the imperatival motif, no explanation is yet apparent why platonist and intuitionist philosophies of mathematics are not both pre-empted thereby.[2] But an argument would still be needed that this strand in Wittgenstein's philosophy of mathematics was revisionary; it is just that the argument of §§ 7 and 8 would no longer be available for the opposite view. However, there is a more important point. If the admission of unacknowledged internal relations is indeed inconsistent with the idea that all necessary 'statements' may legitimately be construed as imperatives, then the latter idea can be correct only if it is right to repudiate all unacknowledged internal relations; for which step an argument is then required. In Wittgenstein's thought, as we have so far interpreted it, this repudiation derives essentially from the ideas on rule-following. The suggestion that *all* necessary 'statements' can be

---

[1] The proposal is reminiscent of Bennett's manoeuvre in 3, discussed in Chapter XVIII below.
[2] A purported such explanation will be considered in Chapter IX.

seen as rules, or commands, is thus plausibly interpreted simply as a way of capturing a key implication of those ideas while simultaneously emphasising the undoubted normative rôle of these sentences. If it were to prove a revisionary suggestion, its revisionary force would thus be that of the rule-following considerations themselves; it would be the revisionary force, if any, of the conception that there is no domain of determinate non-empirical fact for undecided logical and mathematical sentences to describe.

IX

Quasi-platonism

1.   Of the five prima facie potentially revisionary themes in *RFM* listed at the start of Chapter VII, it has now been argued:

(i) that the stress on the essential surveyability of proof allows but does not require interpretation as revisionary;

(ii) that the idea that a mathematical problem is indeterminate in content before a solution is achieved, needs support—if it can be made good at all—from the proof-conditional conception of the meaning of mathematical statements which Wittgenstein shares with the intuition-ists, and appears to have no revisionary implications not already implicit in that conception;

(iii) that the imperatival motif is non-revisionary unless—and perhaps even if—its scope is taken to include all necessary 'statements'; that, so interpreted, it is essentially a way of expressing the rejection, imposed by the rule-following considerations, of all unratified necessity while simultaneously paying heed to the normative rôles of logic and mathematics; and that its revisionary implications, if any, are whatever are involved in that rejection.

We thus arrive at a position in which we are entitled to concentrate on the remaining two themes: Wittgenstein's points of sympathy with the intuitionists, and the rule-following considerations themselves. If Wittgenstein's own philosophising about mathematics is to be consis-tent with his non-revisionary view of the philosophy of mathematics, then it is these two themes which, unpromising though the prospects may look, must somehow be spared revisionary implications. But of course the interest of the matter is not merely exegetical; a host of deep-reaching philosophical questions converge on this area, and I shall allow the ensuing discussion to follow up ramifications to which Witt-genstein, in his published writings at least, gives little or no attention. We shall take the rule-following considerations first.

2.   Granted the idea that it is already determined of any particular finite initial segment of π whether it contains the sequence '550', the platonist then takes the step, rejected by an intuitionist, of concluding that that *is* for it to be determinate whether or not '550' occurs in the

whole expansion. But Wittgenstein's view is that it is not 'already determined' of any particular initial segment of π whether it contains '550'; for it is not predeterminate what, if we judge correctly, we shall *count* as the correct expansion of π up to a particular place. And if the truth-values of its instances are not yet determinate, how can that of a generalisation be so?

In IV. 9 Wittgenstein presents this point by means of an analogy with fiction:

> But what are you saying if you say that one thing is clear—either one will come upon ϕ in the infinite expansion or one will not? . . . What if someone were to reply to a question, 'so far there is no such thing as an answer to this question'. So e.g. a poet might reply when asked whether the hero of his poem had a sister or not; when, i.e. he has not yet decided anything about it.

The suggestion is that with fictional statements we are ready to allow that matters are simply up to the author; fictional characters are as by their authors they are decided to be, so there is no sense in the idea that one of the alternatives about the hero of the poem is determinately true and the other determinately false before the poet decides anything on the matter. And Wittgenstein is, apparently, commending a similar attitude towards mathematical statements before they are 'decided' by doing mathematics. On such a view the platonist loses the premise in his intuitionistically contentious transition from 'at every stage it is determined that either . . . or . . .' to 'either it is determined that at every stage . . . or it is determined that at some stage . . .'. But the intuitionists' attitude to finitely quantified decidable mathematical statements now appears to be under pressure also. If we have not settled whether or not F applies to each object in the range of quantification, then—since settling the matter is now to be viewed as making it so—it appears that the truth-values of finite '$(x)Fx$' and '$(\exists x)-Fx$' must be thought of as indeterminate too. The temptation to think that, on the contrary, the matter has already implicitly been settled, by establishing criteria for membership in the range of 'x' and for being F, cannot survive rejection of the idea of unacknowledged internal relations.

We think that where we are concerned with an effectively decidable mathematical predicate, it is settled already in any particular case whether or not it applies (for there is no question of *change* in a mathematical object). Similarly, where a proof of a statement is available, we think that its constructability was likewise settled in advance by our understanding of the content of the statement and our notion of a valid proof. So in testing a particular case, or constructing a proof, we merely bring to light circumstances which the character of our understanding of the concepts involved had already legislated should obtain. The judgments, for example, that n is prime, or that the proof is correct, are the only *possible* judgments if judgment is to be made in accordance with our understanding of the relevant concepts—that is, if we are to judge these cases in the same way as previous ones. Wittgenstein disputes that we know, in any substantial sense, what way this is. There is, indeed, nothing to be known. For we cannot in principle, it seems, unambiguously explain an intended pattern of use; and, as contended in Chapter II, no respectable sense remains in which we can claim ourselves to know more than we can explain.

The rule-following considerations thus appear to be more than potentially revisionary; they seem to bear against a wider class of accepted applications of excluded middle than any other argument yet considered. Even applications of that principle acceptable to a strict finitist may no longer be so, for decidability in practice of the disjuncts is no longer decisive in its favour. The truth of a disjunction requires the truth of at least one of its disjuncts; so where we are concerned with statements which, for whatever reason, can be regarded as true only if expressly decided to be so, it would seem to be ruled out that a disjunction of such statements could be true—*a fortiori* necessarily true—in advance of any ruling about its disjuncts. And that, it appears, is a class of statements of which a Wittgensteinian should regard every statement as a member whose truth would ordinarily be thought to consist in the obtaining of an internal relation.

3.   Conclusive as these reflections may seem, there is a route which seems to offer a prospect for getting around them and which I propose to consider 'n some detail.

Consider what is the effect of the rule-following considerations upon the distinction between a rule-generated series of natural numbers and one generated by successive free choices.[1] Clearly the distinction is not destroyed; there remains all the difference in the world between a series in which one's nth choice may in certain circumstances be open to correction by other people and a series where that is not so. But what does appear to be threatened is an aspect of the significance which we tend to attach to this distinction. We would, I think, ordinarily suppose that the platonist's idea, that any effectively decidable mathematical question concerning an initial segment of a rule-generated series has a predeterminately correct answer, would be quite inappropriate if a free-choice series was involved; for the question whether, for example, the nth member of a free-choice series has a particular property would seem to *have* no correct answer if the identity of the nth member has not yet been decided. Wittgenstein's argument would then seem to be that while there is, of course, a distinction between rule-generated and

---

[1] In what follows, the series considered are always to be understood as series of natural numbers; and ϕ always as an effectively decidable property of natural numbers. Series considered as generated by free-choice are, of course, fundamental to the intuitionists' mathematical treatment of the continuum; but the use made of the notion in this chapter is entirely intuitive and self-explanatory.

free-choice series, it is one which, properly understood, will not bear the weight of these contrasting attitudes. For the rule does not *determine* the series in the sense needed to sustain the contrast; in neither case is the identity of the nth member fixed before we settle it—it is just that settling it is in the one case *computing*, and in the other *choosing*; and that one's computation can be judged wrong, whereas one's choice cannot.

This contrast in attitude, however, is open to a challenge of another sort. If it were a matter of choosing a reference, and hence a sense, for a *name*, then the truth-conditions of statements involving the name could, of course, not be regarded as determinate before the choice was carried out. But that is not the situation here. If α is a free-choice series, then we have a uniform mode of description, 'the nth member of α', whose instances there is no reason to regard as indeterminate in sense just because their references are a matter of subsequent choice. Indeed, is there any reason to regard them as indeterminate in *reference*? A man who, before he plans to marry anyone in particular, talks in terms of 'my future bride' could expect what he says to be held accountable, after he marries, to the qualities of his then wife, notwithstanding the fact that the reference of the expression was, when he used it, to be settled by a subsequent act of choice.

Suppose we are concerned with a specific finite series of free choices, α, which is actually going to be completed. Here although the identity of the individual places, and hence the character of the whole series, is still a matter for future decision, we ought not —the argument goes—to take that as a reason for regarding statements involving reference to its members (or indeed quantification over them), as taking on a determinate truth-value only when the relevant place in the series has been decided (or when all the choices have been made). Because the series is not determined by rule, the references of expressions of the form 'the nth member of α', are in one clear sense not decided in advance. But that is no ground for regarding '$\phi\alpha_n$' as indeterminate in truth-value. The choice which counts is the fixing of a *sense*, and hence a criterion of identity for the reference, for each term 'α_n'. When this choice has been made, the world, as it were, takes care of the reference; and in this way the reference is decided as soon as the sense is. It is merely that to satisfy the criterion of identity involves being the subject of a further choice. So statements of the form '$\phi\alpha_n$' are no different from any example involving a definite description where we must wait before being in a position to identify its bearer. A bet, for example, on the truth-value of such a statement seems perfectly clear in sense; it is just that the protagonists will have to wait to see who has won.

Once the choosing-agent has made the nth choice—suppose it is 7—do we not have to admit that someone who predicted in advance that $\alpha_n = 7$ spoke *truly*, however unreasonably? Certainly anyone who makes what we should ordinarily take to be the same assertion after α_n is chosen speaks truly; so not to admit this would appear to commit us either to denying that it is the same assertion or to postulating for the predicate, 'true', a significant tense. If neither course is taken, then the fact that α is a series whose exact nature is a matter for future decision presents no obstacle to the idea that statements of the form '$\phi\alpha_n$' are already determinate in truth-value. Hence each disjunction of such a statement with its negation is true—notwithstanding the fact that it would remain appropriate for the choosing-agent, when asked about such a disjunction, to reply like Wittgenstein's poet that he had so far decided nothing about it. For now we are supposing that there are always facts about how he *will* decide. And on such a view there will be nothing objectionable about the validity of the generalised disjunction, '$(x)\phi\alpha_x$ V $(\exists x)-\phi\alpha_x$', since all the instances of each disjunct are determinate in truth-value.

The emergent suggestion, then, is this. Our ordinary conception of the truth-conditions of a decidable but so far undecided mathematical statement S, about a finite, rule-generated series L, is that it is made true or false by the obtaining of a so far unrecognised internal relation. And the reason why the rule-following considerations seem so glaringly revisionary is because they require that there are no unrecognised internal relations; so nothing to make S or its negation true, nor therefore 'S V $-S$' true, before we investigate the matter. What is now being proposed is that it is still open to us, at least in cases where we are actually going to complete the series L, to regard the truth-conditions of S as comparable to those of any statement made before the completion of a finite choice-series, α, of which we can recognise that it will be decidably true or false of α *after* the series has been completed. Indeed the comparison is now specially appropriate. And, at least according to our ordinary conception that statements and their truth-values are timeless, such statements may be regarded as determinate in truth-value before the series which they describe are developed.

The most obvious drawback with the foregoing is the assumption that the series will actually be completed. What if it is infinite? If we are concerned with a free-choice series, β, of potentially infinite allowable extent, no comparable reason seems to be available for holding that every expression of the form, 'the nth member of β', has a determinate reference. On the contrary, that some do not is part of what is meant by regarding such a series as only *potentially* infinite; there is no presumption that the choosing-agent will reach any particular nth stage but only that, in some sense, he *could* do so. If he fails actually to make n choices, that—on one traditional line at least—will rob statements of the form '$\phi\beta_n$' of determinate truth-value.

Suppose, however, that we consider not categorical statements, 'the nth member of β is ϕ', but conditionals: 'the nth member of β, if the choosing agent advances that far, will be ϕ'. It would be awkward to suppose it sufficient for the untruth of such a conditional that there will actually be no nth choice; for in that case its conditions both of truth and untruth will respectively coincide with those of the corresponding categorical statement. But if it is granted that such a conditional's truth does not require that of its antecedent clause, what stands in the way of the view that either it or its contrary[1] must be true irrespective of how many choices are actually made?

If that view is accepted, then an argument is going to be available for statements concerning infinite free-choice series exactly comparable to the platonist's original considerations about Goldbach's conjecture, or the example about π. Let '$\phi\beta_n$' express a conditional of the sort in question. Then '$(x)\phi\beta_x$' will be regarded as the truth-functional conjunction of the infinite set of such conditionals: if each of them is determinately true or false, then that *is*—the reasoning goes—for '$(x)\phi\beta_x$' to be determinate in truth-value also, even if we have no way of finding out what its truth-value is. And, since the effect of the rule-following considerations is to construct the indicated analogy between rule-generated and free-choice series, an analogue of *this* argument will now in turn be available, notwithstanding a rejection of unratified internal relations, for supposing both that all its instances and Goldbach's conjecture itself are determinately true or false.

4.   What is developing here is a kind of *quasi-platonism*: a platonism whose ploys and gambits are essentially those of true platonism save that it rejects the possibility of unratified internal relations. But it is clear that the position needs further refinement. For Wittgenstein's idea of the non-automony, as it were, of the rules which generate rule-generated series applies equally to the decision procedure associated with ϕ. It is equally up to us to settle in each case whether an integer has ϕ or not; the decision procedure does not go ahead of us and settle the matter in advance. So even in the finite, actually-to-be-completed case, the rule-following considerations still potentially provide an obstacle to the idea that each '$\phi\alpha_n$' is determinate in truth-value. It is not enough to propose to regard each 'α_n' as determinate in reference before the nth choice is actually made; unless the application of ϕ has already been settled for all the objects from which the choosing-agent is to select, the application of ϕ to any particular α_n still awaits our ratification.

Applied to series of natural numbers, the effect of the rule-following ideas, as remarked, is to attenuate, though not to destroy, the distinction between those generated by rule and those generated by free choice. So, applied to predicates, they will have a similar effect on the contrast between any predicate associated with an effective test of application and a predicate—if it could rightly be regarded as such—whose application was understood to be a matter of arbitrary, though

---

[1] By the 'contrary' of a conditional, C, is to be understood that conditional which results from appending to C's antecedent the negation of C's consequent.

perhaps collective, choice. In order to develop and assess the quasi-platonist standpoint, we precisely need a notion of a *choice-predicate*. Such a predicate ought otherwise to be as like as possible to ordinary, effectively decidable arithmetical predicates; for its rôle here is to serve, from the point of view of one who accepts the rule-following consider-ations, as a prototype for their logical characteristics. This suggests the following conditions:

(i) The application of the predicate is to be determined by a certain procedure; if the procedure has a certain outcome, the predicate applies—otherwise not.

(ii) Like arithmetical predicates, if the predicate ever applies to a number, then it always does. More specifically, at no time may it truly be said of a number that it was ϕ but is no longer, or that it is ϕ but was not formerly.[1]

(iii) The application of ϕ to a number is not to be thought of as determined solely by the nature of the procedure and the identity of the number; there is no outcome which *must* result if the procedure is properly applied to the number—rather, part of the procedure is pre-cisely to be an element of free choice, or random decision.

The impact of the rule-following considerations upon effectively decidable mathematical predicates can now be described in terms paral-lel to what was said above of rule-generated and free-choice series: it is not that there is *no* generic distinction between such predicates and choice-predicates, but that we are inclined to mistake its nature. In both cases *we* are needed to settle whether the predicate applies to a particu-lar object; it is just that, again, settling this is in the one case computing, and in the other choosing. The logic appropriate to effectively decidable arithmetical predicates thus arguably coincides with that appropriate to choice-predicates of the natural numbers.

However, must this reflection make any difference to our assessment of what logic is appropriate? Consider again the case of a finite choice series α, whose construction is actually going to be completed; and suppose that a ruling about the application of a choice-predicate ϕ will be given with respect to each of its elements. According to the orthodox view that statements, and truth, are timeless, anything which may truly be stated at any particular time may, by suitable transformations of tense and mood, be truly stated at every time. On this account, as we saw, the reference of 'α_n' is determinate as soon as the context and agent of the choosing is settled; 'α_n' then stands for whatever number will as a

---

[1] The imposition of this condition is less than a commitment to the orthodox view of statements and truth as timeless. For example, at any time t_2 at which it is true to say that n is ϕ, it will be true to say that it was ϕ at an earlier time t_1; but without appeal to the orthodox view, nothing follows, so far as I can see, about whether what at t_2 we state by 'n was ϕ at t_1,' could truly have been stated at t_1—or, indeed, could then have been stated at all. But for our present purpose it will not matter if this is a mistake; for in what is to come the quasi-platonist strategy will involve appeal to the orthodox view in any case.

matter of fact be the nth choice—we supposed it turned out to be 7. There may be no way that its identity could reasonably have been predicted; but the number that is chosen was all along the number that was going to be chosen, and someone who had previously asserted the identity of α_n with 7, with however little justification, would thus have spoken the truth. Comparably, it seems likely that where we are concerned with a choice-predicate, the orthodox view will allow that if we anticipate the ruling in a particular case, we can correctly, if unreasonably, describe an object in terms of it. Suppose a machine which when fed with a positive integer issues randomly either one of its prime factors or a multiple of it; and let ϕ be a choice-predicate whose application is decided by which—say the integer is ϕ if a prime factor is issued when it is first tested. (Variations after that do not affect the application of ϕ.) Suppose also that none of the integers constituting the range of selection for α has yet been subjected to the test. There now seems no reason why, on the orthodox view, someone may not truly or falsely state before the series is chosen, or any testing is done, that $\phi\alpha_n$. The orthodox view gives 'α_n' a determinate reference before any choices are made; the integer it stands for is ϕ or not depending on what will happen when the test is carried out; and there is something, for example the issue of a multiple, of which it may truly be said that it will happen, because after the test it will be true to say that is has happened. On the orthodox view, there are at any time true things to be said about the identity of the choices and the verdicts of the machine.

More exactly: let t_1 be earlier than t_2 be earlier than t_3; and let 7 be chosen as α_n at t_2, and be assessed as ϕ at t_3. Then by the changlessness of ϕ—condition (ii)—'7 was ϕ at t_1' is true at t_3 and so, by the orthodox view, '7 is ϕ' can truly be asserted at t_1. Likewise, by the orthodox view, it is true at t_1 that 7 is the integer which will be chosen as the nth element of α; so, 'α_n is ϕ' is truly assertable at t_1. (Here, of course, we have done more than assume that what may truly be asserted at a particular time may, somehow, be truly asserted at any time; we have made specific, though natural, assumptions about how this can be achieved using the tenses.)

Of course, the actually-to-be-completed series is still a special case. But now it appears that the original tactic for coping with infinite series remains available in essentials, even when allowance is made for the bearing of Wittgenstein's ideas upon the predicate involved. The crucial difference acknowledged to be involved in the introduction of infinite free-choice series was that it can no longer be assumed that a reference for each 'β_n' will actually be chosen. It is the same with the predicates; it can no longer be assumed that the application of ϕ will, for each n, be assessed for β_n. Presumably, then, in order to specify the truth-conditions of '$(x)\phi\beta_x$', ϕ a choice-predicate and β an infinite free-choice series, it still suffices to supply an infinite conjunction of conditionals; it is merely that their form will not be exactly as earlier envisaged. For each integer n it will have to be hypothesised not merely that a β_n will be chosen but that the application of ϕ to it will be settled. '$(x)\phi\beta_x$' is thus true if and only if each member of an infinite conjunction is true of which the nth member is:

> if an nth element of β is chosen, and the application of ϕ to it is properly settled by the appropriate procedure, it will be decided to be ϕ.

It appears, then, that if someone is prepared to allow that one of any pair of contrary such conditionals must be true, Wittgenstein's considerations about rule-following will leave available to him a quasi-platonist attitude to excluded middle as applied, for example, to the example about π. The account given of the truth-conditions of ' "550" does not occur in π' can, indeed, remain formally the same as that of the platonist: the statement is true if and only if every member of the infinite conjunction is true of which the nth member is:

> the nth initial segment of π, if the expansion is taken that far, will be decided not to contain '550', if the matter is properly settled in the appropriate manner.

The appropriateness of such an account both for examples involving rule-generated series and effectively decidable predicates, and for examples involving free-choice series and choice-predicates, would ordinarily be held to require a kind of pun on the notion of 'decision'. Not, however, on Wittgenstein's view; for now we surrender the idea that the nature of the concepts involved predetermines the truth-value of each such compound conditional in the former kind of case. In both cases the conditionals involved have to do with what will happen if certain decisions akin to free choices are taken; for that is how the community's verdict on the truth of a hitherto undecided mathematical statement is now to be viewed—it is just that its members are likely to have firm *reactions* to the question what the verdict should be. But even without a realm of determinate but unrecognised internal relations to draw on, it remains open to us, according to the quasi-platonist, to suppose that such conditionals, like future-tense contingent conditionals in general, are determinately true or false more simply just in virtue of the way the world will be if their antecedents are fulfilled.

5.　　Quasi-platonism is apt to seem perplexing in a crucial respect: what sort of account can it give of what—to put it crudely—mathematics *is*? For, interpreted as above, ' "550" does not occur in π' is merely a *contingent* generalisation. How, then, is its truth amenable to discovery by *proof*?

That is not the way to formulate the problem; the whole strategic point of a quasi-platonism is to leave room for a rejection of the special category of fact of which we incline to think of proofs as an instrument of discovery. Rather the question is: what on this view makes that statement, or any of its instances—in contrast with a suitable analogue concerning a free-choice sequence and a choice-predicate—a *mathematical* statement? And what account is a proponent of the strategy to give of the rôle of proof in our acceptance or rejection of such statements?

Central to Wittgenstein's whole attitude to necessity is the idea that the distinction between necessary and contingent statements is not to be seen in terms of contrasting kinds of truth at all. Rather the contrast is to be sought in the contrasting kinds of use made of such statements: crudely, necessary statements have a *normative* rôle. A quasi-platonist could be expected to pursue this suggestion. The statement that '550' does not occur in the first n places of π is indeed to be seen—when so far undecided—as a contingent, implicitly conditional statement, true or false in virtue of the decisions which we will, or would, take in the course of implementing its implicit protases. To that extent it is properly compared to a contingent conditional concerning a free-choice series and a choice-predicate. But the difference is that when we (quasi-) experimentally check out the former, we do not merely take the decisions in virtue of which it is true or false; we lay down the obtained outcome as a condition that these decisions have been *properly* taken—that the matter has been 'properly settled in the appropriate manner'. And this stipulation remains binding for subsequent tests (whereas, with statements concerning free-choice series and choice-predicates, there would be no need to provide for subsequent tests at all). By taking this decision we effect a transition in the status of the conditional from that of contingent statement to that of mathematical *rule*. On this view, the substance in the idea that ' '550' does not occur in the first n places of π' is already a mathematical assertion, before its truth-value is known, is just that here, for reasons of which a Wittgensteinian owes an explanation in any case, we are in the market for such a normative stipulation. But before such a stipulation is made, the statement is not yet used normatively and can only be regarded as a contingent one; the concept of truth, and hence the logic, appropriate to it are thus indeed those appropriate to a suitably corresponding example concerning future free-choice, even though in such a case no such stipulation is in view. To adapt IV. 9: the conditional about the first n places of π 'changes its status'—to that of rule—when the decisions are made in virtue of which it was (all along) contingently true or false; for we then construct a connection, between the outcome and the matter's having been properly settled in the appropriate manner, which 'formerly was not there'. It is in terms of our readiness to effect such a connection that an account of the mathematical character of the example is to be given. And the effect of a proof, should we get one, that '550' nowhere occurs in π is to be seen not, bewilderingly, as a discovery of the truth of an infinite set of contingent conditionals, but as persuading us simultaneously to confer on each of them this normative status.

It appears, then, that the quasi-platonist is in a position to echo very much the sort of account of proof and necessity which is adumbrated in *RFM*. What he is proposing is an amalgam of a thoroughgoing realism about a large class of contingent statements with a denial, motivated by the rule-following considerations, that the rôle of necessary statements is to record a species of necessary fact. If quasi-platonism is a coherent position, then we have an argument that the rule-following considerations are non-revisionary which is of the same kind as that developed in the previous chapter for Wittgenstein's imperatival motif: they are non-revisionary because, at least for the kind of examples on which we have been concentrating,[1] they leave intact the possibility of a realist/anti-realist dispute about what logic is appropriate, neither helping nor hurting either side.

Even if it can be generalised for a much wider class of mathematical statements, however, the argument may seem rather frail. It says: do not imagine that the rule-following considerations call into question aspects of classical number-theory, for example, for they leave open the possibility of a quasi-platonist view of undecided number-theoretic statements. But then the fact is that all that seems to stand in the way of a revisionary harvest from these ideas is a species of realism which Wittgenstein would surely have found repugnant on other grounds. For one who takes the rule-following considerations to heart, undecided but effectively decidable mathematical statements are to be viewed as of a species with future-tense contingent conditionals concerning the outcome of free choices, or random processes—what I shall henceforward call *C-conditionals*. The realism in question takes it as legitimate to suppose that one of any pair of contrary such conditionals must be true. Such a view is not only nowhere endorsed in Wittgenstein's later writings; it appears totally at variance with their general anti-realist temper. And it seems, besides, likely to be objectionable in a variety of ways.

To be sure, the situation seems quite unsatisfactory if all that is preventing a collision between Wittgenstein's ideas on rule-following and his non-revisionary conception of the philosophy of mathematics is the bare possibility of a standpoint which wars with other aspects of his philosophy and which is objectionable in any case. We had therefore better make sure that quasi-platonism has both those characteristics. But there is a problem with the idea that a quasi-platonist can straightforwardly take over the fabric of classical number-theory which we can usefully consider first.

[1] That is, effectively decidable number-theoretic statements, and their universal and existential generalisations. Again, it seems a reasonable conjecture that the argument could be generalised to cover a wide class of statements.

6.   The conception that C-conditionals are determinate in truth-value requires an *account* of their truth-conditions, in particular for the problematic case where the antecedent clause, or clauses, are false. But actually it seems that there is no account to give of this case which sustains the suggestion that classical logic remains valid for the alternatives about π when they are seen as infinitary truth-functions of such conditionals. If we rule that these conditionals cannot be true unless their antecedents are true, we can no longer suppose that the disjunction of one with its contrary is valid. In fact, of course, it is certain that all the conditionals after a certain place would turn out untrue on this account. But if we ruled that such conditionals are true if their antecedents are false, then it would be truly assertable of any sufficiently large initial segment of π *both* that if it is written out and the matter properly settled in the appropriate manner, it will prove to contain '550', *and* that if it is written out and the matter properly settled in the appropriate manner, it will prove not to contain '550'. And if such conditionals become co..npatible, then their respective universal and existential generalisations will do the same, despite being represented as of the forms '$(x) Fx$' and '$(\exists x)-Fx$'. So, either way, this strategy for avoiding the concession of revisionary implication to the rule-following considerations is itself going to engender revision; either way, classically valid principles are going to fail.

   This point is not as powerful as it looks. For it would independently be a mistake on the quasi-platonist's part to give any ruling for C-conditionals with false antecedents. When we are concerned with an infinite free-choice series, we can of course be certain in practice that, for some suitably large n, n choices will not in fact be made. But to entertain the concept of a potentially infinite series of choices is precisely to abstract away from such practical certainty. We are entertaining the idea that it is *possible* to carry on a series of choices arbitrarily far; and the same goes for a predicate whose procedure of application becomes increasingly complex and/or time consuming. So long as we are prepared to countenance this idealised notion of possibility, implicit in the idea of potential infinity, we ought simultaneously to entertain the possibility of the truth of the antecedent clauses of any conditional of the relevant kind. In whatever sense the series is potentially infinite, in that sense the antecedents of the conditional are potentially true. Where, therefore, an attempt is being made to construe such conditionals as determinately true or false of a potentially infinite free-choice series considered *as such*, it would be an irrelevance to stipulate for the supposition of falsity of the antecedents; indeed, it would be in conflict with the fiction of potential infinity. Rather a notion of truth is required for such conditionals such that the actual falsity of their antecedents would neither rob them of truth-value nor settle it. The quasi-platonist requires for C-conditionals precisely a notion of truth such that their truth-value is *unaffected* by that of their antecedents, and such that either they or their contraries must always be true.

Now, ought we to repudiate any such notion of truth? The orthodox view, that truth is timeless, seems to afford some impulsion in the opposite direction. For one thing, we should ordinarily accept as valid the conditional 'If n choices are made, then β_n will either be 7 or it will not'; all, it seems, that can then prevent a valid transition to the disjunction '*either* if n choices are made, β_n will be 7 *or* if n choices are made, β_n will not be 7', is the notion that while the disjunction of consequents will be true in those circumstances, it is not yet true of either consequent individually that it will be true if n choices are made. Rather one or the other will become true if n choices are made, but there is so far nothing true to be said about which. But surely, on the orthodox view, there is in the relevant sense no *becoming* true.

What is absolutely clear is that the orthodox view commits us to *a* notion of truth for these conditionals. Make the hypothesis that as a matter of fact n choices will be made. When they have been, either that $\beta_n = 7$ will be truly assertable, or its negation will be. Whichever it is, it could, on the orthodox view, have been truly asserted at any previous time; and in that case it can scarcely be denied that the weaker, conditional statement, for example 'If n choices are made, β_n will be 7', could likewise have been truly asserted at any previous time.

This argument falls short of the quasi-platonist's requirements just because it involves supposition that the conditional's antecedent is actually true. It is when we consider more carefully the implications of the required *irrelevance* of the actual truth-value of the antecedent that we begin to have doubts about the legitimacy of the notion of truth which he needs. Suppose we are concerned with a finite series of choices, specified as being selected within a particular period of time; and that there is no presumption that the selection will actually be carried out. Let it be the case, in fact, that as things will turn out, the choosing-agent will not get beyond the $n-1$th place. Still, according to the conception of truth in question, if someone asserts 'α_n, if n choices are made, will be decided to be ϕ, if the matter is properly settled in the appropriate manner', he makes an assertion with a determinate truth-value, and one that is not settled by the actual falsity of its antecedents. Suppose, then, that what he says is true. Now, if a conditional in the future-indicative is truly asserted at some time, and if it is allowed to be consistent with, though insufficient for, its truth that a period should then elapse after which its antecedent can no longer be fulfilled and during which it was not fulfilled, then anyone who asserts *after* that period a corresponding conditional in the perfect subjunctive must also speak truthfully. If there is a legitimate notion of truth of the sort which the quasi-platonist requires for these conditionals, it is precisely a notion which, by the orthodox view and natural assumptions concerning the rôles of tense and mood, may legitimately be extended to the

corresponding counterfactuals if the opportunity to fulfil the antecedent clauses of the original conditionals is something that can lapse.
But this seems too much. It is one thing to claim to understand what it is
for a conditional to be true, when the truth of its antecedent, and so
what would ordinarily be regarded as a verification or falsification of it,
is still a possibility; but if a free-choice series is discontinued, in what
can the truth of the residual counterfactuals about the further choices
that might have been made consist? We can suppose that the choosing-
agent was making his selections in a quite spontaneous and unpremedi-
tated way; in that case, neither he nor we can possibly know such
counterfactuals to be true or false—indeed, it is unclear that we could
have even the weakest grounds for believing or rejecting them.

Our doubts about the legitimacy of thinking of such counterfactuals
in the manner described are expressions of an intuitive anti-realism
about them: lacking any clear concept of the kind of information which
would justify their assertion, we are at a loss to explain to ourselves what
kind of state of affairs would make them true. But, granted the orthodox
view, the permissibility of a realist attitude to the counterfactuals seems
a straightforward corollary of the quasi-platonist's realism about their
C-conditional counterparts. So, it is natural to argue, something must
already be amiss with the notions of truth and falsity which the quasi-
platonist applies to the latter; and what is wrong remains wrong, even if
we are concerned with a case, like the decimal expansion of π, where the
opportunity to continue the series and assess the application of ϕ to the
resulting initial segments remains with us for ever, so that the perfect
subjunctive can never be the *uniquely* appropriate form of expression
for the intended statement.

The argument, then, is that the unacceptability of a quasi-platonist
view of C-conditionals in general shows itself when we consider the
subclass of such conditionals for which a counterfactual formulation
can in certain circumstances become necessary. For, on the orthodox
view of truth as timeless, an unacceptable realism about such counterfactuals is then imposed. To be sure, this subclass includes none of the
C-conditionals which have come to be regarded as fixing the content of
effectively decidable but undecided mathematical statements; for in
these cases the opportunity to fulfil their antecedent clauses cannot
lapse. But it would be a strained gesture to try to make anything of that.
Intuitively, it is most implausible to suppose that realism about a
statement like 'If a series of nine positive integers is ever chosen at
random in this place, the seventh will be prime' might be more acceptable than realism about 'If Leonid Brezhnev ever chooses at random a
series of nine positive integers in this place, the seventh will be prime'.

Exactly what sort of inchoate anti-realism does this argument invoke?
Whatever it is, it seems it must have these two features: it must sustain
our intuitive reservations about such counterfactuals, and it must allow
us to refer the doubt back to the future-indicatives. The latter require-

ment, however, involves sustaining the orthodox view, or at least this much of it: that what may truly be asserted at a particular time may always, by appropriate mood and tense transformations for example, be truly asserted at any later time. It is the contrapositive of that principle which the intuitive argument requires. But now, are these two requirements mutually coherent? We are brought up against the question of the overall effect of anti-realism upon our ordinary preconception that truth is eternal, that, while our vantage point changes, there is a single unchanging overall stock of true judgments to be made about the world. On the correct answer to this question turns the kind of anti-realist complaint, if any, which should be brought against the quasi-platonist's attitude to conditionals of the relevant sort. That is a question for the next chapter.

I should stress, in conclusion, that even if realism about C-conditionals proves, contrary to appearance, to be both acceptable and coherent with Wittgenstein's later philosophy as a whole, that will of course be no vindication of the quasi-platonist's treatment of quantified generalisations of such conditionals. But it is in any case already unclear how Wittgenstein's sympathy with the intuitionists' complaints about the classical interpretations of the quantifiers can spare him a revisionary commitment. What we are trying to see at the moment is whether the rule-following considerations are revisionary independently of back-up from Wittgenstein's intuitionistic leanings. If they are not, then out of the original five themes potentially in tension with Wittgenstein's laissez-faire attitude to mathematicians, only the latter, assuming the soundness of our conclusions about the others, will remain. But if, on the other hand, only an unacceptable standpoint comes between the rule-following considerations and rejection of a class of applications of excluded middle acceptable to platonist, intuitionist and strict finitist alike, then—since a *false* view of something or other will always be available to spare us the consequences of a particular thesis—we will have to conclude that the rule-following considerations cannot be reconciled with Wittgenstein's non-revisionism, and that one or the other, at least as here interpreted, must be in error.

X

Anti-realism and Future Choice

Sources

PB: v

1.   Any proposal to the effect that the concept of truth may be intelligibly applied only to, and to all, statements which we are capable of verifying can properly be regarded as anti-realist.[1] Obviously, however, the force of a particular such proposal turns on the intended sense of 'what we are capable of verifying'. Suppose, to begin with, that this is understood as involving a double idealisation: 'what we are capable of verifying' is to mean what we are capable *in principle* of verifying—that is to say, what some being with limitations differing from our own only in finite degree is capable in practice of verifying—at *some* time. Thus any statement of which we can recognise that there is a point in world-history at which its truth-value could, in the relevant sense, in principle be decided may be regarded at all times as determinately either true or false.

The picture of the world appropriate to this idea need not be less than this: it is external to us—we do not create it; we journey through it, temporally speaking, as observers. But it is a world without properties other than those which we can in principle recognise; and for each of its properties an opportunity arises when it can, at least in principle, be recognised to apply—like the number of sheep in a field seen from a moving train. If the opportunity is not taken, still what could have been verified remains so.

Clearly this liberal anti-realism—*I/S anti-realism*[2]—entails the orthodox view that if a statement is truly assertable at a particular time, it may, within a language sufficiently rich to permit the appropriate transformations of mood, tense, and pronoun, be truly asserted at any time. The opportunity for verification can be lost; but the opportunity for truly making the statement can neither arise nor be lost. The facts are timelessly as they would have been determined to be if the opportunity to test them had been taken; they are eternal. But it seems

---

[1] Since to falsify a proposition is to verify its negation, any such proposal eo ipso restricts the concept of falsity to what we are capable of falsifying.

[2] In principle/Sometime.

equally clear that I/S anti-realism is impotent to criticise the quasi-platonist's conception of the truth-conditions of C-conditionals. Indeed, it does not even appear to sustain our intuitive doubts about their counterfactual counterparts. An agent, for example, who stops after n-1 choices, could in principle have gone on to make an nth choice; and we could in principle then have determined the truth-value of the previously asserted conditional: 'β_n, if n choices are made, will be decided to be ϕ, if the matter is properly settled in the appropriate manner'. As it happens, the opportunity to do these things was lost; but for an I/S anti-realist the relevant fact will be for ever as we should have found it to be if the opportunity had been taken, and an appropriate counterfactual now gives it expression.

It seems, then, that if a strain of anti-realism exists which can validate the intuitive argument against quasi-platonism outlined in §6 of the last chapter, it must impose some severer temporal restriction on what we may legitimately deem ourselves 'capable of verifying'. An obvious candidate for consideration is the restriction imposed by *I/N anti-realism*: what we are capable of verifying is what we are in principle capable of verifying *now*. Someone might protest immediately that of no statement which we have not yet verified can it be said that it is capable of verification 'now'; for it we start now, we shall finish in the future. But fortunately we need not waste time on this. Intuitively we can recognise at any time a threefold division among possible assertions: those for which if we have not already tested them, it is too late; those for which if we are to test them, it is too soon; and those for which the opportunity for a relevant test is current. Usually, though not without exceptions, these are respectively statements in the past, future and present tenses. What an I/N anti-realist will propose is to restrict the concepts of truth and falsity at any given time to statements which at that time come into the third category.

The general effect of such a view seems to be that the sufficiently distant past and future have no reality—for there is nothing true to say about them. In general, only the present may be truly described (though it may 'contain' facts about the distant past and future; we may, for example, have no criterion for saying of something that it *is* F, but only that it was). In terms of the railway analogy, the picture would be that nothing true may be stated at any given time save as may in principle be verified by an observation from the carriage window at that time.[1] The totality of facts thus shifts constantly and continuously; there is nothing ahead on the track save what we can see coming, and nothing behind save what we can see receding into the distance. Of course, the relative intelligibility of this picture is feeding upon our having given the period of relevant observation a *span*; things take time

[1] Or may at any time be verified without looking out of the window. Since it is, in the relevant sense, never too soon or too late to seek a mathematical proof, both I/S and I/N anti-realism will generate an intuitionistic attitude to mathematical truth; the same holds for the other I-anti-realisms distinguished below.

to pass by the carriage window—there is a period within which it can be true to say that though we could have tested S already, still it is not yet too late to do so for the facts relevant to S are still in view. Something of this sort, it seems to me, is essential to the anti-realism for which this is a picture, though fully to elucidate the matter looks difficult. (Notice that there would, at any rate, be no constraint to think of such a span as uniform; it could be allowed to vary, depending on the statement concerned.)

Now, how does the quasi-platonist attitude to C-conditionals fare on this view? Rather unsatisfactorily, its fate turns out to depend on what the spirit of the intuitive argument of IX. 6 would surely regard as an immaterial detail concerning the exact character of the conditionals in question. If, for example, we are concerned with a series of choices stipulated as being executed next month, then presumably no C-conditional about it will yet be regarded by an I/N anti-realist as determinately true. But if it is allowable for the series of choices, and the ruling about the predicate, relevantly to take place at any time, then it appears that the opportunity for verifying or falsifying such condition-als is, in the germane sense, *current*. A conditional like 'If n choices are made, the nth digit chosen will be ϕ' will be currently in principle decidable, and thus assertable now with determinate truth or falsity. And this will be the situation of the conditional equivalent of ' "550" does not occur in the first n places of π', when π and the predicate are thought of as comparable in essential respects to a choice-sequence and choice-predicate respectively.

I/N anti-realism will presumably sustain our intuitive reservations about the counterfactual counterparts of C-conditionals. But we can no longer build on those reservations in the way which the argument of IX. 6 proposes; for the presupposed element in the orthodox view, that if a statement is truly assertable at a particular time, it is, in some form, truly assertable at any later time, is now apparently ruled out. But might there be a view intermediate between I/S and I/N anti-realism in terms of which the IX. 6 argument can fare better? What, for example, of an anti-realism which argues that any statement which we *will* be in a position (in principle) to decide is one which we *are* in a position to decide now; it is merely that the 'decision procedure' is a lengthy one?[1] On this view —*I/NF anti-realism*—a statement may at a particular time be regarded as determinately either true or false if and only if it can then be recognised either that an opportunity for determining its truth-value is current or—what for the I/NF anti-realist comes to the same thing—that such an opportunity will arise in the future.

Clearly this is no help. Certainly a realist view of the counterfactual counterparts of C-conditionals is now ruled out; but again so too, it seems, is the element in the orthodox view which the IX. 6 argument needs—the thesis that anything truly, or falsely, assertable at a particu-

[1] Cf. Dummett 25, p. 469.

lar time always remains so. And things are no better, finally, on an *I/NP*—or *Aristotelian*[1]—*anti-realist* standpoint; for now exactly the reverse situation obtains. If a statement may be correctly regarded at a particular time as determinately either true or false if and only if an opportunity for deciding its truth-value is either current or has already occurred, then the needed element in the orthodox view is safe enough; but no support whatever is now forthcoming for our intuitive anti-realism about the relevant kind of counterfactual.

It begins to look as if the strain of anti-realism which the intuitive argument demands is chimerical. The intuitive argument was that if a realist attitude is granted to C-conditionals in general, if it is allowed that either they or their contraries are determinately true irrespective of the truth-values of their antecedent clauses, we have to be prepared to say the same about the corresponding counterfactuals in any case where the perfect subjunctive can become the appropriate form of expression; and this, on intuitive anti-realist grounds, we were reluctant to do. But now it seems that any anti-realism which backs up this reluctance cancels out the element in the orthodox view appealed to in the argument. Moreover, for all we have so far been able to see, each of the four types of anti-realist described seems to be in position to *accept* a realist view of, for example, the conditionals involved in the quasi-platonist account of the truth-conditions of ' "550" does not occur in π'—and indeed a realist view of all C-conditionals which serve, under the umbrella of the rule-following considerations, to express effectively decidable but so far undecided mathematical statements. For each type of anti-realist allows that a statement may correctly be regarded as determinately true or false if an opportunity for determining its truth-value is current; and the opportunity for taking the 'decisions' involved in the solution of an effectively decidable mathematical question is ever-present.

We could, of course, have considered instead species of anti-realism for which the idea of what we are capable of verifying would be very much the idea of what *we*, limited as we are, are capable of verifying. The result would be four corresponding types of 'strict finitist' anti-realism. But no importantly different conclusions would be entailed. Certainly, under each strict finitist view, certain C-conditionals could not be regarded as candidates for truth; those, namely, implementation of whose antecedents was not ever in practice feasible. So a holder of any of these four views would be well enough equipped to criticise the quasi-platonist belief of every nth C-conditional constituent in his account of the truth-conditions of ' "550" nowhere occurs in π' that either it or its contrary must be true. But we should be no nearer to elucidating the intuitive impressions that, even if implementation of the situation depicted by its antecedent clauses was once feasible, something is amiss with a realist view of a counterfactual about free choices;

---

[1] *Locus classicus:* Aristotle, *De Interpretatione* 9, sections 18a–19b.

and that what is amiss is also amiss in an anterior realist view of the corresponding future-indicative. It seems that neither by giving the notion of verifiability a temporal restriction, nor by leaving it unrestricted, can this impression be elucidated; and it thus becomes doubtful whether there is anything coherent there for elucidation.

2.   This conclusion rests on the presupposition, which seemed plausible enough, that N and NF types of anti-realist will have to jettison the orthodox view that truth is timeless; for neither can hold, it seemed, that whatever is truly assertable at a particular time is truly assertable at any later time. Likewise it seems that N and NP anti-realists will have to reject the corresponding thesis with 'later' substituted for 'earlier'. But is it certain that this is so? Or is it perhaps something which seems so only from a standpoint *exterior* to anti-realism of the relevant kinds? Consider some of the standard truth-value liaisons between tensed assertions, the kind of things that would, presumably, feature in a detailed account of the orthodox view. For convenience, imagine them in relation to any language differing from English just in respects enjoined by the circumstance that all tense indications are achieved by inflexions of the predicate, 'to be true'. Let 't_1' and 't_2' range over times, t_1 earlier than t_2; and let 'Q' range over statements of the language. Then, taking the same language as metalanguage, we want to say that the following equivalences will hold:

A: (t_2) ['Q was true' is true at t_2 if and only if $(\exists t_1)$: 'Q is true' was true at t_1.]

B: (t_1) ['Q will be true' is true at t_1 if and only if $(\exists t_2)$: 'Q is true' will be true at t_2.]

C: (t_2) ['Q is true' is true at t_2 if and only if $(\exists t_1)$: 'Q will be true at t_2' was true at t_1.]

D: (t_1) ['Q is true' is true at t_1 if and only if $(\exists t_2)$: 'Q was true at t_1' will be true at t_2.]

It may very well be that an anti-realist (of one of the relevant six kinds) would want to reject some or all of these equivalences; but is it really clear that, as the positions have so far been characterised, there is cogent reason why he *must*? For each of these equivalences has to be understood as providing for what may be correctly asserted at any particular *present* time: t_2 for A and C, t_1 for B and D. (Otherwise there is no way of understanding the tense-indications in the non-quoted expressions which they contain.) Whereas there is a suspicion that, in order to appreciate the putative changes in truth-status which certain statements will undergo if an N, NF, or NP anti-realism is espoused, we need, so to speak, to straddle *distinct* present times simultaneously.

Let us try to chase down this suspicion. Consider equivalence *A* from the point of view of an I/N anti-realist. (For if an I/N anti-realist can avoid revision of the orthodox view, so can the others.) Let T be a tenseless description in the language in question of an event on a star one hundred light years away, which we have just now, at t_2, verified to have taken place. Our inclination was to think that an I/N anti-realist is bound to hold that someone who one hundred years ago, at t_1, asserted 'T is true' could not then have correctly been regarded as stating anything true; for what he said was not, at the time he said it, currently decidable. How, then, can such an anti-realist, having verified at t_2 that T was true, accept the equivalence, A?—for it precisely allows us to infer that 'T is true' expressed a truth one hundred years ago, at t_1.

We can envisage this answer: if an I/N anti-realist accepts the equivalence A, then ' 'T is true' was true' will indeed enter the ranks of currently decidable statements; and our astronomic observations will force him to regard it as now true at t_2 that 'T is true' was true at t_1. But then all that follows is that, although someone who one hundred years ago asserted 'T is true' could not then have been regarded as making a true statement, *we* are now in a position to know better. And is not that just our ordinary view of the matter?

One interpretation of this reply would be as follows: the statement now expressed by ' "T is true" was true at t_1' is a statement which has now, at t_2, *become* true. The essence of I/N anti-realism is that there is no totality of statements true for all times; but the totality of statements true at any particular time includes everything required of it by the equivalences A–D. This interpretation, however, provides too easy a vindication of the incompatibility of I/N anti-realism with the timelessness of truth. If it has only now come to be true, at t_2, that ' "T is true" was true at t_1' expresses a truth, the same statement could presumably not truly have been made *before,* for example at t_1; but at t_1 the appropriate form of expression for the statement was ' "T is true" is true', and that this sentence expressed a truth at t_1 follows from the equivalence A, by substitution of 'T is true' for 'Q'.

If that is an unfairly hostile interpretation, it nevertheless remains quite unclear that an I/N anti-realist can legitimately simply help himself to A–D. We want to agree that our 'ordinary view of the matter' is indeed that an assertor of 'T is true' one hundred years ago could not have been regarded as making a true statement—that is, nobody could reasonably have believed him; but we also want to insist that it was still a *possibility* that he was right, whereas the I/N anti-realist ought to hold that his assertion was not the kind of thing which *could* be true at that time. But that, of course, is just to beg the point in question; for *if* an I/N anti-realist can accept the equivalence A, then he will hold not merely that the assertion of 'T is true' one hundred years ago could have been true, but that it was!

How can we sharpen the issue? In order to demonstrate to an I/N anti-realist that he cannot endorse the equivalence A, it suffices to find a statement of which he must say that it can now be truly asserted that it was true at some time in the past although at no time in the past could it have been truly asserted. And T is prima facie such a statement. Intuitively there seems a clear contrast between the situation of 'T was true in 1879' asserted in 1979, the earliest time at which anyone can know anything about the matter, and the situation of the same statement made at any earlier time, in particular in 1879. We want to say that an I/N anti-realist ought to accept that this statement can be truly asserted in 1979 but at no earlier time. But *when* are we when we attempt to foist this contrast on him? If 'now' is 1979 or later, then our observations have taught us both that 'T was true in 1879' expresses a truth and that 'T is true' expressed a truth in 1879—that someone who then affirmed that statement spoke truly. But if 'now' is 1978 or earlier, then we are not *now* in a position to make either claim. Either way there seems to be nothing in the situation to force an I/N anti-realist to acknowledge a threat to A.

An argument thus emerges according to which the suspicion alluded to above was correct. The changes in truth-status which an I/N anti-realist will allegedly have to acknowledge cannot, the argument runs, be given expression save from an atemporal viewpoint. So long as we take seriously the fact that the equivalences $A–D$ are tensed, that we have to consider them as affirmed at a particular point in time, we cannot identify any particular statement as a counterexample.

The argument will bear a little elaboration. The orthodox view is a conjunction of two theses: that any particular statement can be made by anyone at any time and that its truth-value is invariant. That is:

For any statement P, and times t_1 and t_2, there are locutions, S and R, such that S expresses P at t_1 and R expresses P at t_2 and S expresses a truth at t_1 if and only if R expresses a truth at t_2.

Why did it look certain that I/N anti-realism would be forced to acknowledge counterexamples to this? Merely because, taking 'P' as an example like that about the star, and the times involved as 1879 and 1979 respectively, it seems plain that an opportunity for assessment of 'P is true' is not current at t_1, so that that sentence cannot, for an I/N anti-realist, express a truth at that time; whereas 'P was true at t_1', which expresses the same statement at t_2, is assessable, and has indeed been verified, at $t2$. But the clear-cut status of the counterexample depends upon the legitimacy of the implicitly *tenseless* modes of expression used in its formulation and in that of the orthodox view: 'expresses a truth at t_1', 'expresses P at t_2', etc. In order to force the counterexample on the I/N anti-realist, we have to impose on him a point of view from which he is to describe in an eternal present tense, as it were, the respective situations at t_1 and t_2; rather as the ordinary present tense serves to describe the respective situations at different *places*. But the primary point of I/N anti-realism will be to repudiate the fiction of such a standpoint —this 'spatialisation' of time. We are inextricably stationed in time; it is only from our current specific temporal point of view that the world can be described. And as soon as we pick on a temporal location—a 'now' —relative to which t_1 and t_2 are to be understood, and accordingly tense-inflect the above expression of the orthodox view, there is no demonstrable incoherence in the position of an I/N anti-realist who insists on his right to endorse that view.

It is *current* in principle decidability which, for the I/N anti-realist, puts bounds on the domain of intelligible application of the concept of truth. To suppose it follows that a particular statement must be viewed as having lain outside those bounds at any time at which it was undecidable is to do one of two things: it is either to slip into a point of view from which the truth-status of that statement at a particular time is thought of as settled by something *other* than its status in point of decidability *now*—which is just to part company with the I/N anti-realist. Or it is to suppose that the I/N anti-realist's conception of current in principle decidability can have no place for the inferential powers of *A–D*; and that supposition was meant to be established by the counterexample, not assumed by it.

3.   The foregoing is no final demonstration that an I/N anti-realist can retain the orthodox view; it merely brings out that we should not too readily assume that he cannot. But if I/N anti-realism—and a fortiori NF and NP anti-realisms—really are non-revisionary of the ordinary apparatus of tense and mood liaisons of which *A–D* are components, will it not follow that the Chapter IX, §6 argument ought to be regarded as sound from an I/N anti-realist viewpoint?

That, admittedly, would be an unexciting result so long as our concern is merely to explicate our original intuitive reservations about quasi-platonism; for there is no doubt that the qualms we feel concerning realism about the relevant kind of counterfactual do not *seem* to us like expressions of anything so radical as I/N anti-realism. The interest of the situation would rather concern its bearing on the *coherence* of I/N anti-realism: if the intuitive argument of IX. 6 is sustained, how is that to be reconciled with the current in principle decidability of the sort of C-conditional which corresponds to an effectively decidable mathematical statement? The prospect opens that I/N anti-realism will have no consistent response to the question whether such conditionals can legitimately be treated in the quasi-platonist manner; and if that prospect is realised, a very sharp edge will be put on the question how best to adapt to ordinary statements the ideas of the intuitionists about mathematical truth.[1]

Let us follow the intuitive argument through wearing I/N anti-realist

[1] For I/NF anti-realism will confront the same difficulty.

hats. Since, to begin with, nothing counts as a current opportunity to determine their truth-value—or even, plausibly, to arrive at a reasonable belief about the matter—we shall reject the idea that one of any pair of contrary counterfactuals of the relevant sort must be true. Thus, at t_2—now—we will not accept that a disjunction of such a counterfactual (about, say, how I would have continued a series of arbitrary choices broken off five minutes ago) with its contrary must be true. By an analogue of equivalence A, then, and the necessary further assumptions about translations of mood, we will not regard the assertion of the disjunction of the corresponding future-indicatives made at t_1—six minutes ago—as having had then to have been true. A quasi-platonist view of those conditionals was therefore mistaken.

We seem to be well on the way both to sustaining the intuitive argument and to uncovering the threatened incoherence. Arguably, however, we are as yet in no position to do either; for there is so far no commitment on the question what our attitude should be to comparable conditionals which it is appropriate to express in the future-indicative *now*. We have the example of such a conditional, call it A, asserted in the past, of which we are prepared to argue that it did not have to be the case that either it or its contrary was true; so it might seem that we must be in a position to reject the quasi-platonist concept of truth for such conditionals in general. There is a temptation to argue that, since the quasi-platonist's claim that one of that pair of conditionals had to be true was, as we now see, incorrect, why should it be any different with any comparable conditional, B, now assertable in the future indicative? But the question betrays how easy it is to slip out of the I/N anti-realist view of these matters. The suggestion is that we should attempt to see C-conditionals asserted now and similar conditionals asserted six minutes ago as on an equal footing; that we ought to be ready to transfer our verdict about the validity of any particular logical compound of the latter to any structurally identical compound of the former, and *vice versa*. And this is just an expression of the traditional view that it is appropriate to essay a general account of what sort of compounds of a particular kind of statement are validly assertable, irrespective of the relation between the times at which the assertion is envisaged as made and the question of validity raised. An I/N anti-realist, however, will insist that matters of validity be addressed from our specific temporal viewpoint. So the questions we are confronting are: ought we now to suppose that the disjunction of A with its contrary was validly assertable six minutes ago? and: ought we now to suppose that the disjunction of B with its contrary is validly assertable now? No reason is apparent why the same answer should be given in both cases. If B is currently in principle decidable, the answer to the second question is affirmative. But if, as seems clear, in order for it to be true now that a disjunction was true at some particular past time, is has to be true now that one of its disjuncts was then true; and if it is currently in principle verifiable of neither disjunct that it was then true—the situation of A and its contrary—then an I/N anti-realist is currently in no position to hold that the disjunction was, let alone had to be, true. So the fleeting prospect of support for the intuitive argument of IX. 6 from an I/N anti-realist standpoint—and the attendant threat of internal incoherence in that standpoint—now seem to be empty.

4.   The feeling may persist that somehow or other it ought to be possible to demonstrate that I/N anti-realism and the orthodox view that truth is timeless simply will not marry. For surely there is *some* sort of tension between the orthodox view and, what is surely evident even from an I/N anti-realist standpoint, that the class of currently in principle decidable statements varies over time: that it can, for example, be true that S was, or will be, decidable at some past or future time whereas T which now expresses what S used to, or will, express is not currently decidable.

We could actually generate a contradiction if we could soundly reason as follows:

(a) ' "Jones is brave" is currently decidable' was true at t_1.
(b) ' "Jones is brave or Jones is not brave" is true' was true at t_1.
(c) ' "Jones is brave or Jones is not brave" was true' is true at t_2.
(d)   'Jones is brave or Jones is not brave' was true.
(e)   'Jones was brave or Jones was not brave' is true.

For, where Jones was alive and well at t_1 but is now, at t_2, dead, and never had the opportunity in his life to show whether he was brave or not, the anti-realist has to hold that (a) is now correctly assertible whereas (e) is not.[1] So what is wrong with the reasoning? The transition from (b) to (c) is a straightforward application of equivalence A above,[2] and that from (c) to (d) a consequence of the principle that if 'P' is now true, we may now truly assert that P. So either the I/N anti-realist must discard some of the tense-equivalences, or he must find fault with one or both of the other transitions.

Now the transition from (d) to (e) is not, to be sure, a direct application of the equivalence A. It appears to require that we can validly distribute the past tense across the disjuncts in (d), so arriving at:

(d)* 'Jones is brave' was true or 'Jones is not brave' was true. However if that step is in order, the reasoning can then proceed in the pattern of a disjunction elimination: we apply equivalence A to each disjunct of (d)* in turn, inferring (e) in both cases by a disjunction-

---

[1] Cf. Dummett 22 (Pitcher pp. 107-9).
[2] Save that, irrelevantly to the present context, the statement variables in A–D were stipulated to be ranging over tenseless expressions.

introduction. And the validity of the step from (d) to (d)* is a consequence of the principle enunciated towards the end of the preceding section that if a disjunction is now truly regarded as having been true at some time, it has to be true now of one of its disjuncts that it was true at that time.

These matters are so slippery that it is impossible to be confident that nothing is being overlooked. But I believe that the anti-realist has no good reason to discard the principle in question—save the prospect of an emergency if he accepts it!—and that he ought therefore to accept the transition from (d) to (e). (And if he does not accept it, he owes an explanation of how, if he professes allegiance to the timelessness of truth, we can *now* give expression—without mentioning any sentences—to the state of affairs necessary and sufficient for the truth of 'Jones is brave or Jones is not brave' at t_1.)

Where the reasoning seems certain to be frustrated is in the step from (a) to (b). There would be no objection to the step if the essential thesis of I/N anti-realism could be fairly expressed as that the class of statements determinately true or false at any particular time coincides with the class of in principle decidable statements at that time; it would follow that any statement may be regarded as having been determinately true or false at any time at which it was in principle decidable. But that is not a fair expression of the essential thesis; it is merely a description of the effect which it has from a viewpoint *sub specie aeternitatis*. Whereas for the I/N anti-realist it is, to stress the point again, *current* in principle decidability which limits the domain of application of the concept of truth.

Again, then, the attempt to elicit an inconsistency between I/N anti-realism and the standard tense-equivalences founders on a lapse into a realist interpretation of the anti-realist's view. But we are at least in a position, I think, to locate more specifically the sense of *mauvais foi*, if not actual inconsistency, which that view provokes. For it seems clear that the I/N anti-realist could have no objection to the move from (a) to (b) if *now* were t_1 and 'was true' was replaced by 'is true' in each occurrence. So at t_1, he would presumably have been willing to accept an inference which he now rejects (for what (a) and (b) respectively state now—at t_2—is, by equivalence A, what the sentences which result from that substitution respectively stated at t_1). But he cannot hold, save at the cost of finally surrendering the timelessness of truth, that his present judgment about the transition from (a) to (b) is true *and* that his hypothetical judgment at t_1 about the transition between their equivalents at that time would then have been true *and* that the content of those judgments is the same. Perhaps he would jettison the last of these; but, since a judgment about the acceptability of an inference is a judgment about the acceptability of a conditional, that would involve abandoning the equivalence of (i) 'If "P" was true at t_1, then "Q" was true at t_1' is true at t_2, with (ii) 'If "P" is true, then "Q" is true' was true at t_1. For, substituting for 'P' the largest quoted part of (a) and for 'Q' the largest quoted part of (b), the anti-realist would be in the position of rejecting (i) but accepting (ii). And that would be to safeguard his loyalty to the timelessness of truth at the price of discarding a standard part of the very apparatus which enables us to make the same statement at different times.

Otherwise, there appear to be just two courses open to the anti-realist. One is heroically to insist that the judgment he would have made—perhaps actually did make!—at t_1 is wrong; wrong despite its irreproachability at that time from an I/N anti-realist viewpoint. But he is bound also to insist that any similar judgment which is irreproachable from that viewpoint at the present time is most certainly right; so how is he to blinker himself from the manifest possibility that he may later be constrained by his views to revise that assessment in any particular case? For nothing is more certain than that not every currently in principle decidable statement will actually be decided before it becomes too late. Perhaps we can at no time convict an I/N anti-realist who retains the orthodox view, and all the attendant apparatus of conversions of mood and tense, of holding inconsistent beliefs; but how can he put his heart into the claim that any inference from the current in principle decidability of a statement to its being determinately true or false is valid, when he must know that some of the conclusions which he thereby grants will have to be disallowed in the future?

The other possibility open to the anti-realist is to refuse to grant that at t_1 he would have accepted an inference which he now rejects. Such a refusal could rest on a denial that the inference he would have accepted at t_1 is the same as that which he now rejects; in that case the move is the one we considered first—it amounts to rejecting the equivalence of (i) and (ii) above, and we continue to await an explanation of how to express now the acceptable inference which he would have accepted at t_1 and of how we might have expressed at t_1 the invalid inference now rejected. But it could be that the anti-realist meant to deny that he would have accepted—or, indeed, did accept!—at t_1 the inference from the t_1-equivalent of (a) to the t_1-equivalent of (b). And such a denial seems intelligible only as an act of insincerity.

I/N anti-realism is no mere eccentricity, but is arguably the most natural extension to statements in general of the intuitionists' ideas about mathematical statements.[1] Such an extension must be possible if, as Dummett has so forcefully argued, an impressive case for (at least) intuitionistic views on mathematics can be based on quite general

---

[1] It is, at any rate, a nice question why the conception of the truth-conditions of erstwhile but no longer in principle decidable statements with which both I/S and I/NP anti-realism would credit us is not open to reproach in just the manner in which the platonist's conception of the truth-conditions of Goldbach's conjecture is open to reproach: that nothing serves to distinguish the responses of a man who has grasped those truth-conditions from those of a man who merely understands when he is entitled to assert or deny the statement. And I/NF anti-realism, if it escapes that reproach, does not escape the difficulties facing I/N anti-realism which this section has been concerned to bring out.

considerations in the philosophy of language. It is therefore of importance to determine at what price such considerations can coherently be
extended to non-mathematical statements, and the foregoing is some
indication of what that price may be: the choice, it appears, is between
(1) discarding the orthodox view that what may truly be stated at some
time may truly be stated at every time;[1] or (2) discarding some of the
equivalences which make up our ordinary conception of *how* to make
the same statement at different times; or (3) a kind of blinkered 'bad
faith'; or (4) insincerity about which judgments one would consider
correct at times other than the present.

5.   We have, after all, been able to make absolutely nothing of the
intuitive argument of IX. 6. We have succeeded in describing no strain
of anti-realism which would endorse that argument or, indeed—so long
as we confine our attention to 'in principle' varieties—which would
reproach the quasi-platonist attitude to C-conditionals on any other
grounds. In the last chapter I envisaged the complaint that quasi-
platonism views these conditionals in an unacceptable way and that
Wittgenstein would certainly have thought so. But the fact is that we
have so far found no basis for either part of the complaint; and while
that remains so, it has still to be shown that to endorse the rule-
following considerations does not leave one free to do mathematics in
classical, intuitionist, or even —whatever it may be —strict finitist style.
     Nevertheless a case can be made for the first part of the complaint;
and, in a way whose ramifications are important for the interpretation
of Wittgenstein's conventionalism, for the second also.
     Throughout the discussion we have been conceiving of the process of
implementing the antecedents and settling the truth-values of the consequents of C-conditionals merely as a way of *finding out* their truth-
values; so wherever it is at least in principle in our power to do both
things, we have been constrained to think of the conditionals as in
principle effectively decidable. And this makes it seem that there can be
no cogent reservation about excluded middle here which does not
emphasise that the conditionals are not in general *in practice* decidable.
(An anti-realism concerning the future might have seemed another
way; the conditionals, after all, are expressed in the future tense. But in
key cases they are not such that we have to *wait* before we can relevantly
assess them, and it is unclear how to formulate a coherent anti-realism
about the future for which this would not be the crucial point.) But
there is another possibility: to reject the idea that implementing the
antecedents and settling the truth-value of the consequent is just a way
of finding out the truth-value of a conditional of this sort. Perhaps it is
an instinctive doubt about this which underlies our attitude to the
corresponding counterfactuals; for, where arbitrary choices and

---

[1] In 24 Dummett concludes by having his anti-realist jettison the timelessness of statements.

random processes are involved, our feeling is not merely that it is mistaken to suppose of such a counterfactual that either it or its contrary *must* be true but, more, that neither can be, that there is nothing to make them true. So the suggestion is that implementing the antecedents of these conditionals is not to be conceived as the first stage in a decision procedure but as bringing about a *precondition* for their being true.

One argument for this view would be the following. Nothing counts as knowing, even by the most geneous criteria of knowledge, what, for example, the *n*th element in a free choice series will be if *n* choices are actually made; the possibility of any knowledge here is itself conditional on implementation of the antecedent. But the idea of the truth of any particular statement is intelligible only if its being true is taken to require the existence of a state of affairs which a man could come to know about and which would then justify him in asserting the statement's truth; truth cannot be *symptomless,* so to speak. This much 'anti-realism' would surely be quite acceptable to most realists. It follows that any necessary condition for the existence of all symptoms of a particular statement's truth is a necessary condition for its truth. Thus in the case of C-conditionals the truth of the antecedents is a necessary condition of the truth of the whole; and the quasi-platonist idea, that of each pair of such a conditional with its contrary one must be true and the other false, is unacceptable. (This suggestion, of course, would not apply to future-indicative conditionals as a class; it would be restricted to those about which a reasonable knowledge-claim could not be made except in the circumstances depicted by their antecedents.)

The force against quasi-platonism of this train of thought, however, has obviously to be qualified by the extent to which, even under the umbrella of the rule-following considerations, disanalogies remain between rule-governed and free-choice procedures. And the fact is that to regard effectively decidable but so far undecided mathematical statements as a species of C-conditional by no means clearly precludes the possibility of a reasonable view about their truth or falsity in advance of implementation of their antecedent clauses. Whatever sort of reason Goldbach might have had for believing his celebrated conjecture would eo ipso be reason for believing any conditional of the form:

the *n*th even number will be determined to be the sum of two primes if the matter is properly settled in the appropriate manner,

even if it is regarded merely as a contingent prediction about a decision which we will take when we are satisfied that the appropriate procedures have been properly observed. There is thus a substantial question whether the very gentle anti-realism described poses a threat to any feature of the quasi-platonist's position save his realism about absolutely prototypical C-conditionals.

Whatever the answer to that question, Wittgenstein would surely have had to reject quasi-platonism utterly. Nowhere, as far as I know, does he himself explicitly draw such a consequence; but it seems a clear implication of the rule-following considerations that we ought not to regard any *contingent* statement as determinate in truth-value whose truth or falsity has not been expressly settled. This is not straight-forwardly because we have to think of the world as somehow indeterminate in unassessed respects; it is rather that the open character of our understanding requires that it does not predetermine what aspects of the world we shall *count* as showing such a statement to be true or false. The picture is that however determinate the world is, our verdict on a particular contingent statement can still ultimately go either way with-out illegitimacy. The idea of a rigid, objective division of possible states of affairs into those consistent with the statement and those which are not is no longer in order. World determinacy, even if somehow still intelligible, is insufficient to secure determinacy in truth-value for unassessed statements, however precise we should ordinarily regard their senses. For one who accepts the rule-following considerations there seems to be exactly the same objection to the idea of unratified contingent truth as there is to that of unratified internal relations; for each idea requires an illicit conception of what it is to use an expression in a manner which conforms to the way in which one understands it.

Quasi-platonism is thus in the end unavailable to intervene between the rule-following considerations and the wholesale revision of estab-lished schools of logic and pure mathematics which they threaten to require. In the next chapter we shall consider, by way of an apparent digression, what, if any, idea of the objectivity of our system of beliefs is available to a global anti-realism of the kind which Dummett has expounded. The digression, however, is only apparent. We need to consolidate our grasp of the notion of objectivity as it has featured in the discussion to this point, and to elaborate the thought of the preceding paragraph. And there is waiting in the wings an argument to the effect that there is no clash between intuitionistic mathematical practice and the rule-following considerations which demands consideration.

XI

Anti-realism and Objectivity

Sources

RFM: i. 20–4
  PB: 192
LFM: lectures xix, pp. 181–4; xxviii, p. 275

1.   What exactly is it to believe in the objectivity of mathematics, or in that of any seemingly statement-making region of discourse? The idea with which we worked in Chapter I was that a belief in the objectivity of the statements within such a region is expressed by allowing a distinction for its statements between their meeting our most refined criteria of acceptability and their actually being true. To see the point of this idea, it is enough to reflect that by admitting such a distinction we commit ourselves to regarding truth for these statements as not simply a matter of our —human —judgment; to supposing that for such statements it is not human judgment which *constitutes* truth, but something else: precisely, 'objective reality'.

Obviously, we can make such a distinction in either or both of two ways: we can allow that statements can meet our most refined criteria without being true, or conversely. Either way, we have a purchase on the central intuitive idea of objectivity: being true is not the same thing as being judged to be true by human beings, even when their judgment is at its most sophisticated.

Ordinarily we are ready to grant the distinction the first way round for a wide variety of statements: statements about remote regions of space and time, general scientific laws, counterfactual conditionals, and oddments like conjectures about animals' mental states and the thesis that our phenomenology of colour is the same. So, in fact, for all statements where we have no working notion of decisive verification; that is to say, where the claim 'I have conclusively verified *Q*' is certain to be immediately objectionable, irrespective of the evidence which I can go on to adduce. To be sure, to allow that there are statements of this kind, for which the satisfaction of ones most refined criteria of acceptability would be insufficient for truth, would not yet be to commit oneself to thinking of them as candidates for *truth* at all. The connection with the intuitive idea of objectivity adumbrated only arises when such statements are thought of as potential truth-bearers. In particular, a belief of each member of a class of statements of this kind that either it or its negation must be true is eo ipso a belief in their objectivity.

By describing what we possess in contrasting cases as merely a 'working' idea of verification, I have in mind only the point that the claim 'I have conclusively verified Q' is always defeasible. Even in the case of a Q where it is not an immediately objectionable, because quite inappropriate, claim, we always allows that at any particular stage things can seem in all respects as though the relevant statement is true without its actually being so. What makes the difference between the two kinds of case is the rôle played by perceptual error. Where we have no working notion of verification, incorrect assertion of a statement in what appeared to be optimal conditions need involve no perceptual error; no error, indeed, of any kind save barely that the statement is incorrect. To have a working notion of verification for a statement, on the other hand, is just for its incorrect assertion in seemingly optimal conditions to be possible only if they are not really optimal.

For statements for which we have no working notion of verification, then, it seems we have a clear account of what it is to believe in their objectivity: it is to accept that they may none the less be true. But they are not the only class for which the notion is clear. A belief in objectivity is just as plainly expressed if we merely allow as a possibility what for statements for which we have no working notion of verification we have to allow as a fact: namely, their being, if true, unrecognisably so. And this, of course, is the way in which a belief in the objectivity of pure mathematics typically shows itself: decidability in pure mathematical theories is not, by those who hold to objectivity, supposed to be a necessary condition for truth.

That is a resumé of the notion utilised in Chapter I. So interpreted, however, a belief in objectivity just comes down to an endorsement of realism, in Dummett's sense, for the statements for which the belief is held (or rather, part at least of Dummett's sense, viz. the admission of possibly verification-transcendent truth-values; there need be no implication of bivalence). What we have so far, then, comes to a generalised gloss on Kreisel's suggestion that what is in dispute between platonists and constructivists in the philosophy of mathematics are not essentially any questions to do with the existence or character of mathematical objects but the question of the objectivity of mathematics itself. For one thing which evidently is in central dispute between platonists and intuitionists is the legitimacy of methods of inference whose explanation requires appeal to a proof-transcendent idea of truth.

The main question to be considered in this chapter is in effect a query about Kreisel's proposal. *Are* the intuitionists committed to rejecting the objectivity of mathematical truth? More generally, if a verification-transcendent idea of truth is rejected, does any clear sense remain in which truth can be thought of as something *objective* at all? To repeat, the root idea of objectivity is that truth is not constituted by but is somehow independent of human judgment. Realism gives this independence the obvious interpretation: *logical* independence—the idea that for particular true statements it is either unnecessary or insufficient, or both, to meet our most refined criteria of acceptability in order to be true. So the question is, is there any other interpretation? Can we give content to a belief in objective reality for someone who finds realism unacceptable?

2.   The question appears to concern the relation between anti-realism and old-fashioned *idealism*: the conception, whether or not anyone exactly held it, that we in some sense *create* the world, that it is misconceived if taken as an independent domain autonomously rendering our judgments true or false. Obviously, someone who endorsed such a picture, whatever his motivation, would have no time for the idea of verification-transcendent truth. The issue is whether the rejection of transcendent notions insisted on by the anti-realist in effect provides sufficient motivation for such a picture, or whether some sort of belief in objectivity is still available to him. The notion that we create the world by the concepts we employ and the judgments that we make is, plainly, a metaphor; what is the substance to the metaphor if it is not precisely the case against the intelligibility of transcendent notions which the anti-realist advances?

It is natural to think that if the metaphor *has* any substance, it must be something other than the anti-realist arguments by now familiar to us; that those arguments will leave available some version of the realist's traditional picture. It is true that Dummett has once or twice[1] suggested that the picture of mathematical reality appropriate to intuitionism —and so presumably, that of all reality appropriate to a generalisation of the intuitionist's ideas—is a curious hybrid of the platonist and constructivist images: we are to think of reality as *coming into being as we explore*; particular aspects of reality do not pre-exist our investigations into them, but what comes into being when we investigate is not of our making. On the face of it, though, this idea is so opaque as scarcely to amount to a picture at all. (I shall return to it.) Better, it might well be thought, to hold on to as much as possible of the realist picture—which would seem to be most of it. Thus: the world is independent of us, having the properties and organisation which it does irrespective of whether or not we ever choose to investigate them or even form the concepts in terms of which a particular investigation might be possible. The only modification required to the realist picture is that we must now think of the world as in a certain sense *epistemically transparent*;

[1] 22, (Pitcher, p. 110), and 23 (2, p. 509).

that is, all the variety which it can display is in principle conceptualis-
able by human beings, and all the true statements which there are to be
made about it are by human beings (in principle) recognisably true. We
thus excise from the realist picture all objectionable play with transcendent notions, but our assertions remain answerable to an external
reality in exactly the traditional spirit of realism.

This may seem a reasonable suggestion.[1] We do not normally think
of those aspects of the world depicted by statements which, if true, we
can verify, as being any less *of* the world just on that account. So it
appears that all that need be involved pictorially in the shift to anti-
realism is an endorsement of the idea that such features *constitute* the
world.

In fact, however, matters are less straightforward than that. To begin
with, we ought to enquire into the assumption that the consistency of
such a picture with anti-realism would be sufficient for its *availability* to
the anti-realist. When philosophers put this sort of pictorial conception
into opposition one with another, they certainly intend to be concerned
with very abstract and profound questions: what is the nature of the
relation between human thought and reality in general? in what sense, if
any, is non-mental reality autonomous? and so on. But no one will
contest that images of this kind need to be *given* sense; their proper rôle
is as souvenirs of considerations which are in some sense of greater
philosophical substance, or, at least, of more direct content.
Philosophical imagery is respectable only when it is so related to a
particular group of more down-to-earth considerations that it both
represents everything which is essential to them and adds nothing
inessential. We do not want any ornament in our philosophical pictures,
any more than we want them to miss essential aspects of the thoughts
which motivate them.

These platitudinous constraints are conditions on the *intelligible* use
of philosophical imagery; to ignore them is to risk vacuity. But now it
begins to look doubtful whether *any* of the three above pictures is
genuinely available to an anti-realist. For nothing in his rejection of the
core theses of realism—that the truth of a statement need not require
that we can (in principle) recognise its truth, and that all intelligible
statements are determinate in truth-value—seems to give any content to
the respects in which the three pictures sketched *contrast*.

To rehearse: the anti-realist wants to say that if, per impossibile, we
attempt to associate a statement with transcendent conditions of truth,
nothing will count as a display of grasp of the content which we thereby
try to give it; the best anyone will be able to do will be to show that he
knows under what conditions its assertion is reasonable. So, appealing
to the emptiness of the idea that we may possess concepts which nothing
counts as exercising, it appears that the totality of intelligible *truths*
stops where transcendence begins, that all the facts which we are

[1] Cf. Wright 77, pp. 225–6.

capable of grasping must be facts which we would be capable of ratifying if appropriately placed.

That conclusion, however, is apparently a consequence of *each* of the three pictures. The idealist's image of creation would appear to entail that the whole world, a fortiori the intelligible world, is at any stage exhausted by its already ratified, a fortiori ratifiable, aspects; Dummett's picture confines the world at any stage to those of its aspects which have come into being 'in response to our probing', that is to say, presumably, our recognition of which has been the outcome of our probing; and the 'epistemically transparent' refinement of the realist's picture is precisely designed to confine the world to what, according to the reflections rehearsed, are its intelligible features. Each of the pictures, however, while consistent in this way with the anti-realist's leading train of thought, has aspects which that train of thought does nothing to explain or suggest. In the case of the idealist's and Dummett's pictures, it is the play made with the idea of the world as constantly *becoming*; it seems quite unclear what invites the use of any such image, let alone what favours one side of the 'of our making'/'not of our making' opposition. But the modified version of the realist's picture is no better off. For equally, what is there in the anti-realist's train of thought to suggest a world picture in which truth is incapable of becoming, in which the facts always somehow pre-exist our probing? Of course, this remains the more *natural* kind of thing to say. But that could be viewed merely as testimony to the strong attraction which realism proper has for us anyway —so that we tend to cling to its images even after we have abandoned their motivating substance. Or, if it is not that, then the explanation is that we do not know what to *do* with the other pictures, what it *is* to think of the world in the manner which they recommend; and our impulse to reject them takes the form of a denial. But it is equally fair to say that it is unclear what at bottom it is to think of the world in the realist's way, once we are no longer prepared to grant that statements may be determinate in truth-value without its having to be possible to recognise what their truth-value is.

Perhaps, indeed, a fourth picture is consistent with anti-realism. The anti-realist's arguments focus upon what he takes to be the character of our understanding of language, as determined by the manner in which such understanding can be displayed—the data which a theory of meaning, that is to say, a systematic description of the practice of speaking a particular language, can have available. He disputes the realist's idea that our understanding of statements for which we lack even a working notion of verification is correctly viewed as consisting in an apprehension of their purported conditions of truth. Is a position conceivable, then, which while respecting the anti-realist arguments as points descriptive of the essential character of our understanding of language, disputed their projection, as it were, onto what we talk about? The suggestion would be that while of no intelligibly assertible statement can we suppose ourselves to have grasped possibly verification-transcendent truth-conditions, there is no reason to think that the world falls into line with *our* limitations, that restrictions on the character and scope of our understanding are automatically restrictions on the character and scope of reality. To think that would be a mere anthropocentric conceit. Rather, realism as a world view is objectionable only when it intrudes into the theory of meaning.

According to such a view the realist picture could be maintained by an anti-realist without even the restriction to epistemic transparency. The point would be merely that we cannot in linguistic commerce evince a grasp of any of its possibly epistemically opaque aspects. No doubt such a conception is unattractive. But it is prima facie consistent with anti-realism; and no reason is so far apparent why an anti-realist should have preferences about which of these four pictures he should endorse.

The idea that an anti-realist can evince a belief in objectivity by retaining—perhaps, as it turns out, in one of two ways—the realist's picture thus remains so far totally superficial. Such a picture has to be *given* content; and the fact is that what gives content to the realist's world view—that of an external, mind-independent totality of facts, one for each pair of statements, P and its negation, to which we have attached clear sense—is his acceptance of universal, possibly verification-transcendent determinacy in truth-value; which is just what the anti-realist rejects. Moreover, nothing in the arguments for a global anti-realism seems directly to give content to the respects in which the static, realist pictures contrast with the dynamic pictures of Dummett and the idealist. If, therefore, an anti-realist endorses a picture of the former kind, rejecting the latter, how are we to understand the contrast which he intends?

What is it to understand any image or metaphor? Clearly, it would be too stringent to insist that one be able to dispense with any metaphor which one understands; for metaphor is not in general stylistic ornament but is integral to the expressive power of language. But it seems reasonable to require that we should at least be able to call attention to the kind of thing which would give the metaphor point, and that we should have some concept of the consequences of its appropriateness. If one thinks of any metaphor which is easily understood—'shallow' as applied to someone's thought, 'striking' as applied to a suggestion, 'octopus' as descriptive of the family, 'arid' as applied to Oxford philosophy of the fifties—it is apparent that one's ease with the description rests on a firm conception of how one would debate its appropriateness, and of what would follow from its being appropriate. In contrast, not understanding a metaphor, for example, perhaps 'impudent' as applied to a wine, is always a matter of unclarity in one or both of these respects. (If all you knew was that someone had so described a wine, you would be unable to say whether he liked it or not.)

As far as we have so far been able to see, then, the situation appears to be that, if an anti-realist were to endorse any of the world-pictures described, it would be as if he had described a wine as 'impudent'. It simply isn't clear what he could have in mind which would give any of the pictures a special appropriateness; it is not even clear what features he could *mistakenly* believe his position to have which would do so. And it is unclear what would follow from the appropriateness of any of the pictures. What should we expect of an anti-realist who endorses, for example, Dummett's picture as against one who endorses one of the alternatives? Each will presumably reject classical logic, or at the very least its traditional semantics; but what can they be expected to do *different* from each other, other than besport their favoured pieces of imagery?

I am not claiming that absolutely no kind of consideration is apparent which might motivate an anti-realist to propose a dynamic rather than a static world-picture. On the contrary, it is natural to suggest at this point that both motive for and expression of a dynamic view may be forthcoming if, pressed for more detail about the class of statements for which he would regard a truth-conditional conception of understanding as unobjectionable, the anti-realist proposes an N- or NP- type of view as sketched in the last chapter. In the present context, however, this suggestion is of no use to us. We should want a much fuller account of the reasons which might impel an anti-realist to develop his view in one of those two directions, or the point of his doing so will be no clearer than the point of the contrasting pictures which we already have; and it is, in any case, unclear that an N or NP anti-realist would have any reason to propose a dynamic view unless he thought that the orthodox view of truth as timeless was inaccessible to him—an issue of some complexity, as we saw. What is more important, however, is that the kind of 'dynamism' which might issue along these lines is not that with which Dummett's and the idealist's pictures are meant to be associated; this is clear as soon as one reflects that while Dummett's and the idealist's pictures would be meant to portray, inter alia, the relationship between the mathematician and his field of study, the timelessness of opportunities for mathematical verification will leave N, NP, NF, and S anti-realists with nothing to disagree about as far as the philosophy of mathematics is concerned. The dynamism which Dummett and the idealist intend has to do with the relation between our investigations and the genesis of 'the facts'; the facts spring into being in response to, or are created by, our investigative activities. But the sort of dynamism to which, according to the suggestion, an N, or NP anti-realist might be led would have the facts 'spring into being' not as a result of any investigative activity on our part but as a consequence merely of the passage of time. In short, it is a dynamism which would have nothing to do with objectivity—with the relation between human judgment and truth. But our present question is precisely whether anything properly describable as a belief in the objectivity of a certain class of statements is available to someone who holds an anti-realist view of them. This question will arise for each of the eight kinds of anti-realist distinguished in the preceding chapter: can he intelligibly believe in the objectivity of those statements which he currently regards as determinately true or false? And it remains unclear what the answer should be.

3.   This unclarity affects Dummett's suggestion[1] that, following the intuitionists, it is open to us to reject the objectivity of mathematical truth while retaining that of mathematical proof. Proof, for the intuitionists, is an effectively decidable notion; there is never any question that the correctness of a proof might elude our capacity, at least in principle, to recognise that it is so. So the objectivity of the claim that a proof is correct cannot be the realist's conception of objectivity sketched above. What, then, is it to think of the correctness of a proof as an objective issue? What would vindicate such a conception, and what are its consequences?

Dummett intends a contrast with what he takes to be Wittgenstein's view, that a valid proof does not *command* acceptance. He expresses his opposing idea like this:[2]

> . . . a good proof is precisely one which imposes itself upon us, not only in the sense that once we have accepted the proof, we use rejection of it as a criterion for not having understood the terms in which it is expressed, but in the sense that it can be put in such a form that no-one *could* reject it without saying something which would have been recognised before the proof as going back on what he had previously agreed to.

What, though, does the 'but' clause add to the 'not only' clause? The idea seems to be that it is not merely the case that *we* say that rejecting the proof evinces a misunderstanding, but that it really is not possible to reject the proof without giving some account of one's rejection —since we are not concerned with an inarticulate rejection, or a mere failure to follow the proof —which betrays what would have been counted as a misunderstanding even before this particular proof had been devised. But how are we to get a purchase on the idea that this too is not merely something which *we* say after we have accepted the proof? Dummett wants to urge the view that someone's acceptance of a new proof is not in general a *new* criterion for his having grasped correctly certain relevant concepts; that the sorts of thing which, after we accept the proof, we regard as evincing a misunderstanding—of, say, a particular principle of inference which it uses —are typically all of a kind which we would have counted as symptomatic of a misunderstanding before this particular proof was presented. The trouble is, however, that no clear

[1] 23 (2, p. 507).
[2] ibid.

reason is apparent why his Wittgensteinian opponent should not allow such a claim; what is essentially in dispute, it seems, is the status of the claim—precisely, its *objectivity*.

It is open to both, for example, to endorse a counterfactual like: 'If before we had this proof, Jones had said what he just said, he would have betrayed a misunderstanding of the serial structure of the integers.' But Dummett wishes to endorse the idea that the correctness of such an assertion is something independent of the context in which it is asserted, and of its recognition by us; whereas his opponent will regard it as a misrepresentation of our understanding of the conditional to suppose that it captures a truth about the concept of the series of integers whose status as such is quite independent of the judgments involving that concept which we have actually made—in particular, our acceptance of the proof.

It is difficult to see how someone who sympathised with Dummett could establish his view if it turned out that *everything* he wanted to say by way of expression of it could be endorsed by his opponent. How is content to be given to a belief in, for example, the stable, objective character of the concept of the series of integers if the things one naturally says by way of attempt to give expression to the continuity of that concept through new applications —like the counterfactual conditional above —are all available to someone who rejects one's belief? What is clear is that the legitimacy of transcendent notions is not in play in the dispute; it is internal to anti-realism, and its inchoate character seems to bring out a problem. The cardinal thesis of Wittgenstein's later philosophy of mathematics, as we have so far interpreted it, is a rejection of the objectivity of diachronic conceptual identity. This goes far beyond rejection of the transcendent kind of objectivity espoused by the realist. The difficulty now is that, to the extent that it is unclear what an anti-realist belief in objectivity amounts to, it is arguably equally unclear what is the force of Wittgenstein's negative thesis. Unless some improved account is forthcoming of what it is for an intuitionist to accept the objectivity of proof, or in general of what it is for an anti-realist to accept the objectivity of decidable truths, how are we to interpret the corrective content of Wittgenstein's idea?

Against this, a Wittgensteinian could, surely rightly, protest that things have come to a pretty pass if a philosopher cannot legitimately reject a concept without first supplying a full explanation of it —his rejection, after all, may be based on the belief that a satisfactory explanation cannot be given. (In order to understand the *type* of nonsense someone is taking, we do not need to understand what he is saying.) He might well protest also that the relative clarity which we granted to the realist's conception of objectivity is an illusion. For we ought to ask what account is to be given of the realist conception when effectively decidable statements are concerned—statements for which we do possess, in the sense earlier sketched, a technique which must culminate in a 'working' verification or falsification? Here there is no question of explaining the belief in objectivity in terms of a possible transcendence of truth over in principle accessible human knowledge. So for such statements the realist, too, needs some other account. If none is forthcoming, however, the resulting unclarity will not be confined to effectively decidable statements. The sorts of statement for which a realist allows verification-transcendent possibilities of truth are, typically, compound; and there seems no sense in the idea that a belief in objectivity, while having clear content for certain compound statements, might have no substance in relation to the statements out of which they were compounded.

The point is striking in connection with statements which the realist conceives as infinitary *truth-functions* of effectively decidable ones. If '$(x)Fx$', 'x' of infinite range, is thought of as determinate in truth-value, irrespective of whether we can ever determine what its truth-value is, then so must each of its instances be—at least in case it is true. But it is certain that we will not check each instance, even if all are effectively decidable. So it is a presupposition of the intelligibility of the realist's attitude to '$(x)Fx$' that sense can be made of the idea that each 'Fa' has an investigation-independent, determinate truth-value—is true, or false, independently of our investigation. And is that not in effect just the idea of objectivity that Dummett was seeking after concerning proofs: that *before* we agree about, for example, a proof in an unopened envelope in a trunk which belonged to Fermat, the statement 'The enclosed proof is sound' is of determinate truth-value?

It is the conception of objectivity thus outlined which the Wittgensteinian is opposing: the idea of the investigation-independent determinacy in truth-value of decidable statements—henceforward, for short, *investigation-independence*. He will contend that we have a sufficient grasp of the essential spirit of this conception to evaluate its shortcomings, and that our possession of that degree of grasp is quite consistent with an irremediable unclarity concerning what a belief in the investigation-independence of a particular class of statements really consists in. It is also open to him to argue, of course, that such irremediable unclarity would itself constitute a serious criticism of the conception—a conception of which, for the reasons above, not only the Dummettian anti-realist but the realist too must provide an elucidation and defence.

There is, nevertheless, a sense in which the content of the realist's belief in objectivity continues to enjoy a relative perspicuity. To be sure, it is unquestionably a correct diagnosis of the *motivation* for realism about infinitary quantification that it feeds on the same conception of objectivity as that in which the Dummettian anti-realist believes. It is *because* the realist accepts the investigation-independence of each of the infinitely many instances that he is prepared to regard what he sees as essentially an infinitary truth-function of them as determinately true or false; when, of course, he has no option but to regard its truth-value as possibly recognition-transcendent. A *justification* of realism about quantified statements of this kind would thus presuppose a justification of, and a fortiori an explanation of the content of, a belief in the objectivity of their instances. Then, since the transition to acceptance of the determinate truth or falsity of infinite quantifications is actually a further step, one would expect that an anti-realist of Dummettian sympathies could indeed take over the initial part of any account which the realist might produce. But as far as the *content* of the realist's belief in objectivity is concerned, possible justification is only part of the story. After all, at least the realist's belief in the objectivity of the instances has consequences for him: it leads him to regard their quantified union as determinately true or false. So it is open to us precisely to take his belief in the objectivity of effectively-decidable statements to be expressed in his treatment of non- effectively-decidable compounds of them. However short of justification it may prove, the realist's belief does at least in this way have some non-trivial bearing on his linguistic practice. Asked to explain in what his belief in the objectivity of effectively-decidable statements consists, he can do better than patter on about determinacy in truth-value irrespective of whether or not we ever check them; he can point to his admission of operations involving logical compounds of those statements which intuitively require their—the compounds'—determinacy in truth-value; and these operations are exactly what his anti-realist opponent does not allow.

In order, then, satisfactorily to elucidate the content of an anti-realist belief in investigation-independence, some operational implications of the belief have to be found. But that is just to require that it is to have practical consequences whether one cleaves to this most basic notion of objectivity or accepts the criticism of it which Wittgenstein's rule-following considerations contain; in other words, *either* the rule-following considerations are revisionary of the linguistic practice of a Dummettian anti-realist *or* a belief in objectivity is operationally idle unless it is embedded in a full-fledged realism. Thus if Wittgenstein is right in thinking that his later philosophy as a whole is non-revisionary, then the belief in the objectivity of proof which Dummett attributes to the intuitionists is a vacuous belief.

4.   The crucial question, then, is: what, if any, aspects of intuitionistic mathematical practice *express* a belief in the objectivity of proof, that is, of decidable mathematical truth? The obvious suggestion is that we should look to the intuitionists' willingness to accept excluded middle for undecided but in principle effectively decidable statements. If they are prepared to count such a disjunction true, surely their conception must be that one of the disjuncts is true; and there seems no

sense in that conception unless there is, in advance of any investigation, a determinately correct answer to the question, which disjunct?

In fact, though, it is not entirely clear how such 'disjunctions' are to be interpreted from an intuitionistic standpoint. Any effectively decidable statement is implicitly equivalent to a conditional: if such-and-such a test is correctly carried out, such-and-such will be the outcome. For example, 'n is composite' is equivalent to: if an exhaustive, correct check is made on the predecessors of n, a pair will be located, $x, y, \neq 1$, of which n is a multiple. So the obvious interpretation of 'n is prime or n is composite' is as a disjunction of such conditionals with shared antecedent and contrary consequents: $(P \to Q) \ V \ (P \to -Q)$. But suppose that we were to interpret the intuitionists' assertion as itself a *conditional* with suppressed antecedent and disjunctive consequent: $P \to (Q \ V - Q)$. There is no doubt that it would be validly assertible under that interpretation, since there are only those two possible outcomes of correct implementation of the test. That only two outcomes are possible, however, is quite insufficient to warrant the view that there is an investigation-independent fact about which will result if the test is actually carried out. Under the second interpretation, therefore, the assertion intuitively carries no commitment to investigation-independence for the conditionals which are disjoined in the first interpretation.

However, is such an interpretation viable? Well, Heyting at least gives an informal explanation of disjunction: 'a disjunction may be asserted just in case at least one of the disjuncts may be asserted',[1] which would actually exclude valid assertion of our statement under the *first* interpretation, the disjunction of conditionals. For we need not be in a position to assert either conditional. We know we can *get into* a position when we will be entitled to assert that the statement effected by the assertion of one of them now was true; but, given that it is in our power to bring it about that P is true, *that* knowledge seems adequately expressed by the second interpretation.

Perhaps, though, we ought not to take Heyting's explanation as gospel. After all, if the intuitionist—or anti-realist in general—is to have any *use* for assertions whose main connective is disjunction, he must weaken their conditions of legitimate assertion to something less than those for asserting either individual disjunct. Otherwise it will never be necessary to infer *through* a disjunction—one disjunct will always be available independently—nor, therefore, to derive one. And their assertion in non-inferential contexts would violate Gricean 'conversational' principles.[2] So it seems we must broaden the conditions of legitimate assertion of a disjunction at least so far as is necessary to include the assertor's recognition that he can find out which one of the disjuncts is

[1] 46. p. 102.
[2] See 42—various copies have circulated, and one resides in the Oxford Sub-faculty of Philosophy library in 10 Merton St.—and 43.

correct, even if he does not yet know which. And if we so broaden them, a disjunction of conditionals, equivalent respectively to an effectively decidable statement and its negation, will be validly assertible.

The question now arises, what difference does it make which of the two interpretations is adopted? In particular, is there anything in the intuitionists' practice which requires that they do *not* mean the conditional with disjunctive consequent? We have noted that '*n* is prime or *n* is composite' would be validly assertible under that interpretation. So, taking the question in the context of some appropriate system of natural deduction, what we have to consider is whether analogues of the disjunction-introduction and -elimination rules would survive; that is, whether, supposing we have the standard rules for disjunction, there will be any essential difference in the proof-theoretic behaviour of our two candidates for what the intuitionists mean by '*n* is prime or *n* is composite'.

Since we are concerned with a valid statement, the I-rule is nugatory. But clearly there is no problem with it. Either premise from which the first interpretation follows by Vel-Introduction entails the second interpretation also. What about Elimination? We have to consider whether in

$$\frac{(P \rightarrow Q) \text{ V } (P \rightarrow -Q); \quad (P \rightarrow Q) \rightarrow R; \quad (P \rightarrow -Q) \rightarrow R}{R}$$

we can substitute the second for the first interpretation among the premises without affecting the validity of the inference. Plainly we cannot do so straightforwardly. But we can do something interesting. Note that from the middle and right hand premises we can derive respectively:

$$P \rightarrow (Q \rightarrow R), \text{ and } P \rightarrow (-Q \rightarrow R).{[1]}$$

So if we have as a third premise:

$$P \rightarrow (Q \text{ V } -Q), = \text{ our second interpretation,}$$

we can straightforwardly infer $P \rightarrow R$.

It might be thought that that is not yet good enough, since the intuitionistic reasoning whose possible interpretation we are

---

[1] —granted the usual introduction and elimination rules for conjunction, and the rule of conditionalisation:

$$\frac{\Gamma, B \vdash A}{\Gamma \vdash B \rightarrow A}$$

(Intuitionists, of course, accept these rules, so the intuitionistic conditional is effectively a material conditional.)

considering will culminate in the *unconditional* assertion of R. But consider the actual type of case with which we are concerned: P is the hypothesis of the correct implementation of a certain effectively implementable investigative procedure. Now, what does the conditional, $P \rightarrow R$, mean? For the intuitionists the meaning of a mathematical conditional is fixed by the stipulation that it may be counted as proved just in case we have proved that any proof of its antecedent could be effectively transformed into a proof of its consequent. In the present case, admittedly, there is no question of a *proof* of the relevant P—for whether or not the appropriate procedure has been correctly carried out on a particular occasion is not a mathematical question. But P still admits of verification; and the conditional, $P \rightarrow R$, is thus naturally interpreted as warrantedly assertable for an intuitionist just in case he has verified that a verification of P will result in a state of information in which a proof can effectively be found for R—supposing, as we may, that R for its part admits of no kind of warranted assertion save on the basis of proof. Accordingly, to have proved the conditional is to have proved that there is an effectively accessible state of information in which R can be proved; which is to have proved that R can be proved; which, since there arguably is no material difference between knowing that one can get a proof of a particular statement and actually having the proof, is to have proved R.

The argument, then, is that the intuitionists' acceptance as valid for effectively decidable mathematical statements of what looks like excluded middle admits interpretation, consistently with their mathematical practices, as acceptance of the validity of a certain kind of conditional instead; a conditional which could be accepted as valid by someone who disputed the investigation-independence of effectively decidable mathematical statements. So there is a substantial question whether the intuitionists' acceptance of classical logic within the region of effective decidability has any bearing on any belief they may have in the objectivity of mathematics; and whether, by the same token, the rule-following considerations are indeed, as they threatened to be, revisionary of intuitionist mathematics.

5.   A natural reaction would be to wonder what the foregoing reasoning stands to achieve unless we are supposing that to regard the first interpretation as valid may well serve to express a belief in the investigation-independence of the disjoined conditionals; unless, that is, for any effectively decidable statement S, equivalent to a conditional, $P \rightarrow Q$, to accept the validity of the inference:

$$\frac{P \rightarrow (Q \vee -Q)}{(P \rightarrow Q) \vee (P \rightarrow -Q)}$$

is to accept the investigation-independence of S. And if that is so, then whether or not intuitionistic mathematics can be interpreted so as to involve no expression of a belief in the investigation-independence of effectively decidable mathematical statements, the larger claim that belief in investigation-independence might be totally void of non-trivial operational consequences is wrong. For such a consequence will emerge as soon as, unlike the intuitionists, we make it our policy to express all effectively decidable statements in conditional form.

The intuitionists, of course, would not accept the above transition in general. The 'official' explanation would be something like this: we may have a construction, C, of which we can recognise that had we a proof of P, we could use C, to transform it into a proof either of Q or its negation *without* C's being such that we can recognise in advance *either* that a proof of P could thereby be transformed into a proof of Q, *or* that a proof of P could thereby be transformed into a proof of $-Q$. That is, our state of information does not justify the assertion of either disjunct in the conclusion. This, however, is conclusive only if we are still working with Heyting's (pointless) interpretation of disjunction. Suppose now that we actually know *how* to get a proof of P. Then we thereby know how, using C, to get a proof either of Q or of its negation; so much is guaranteed by the premise. So we know how to get into a situation in which, in virtue of having proved its consequent, we shall be in a position to prove either $P \rightarrow Q$ or to prove $P \rightarrow -Q$. Hence, on the more liberal interpretation of disjunction proposed above and the supposition that we know how to get a proof of P, the transition is sound.

Suppose now that we generalise the intuitionists' readings of 'V' and '$\rightarrow$' to cover both mathematical and non-mathematical statements in what seems intuitively the most natural way:

'$P \vee Q$' is warrantedly assertable in state of information Σ just in case Σ either recognisably justifies the assertion of P, or recognisably justifies the assertion of Q, or—generalising our earlier proposed liberalisation—can be recognised to be capable of effective enlargement into a state of information of one of those two kinds.

'$P \rightarrow Q$' is warrantedly assertable in state of information Σ just in case Σ is recognisably such that any enlargement of it into a state of information recognisably justifying the assertion of P will result in a state of information recognisably justifying the assertion of Q.

Then the transition in question can be seen to hold good for every example, mathematical or otherwise, in which P hypothesises the implementation of an effective investigative procedure. We shall be entitled to assert $P \rightarrow (Q \vee -Q)$ if and only if we have recognised that any enlargement of our state of information into a state justifying the

assertion of P would result in a state of information *either* already
recognisably justifying the assertion of Q or of $-Q$ or recognisably
capable of effective enlargement into one such. But in that case we can
recognise that by carrying out the investigation depicted in P we have it
in our power to enlarge our state of information into one which recog-
nisably justifies the assertion of Q, so of $P \rightarrow Q$; or of $-Q$, so of $P \rightarrow$
$-Q$. And that, by the proposed account for 'V', is to be in a position
justifiably to assert $(P \rightarrow Q) \text{ V } (P \rightarrow -Q)$.

It seems clear, then, that the inference is valid for the described class
of cases. It also seems intuitively clear that the inference is invalid
unless the investigation-independence of the conditionals disjoined in
its conclusion is assumed; for if implementing P is thought of as
actually bringing into being something in virtue of which one or the
other conditional holds true, then it can have been true that the premise
of the inference held good without there having been anything in virtue
of which the conclusion held good. However, before we take it that the
requisite operational content for a belief in investigation-independence
has therefore been supplied, we had better be clear exactly *where* the
reasoning of the immediately preceding paragraph assumed the
investigation-independence of the relevant conditionals; for if it was
sound, it must, apparently, have done so somewhere.

But where? All that appears to have been essential is that P hypothesise implementation of a procedure which recognisably lies within our
powers; the assumptions that it be an *investigative* procedure, and that
the disjoined conditionals be thought of as determinately true or false
independently of carrying it through, seem to have been otiose. If it is
recognisably within our power to bring it about that P, and the premise
of the inference holds good, then it is recognisably within our power to
enlarge our state of information to the point when it will justify either
the assertion of Q or of its negation; and hence either the assertion of
$P \rightarrow Q$ or of its contrary. And to recognise that we are in a position so to
enlarge our state of information is, by the proposed account of disjunction, to be in a position to assert the conclusion of the inference.

The suggestion, in fact, that a belief in the investigation-
independence of an effectively decidable mathematical statement, S, =
$P \rightarrow Q$, might be adequately expressed by endorsement of the validity
of $(P \rightarrow Q) \text{ V } (P \rightarrow -Q)$ seems simply to be at odds with the reasoning
of the preceding section. It was argued that if we can effectively bring it
about that P, and if R is warrantedly assertable only on the basis of a
proof, then to have proved $P \rightarrow R$ is to have proved that a proof of R
can be effectively found—that is, to have proved R.[1] But then what is
there to prevent the substitution of '$(P \rightarrow Q) \text{ V } (P \rightarrow -Q)$' for '$R$' in
that reasoning, given that from $P \rightarrow (Q \text{ V } -Q)$ it follows directly that
$P \rightarrow [(P \rightarrow Q) \text{ V } (P \rightarrow -Q)]$? The substitution would be illicit only if

[1] Strictly, in view of the non- a priori character of the relevant P, the argument anticipated
something close to the generalisation of the intuitionistic conditional since suggested.

something other than a proof could warrant the assertion of the disjunction; but all that the intuitionists allow to warrant that assertion is a proof of one of the disjuncts or—*pace* Heyting—a proof that such a proof can be effectively achieved. It would appear to follow that if to endorse the validity of $P \rightarrow (Q \vee -Q)$ is neutral concerning the investigation-independence of S, then so is an endorsement of the validity of $(P \rightarrow Q) \vee (P \rightarrow -Q)$.

6.   That completes the argument, signalled at the end of the previous chapter, for the thesis that the rule-following considerations may be non-revisionary of intuitionistic mathematics; and, perhaps, non-revisionary of whatever inferential practices would be acceptable in general to a Dummettian anti-realist. If the argument is sound, then the interest of the situation is, to stress, that it becomes open to doubt whether there is defensible middle ground between Wittgenstein and fully-fledged realism; the doubt arises because of the consequent difficulty of giving a satisfactory account of what the distinctive beliefs of one who essayed to occupy middle ground would fundamentally consist in.

Now, however, the boot is on the other foot: before we grant the soundness of the argument, we had better be clear exactly where our above intuition, that the inference in question may fail of validity unless the investigation-independence of the disjoined conditionals is assumed, went astray. But the truth is that it is not at all clear that this intuition did go astray. To begin with, nobody—intuitionist or not—would want to claim that whenever P hypothesises implementation of a procedure within our effective power, to be in a position to assert $P \rightarrow R$ is eo ipso to be in a position to assert R; the claim would have the consequence, for example, that to be in a position to assert 'If I go to the match, I shall be late home' would also entitle one to assert 'I shall be late home'. Recognition that, by implementing P, one can bring about a situation in which one is justified in asserting R entitles one to assert R categorically only if there is no question of the implementation of P *bringing it about* that R. Now, certainly, when R is a mathematical statement, we should not ordinarily suppose that anything we do is capable of affecting whether or not a correct proof of R can be given—though it is, of course, within our compass to affect whether *we* can actually construct such a proof. For this reason we naturally took it that to have verified $P \rightarrow R$, P hypothesising correct implementation of an effective investigative procedure, is to have verified that we have *access* to a proof of R; access, that is, to a construction whose status as a proof of R in no sense awaits our actual recognition. We were, in other words, tacitly conceiving of the provability of R as an investigation-independent matter.

What this means is that the argument of §4 is open to the following counter-attack: we can legitimately pass from recognition that the assertion is justified of a conditional, construed in the generalised intuitionistic manner, to a justified assertion of its consequent only if it is true both that we can effectively bring about the justified assertability of its antecedent *and* that we can rightly conceive of the state of information justifying assertion of its consequent in which we can thereby place ourselves as something at which we should *arrive*, rather than something which we should create. We have to be able to think of that state of information as, so to speak, a place which we have not yet visited but which we now know how to get to. More prosaically, we have to think of the statements whose verification will be involved in our being in that state of information as investigation-independent. The argument of §4 thus implicitly made use of the very notion about which it aimed to nourish a scepticism. If there are doubts about the philosophical respectability of an anti-realist belief in investigation-independence, no anti-realist can accept the play which, it now appears, must be made with the notion in order to pass legitimately, even in the relevantly restricted class of cases, from $P \to R$ to R. But without that passage, an argument is still wanted for the operational neutrality of the intuitionists' treatment of effectively decidable mathematical statements in relation to their belief—if they have it—in the investigation-independence of those statements; no reason is any longer apparent why their acceptance of what appears to be excluded middle can legitimately be interpreted merely as an endorsement of the validity of $P \to (Q \lor -Q)$.

The argument continued by development of the claim that acceptance of the validity of excluded middle for effectively decidable statements, even if taken as an endorsement of the first, genuinely disjunctive interpretation, is available to a sceptic about investigation-independence in any case. To be sure, an answer has now been sketched to the way in which the point was finally presented, when it depended upon an elision of the antecedent in $P \to [(P \to Q) \lor (P \to -Q)]$. But the claim was also argued for by direct appeal to the proposed generalisation of intuitionistic disjunction: if it is within our power effectively to implement P, and one of Q or its negation is certain to be warrantedly assertable if we do so, then it is now open to us so to enlarge our state of information that it will warrant assertion either of $P \to Q$ or of its contrary; and recognition of this state of affairs now justifies us, according to the proposed account of disjunction, in asserting the disjunction of those conditionals, whether they are thought of as investigation-independent or not.

Or does it? The crucial question is what 'enlargement' of our state of information is taken to mean. Consider 'John will be dead by Christmas or I shall be bearded by the New Year' asserted, say, in September. Since, we may suppose, I am able to bring it about that either, or indeed both, disjuncts are true, there is a sense in which it is recognisably in my power to 'enlarge' my present state of information in such a way as to justify the assertion of one or both disjuncts. But it would be quite alien to the intention of the generalised intuitionistic account to conclude that the disjunction is therefore correctly assertable. Clearly we cannot understand 'enlargement' of our state of information in such a way as to have it include *making* things true and simultaneously hope to offer sensible proposals about how to generalise the intuitionistic logical constants which make essential use of the notion of enlargement. The proposed liberalisation cum generalisation of Heyting's account of disjunction cannot possibly be acceptable unless the allowed-for 'enlargement' is taken to exclude effecting changes in those aspects of the world which are material to the correct assertability of the disjuncts. But now, how is the intended exclusion to be explained if the whole idea of investigation-independence is rejected? That is: if the generalised account of disjunction is taken to imply that $(P \rightarrow Q) \vee (P \rightarrow -Q)$ is validly assertable even if the disjoined conditionals are not conceived as investigation-independent, it is—to say the least—unclear what obstructs its implication of the correctness of asserting *any* disjunction the truth of at least one of whose disjuncts it is recognisably in our power to bring about. On the other hand, if 'enlargement' of one's state of information is explained—no easy matter though it might prove to do so in detail—simply as involving discovery of certain antecedent facts, then it is unclear how the account of disjunction can sustain the validity of $(P \rightarrow Q) \vee (P \rightarrow -Q)$ *unless* the disjoined conditionals are thought of as investigation-independent. What the sceptical argument needs by way of supplementation, it appears, is an explanation of how one can reject the concept of investigation-independence without surrendering one's title to the ordinary distinction between finding out that so-and-so is the case and bringing it about that it is; by means of such an explanation, the idea of enlargement of one's state of information will be capable of clarification in such a way as to have the resulting account of disjunction both exclude the correct assertability of spurious cases and imply the valid assertability of $(P \rightarrow Q) \vee (P \rightarrow -Q)$ in the relevant class of cases; but without such an explanation the argument fails.

7.   As at the conclusion of the discussion of quasi-platonism, it should be emphasised that what has failed is one argument for denying the rule-following considerations a certain sort of revisionary implication; it has not been shown that those ideas are unavoidably revisionary. By the same token, it has not been shown that an anti-realist belief in investigation-independence has a clear operational content; it is only that the most natural preconception about the place to locate such content has survived an attack. But, of course, the question whether there is a defensible area of middle ground between Wittgenstein and realism would not be settled affirmatively just by discerning a

distinctive operational content for the beliefs of someone who occupied it Here it is in point, I think, to rehearse the bearing of the rule-following considerations upon the whole idea of investigation-independence, whether it is applied to non-contingent statements or to contingent ones. The effect will be to teach us something about the relation between Wittgenstein's and Dummett's respective objections to realism.

Investigation-independence requires a certain stability in our understanding of concepts. To think, for example, of the shape of some unobserved object as determinate, irrespective of whether or not we ever inspect it, is to accept that there are facts about how we will, or would, assess its shape if we did so *correctly*, in accordance with the meaning of the expressions in our vocabulary of shapes; the fact about the object's shape is a fact about how we would describe it if on the relevant occasion we continued to use germane expressions in what we regard as the correct way, the way in which we have always tried to use those expressions when aiming at the truth. This idea leads us to look on grasp of the meaning of a shape predicate as grasp of a pattern of application, conformity to which requires certain determinate verdicts in so far unconsidered cases. In that sense, the pattern extends of itself to cases which we have yet to confront.

We have to acknowledge, however, that the 'pattern' is, strictly, inaccessible to definitive explanation. For, as Wittgenstein never wearied of reminding himself, no explanation of the use of an expression is proof against misunderstanding; verbal explanations require correct understanding of the vocabulary in which they are couched, and samples are open to an inexhaustible variety of interpretations. So we move towards the idea that understanding an expression is a kind of 'cottoning on'; that is, a leap, an inspired guess at the pattern of application which the instructor is trying to get across. It becomes almost irresistible to think of someone learning a first language as if he were forming *hypotheses*. 'Cottoning on' would be forming the right hypothesis; failing to do so would be forming the wrong, or no hypothesis. And the 'leap' involved would be just that with the best will in the world we—the instructors—cannot do better than leave an indefinite variety of hypotheses open.

What is wrong with that conception? Sometimes we do, no doubt, self-consciously form hypotheses in order to interpret what someone means, for example, if playing a guessing game about the expansion of a series of integers. What is wrong with thinking that, when we are learning a new expression, it is always *as if* we did that? It depends how circumspect we are in our application of the conception. It makes sense to think of someone as having a hypothesis only if *he* knows what it is, that is, what its correctness will require. So the question is whether it is legitimate to think of our trainee as knowing what he himself has come to mean by an expression, what hypothesis about its correct use he has formed. If the 'as if' means what it says, we have to be able to credit the trainee with knowledge about what we will allow in this new case if his hypothesis is right, if his conjectural understanding of the expression coincides with ours. The picture thus encourages our drift into the idea that each of us has some sort of privileged access to the character of his own understanding of an expression; each of us knows of an ideolectic pattern of use, for which there is a strong presumption, when sufficient evidence has accumulated, that it is shared communally.

The hypothetico-deductive picture thus encourages us to accept as a matter of course the rider of certain knowledge of the character of one's own understanding of an expression. And this idea is fundamental to our whole conception of concepts as stable, on which the idea of investigation-independence feeds. But the rider is not a matter of course; it involves an over-dignification. When there really *is* a hypothesis, as in the number-series case, it can be stated and its correct application is open to communal assessment. But when it is merely a matter of an 'as if', the trainee will not have the vocabulary to tell us what it is he thinks we mean. To be sure, the idea of a hypothesis is then psychologically artificial—but it was already conceded to be a mere picture. The real objection is that nothing can be done to defend the description that the trainee *recognises* how, if his conjectural under-standing is right, we will treat the new case, against the more austere account that he simply expects *this* application to be acceptable to us.

The trouble is to prise apart the fact of what his 'as if' hypothesis requires and his expectations about the new case—or, what comes to the same thing, his response to it. *We* cannot tell whether he implements his hypothesis correctly, that is, whether his expectations here really are consonant with the interpretation he has put on our treatment of, say, the samples which we gave him; and *he* cannot provide any basis for a distinction between their being so and its merely seeming to him that they are. It would make all the difference if he could tell us what his hypothesis was; then there could be a communal assessment of the situation. But this is just what he cannot do in the fundamental case when the whole idea of a hypothesis about a pattern is meant merely as an 'as if' idea.

The proper conclusion is not merely that the hypothetico-deductive picture is misleading, but that there cannot be such a thing as first-person privileged recognition of the dictates of one's understanding of an expression, irrespective of whether that understanding is shared. The only circumstance in which it makes sense to think of someone as correctly applying a non-standard understanding of, for example, 'square' is where there is a community of assent about how, given that *that* is what he means, he ought to characterise this object.

Now, how exactly do these reflections bear on the idea of investigation-independence? They bear on the implicit idea that if we do not allow the description of so far uninvestigated objects in particular predeterminate ways, we shall break faith with the patterns of use which we have given to the relevant vocabulary. The question is whether it is legitimate to think of mastery of a language as involving a capacity of *recognition* of what preservation of such patterns requires. Suppose that one of us finds himself incorrigibly out of line concerning the description of a new case. We have just seen that he cannot single-handed, as it were, give sense to the idea that he is at least being faithful to his *own* pattern; that is, that he recognises how he must describe the new case if he is to remain faithful to his own under-standing of the relevant expressions. How, then, does his disposition to apply the expression to a new case become, properly speaking, recognition of the continuation of a pattern if it so happens that he is *not* out of line, if it so happens that there is communal agreement?

If there can be such a thing as a ratification-independent fact about whether an expression is used on a particular occasion in the same way as it has been used on previous occasions, it ought to be something which we can recognise—or at least be justified in claiming to obtain. Otherwise the correct employment of language will become on every occasion radically transcendent of human consciousness. But I cannot legitimately credit myself with the capacity to recognise that *I* am here applying an expression in the same way as *I* have used it before, if this capacity is to be indifferent to whether I can persuade others of as much or whether that is the way in which the community in general uses the expression. How, then, is it possible for me to recognise that I am using the expression in the same way that *we* used it before? How, where it seems that I cannot significantly claim to have grasped an idiolectic pattern, can I claim to have grasped a *communal* one? How does others' agreement with me turn my descriptive disposition into a matter of recognition of conformity with a pattern, recognition of an antecedent fact about how the communal pattern extends to the new case?

To put the point another way: none of us, if he finds himself on his own about a new candidate for ϕ-ness, and with no apparent way of bringing the rest of us around, can sensibly claim to recognise that the community has here broken faith with its antecedent pattern of applica-tion for ϕ; the proper conclusion for him is rather that he has just discovered that he does not know what ϕ means. So there can be no such thing as a legitimate claim to unilateral recognition that here a community's pattern of application of a particular expression is con-travened. So long as the 'recognition' is unilateral, it cannot legitimately be claimed to be recognition, nor can that state of affairs of which it is supposed to be recognition be legitimately claimed to obtain—though if the community can be brought around, it will later be legitimate to claim that the unilateral verdict *was* correct. But how, in that case, can there ever be such a thing as a legitimate claim to unilateral recognition of how any particular expression should be applied in particular cir-cumstances if it is to be applied in the same manner as previously? How can the mere *absence* of the information that the community at large does not agree legitimise the claim to recognition of a fact whose status is supposed in no sense to await our ratification.

Like everyone else, I am tempted to reply that a solicitable community of assent just does make the relevant difference, just does supply the objectivity requisite to transform one's unilateral response into a matter of recognition or mistake. Thus in the absence of the information that the community at large does not agree with one's verdict, standard inductive grounds remain intact for the supposition that there will be solicitable agreement; and one's claim to have recognised how the 'pattern' extends to the particular new case is justified accordingly. But now, what *type* of bearing would the discovery that one was in line with the rest of the community concerning this case have on the question of the correctness of the original unilateral verdict? How does finding out that other people agree with one make it legitimate to claim apprehension of a fact whose status does not require that we acknowledge it at all? Is it that it is somehow less *likely* that lots of people are in error? To suppose that would be to confront the unanswerable challenge to explain how the probabilities could be established.

Assuredly, there is truth in the idea that it is a community of assent which supplies the essential background against which alone it makes sense to think of individuals' responses as correct or incorrect. The question is whether it can be explained why there should be any truth in this idea unless the correctness or incorrectness of individuals' responses is precisely *not* a matter of conformity or non-conformity with ratification-independent facts; for if the contrary account were correct, the rôle of communal assent in determining the correctness of individuals' responses would require the capacity of the *community as a whole* to recognise what conformity to antecedent patterns of use required of us. And now essentially the same problem arises again that besets the attempt to make unilateral sense of correct employment of an ideolect. What is it for a community to recognise that it here continues a pattern of application of an expression on which it previously embarked? What does it add to describe the situation in two-fold terms, of the fact of conformity to the pattern *and* the community's recognition of that fact, rather than simply saying that there is communal agreement about the case? It is unclear how we can answer. We are inclined to give new linguistic responses on which there is securable consensus the dignity of 'objective correctness'; but we have, so to speak, only our own word for it. If 'correctness' means ratification-independent conformity with an antecedent pattern, there is apparent absolutely nothing which we can do to make the contrast active between the consensus description and the correct description. Of course, it may happen that the community changes its mind; and when it does so, it does not revise the judgment that the former view enjoyed consensus. But that is a fact about our procedure; to call attention to it is to call attention to the circumstance that we make use of the notion that we can all be wrong, but it is not to call attention to anything which gives sense to the idea that the wrongness consists in departure from a ratification-independent pattern.

This, I believe, is the fundamental point of the 'limits of empiricism' remarks.[1] None of us unilaterally can make sense of the idea of correct employment of language save by refence to the authority of securable communal assent on the matter; and for the community itself there is no authority, so no standard to meet. It is a different question why we find the ideas so attractive that the communal language constitutes a network of determinate patterns, and that our communal judgment is somehow in general a faithful reflection of the world. Probably it is because we do not pause to consider how much sense we can really give to them; and their opposites seem intolerable.

To put the point in the starkest possible way: how can we penetrate *behind* our consensus verdict about a particular decidable question to achieve a comparison of our verdict with the putative investigation-independent fact? Once we accept the double-element conception of the situation, verdict and investigation-independent fact confronting one another, we feel constrained to say that all we can know is how things *seem* to us to deserve description; to be sure, we may come to revise our assessment—but then again, the revised assessment will be merely what *seems* right. The dilemma we confront as a community is thus essentially that of the private linguist: faced with the impossibility of establishing any technique of comparison between our judgment and the putative objective fact, we must construe the fact either as something we cannot know at all or—the classic choice for the private linguist—as something which we cannot but know. Wittgenstein's response is to urge that both courses are equally disreputable; and that we should therefore abandon the conception which leads to the dilemma. If we do so, we shall reject the idea that, in the senses requisite for investigation-independence, the community goes right or wrong in accepting a particular verdict on a decidable question; rather, it just goes.

So much by way of rehearsal and elaboration of the rule-following considerations. But now, I think, we are in a position to see the following. When Wittgenstein talked of the 'limits of empiricism', it would have got his sense just as well if, anticipating Dummett's later terminology, he had talked of the limits of *anti-realism*. For the points just sketched are shot through with anti-realist strands. The idea that our inability unilaterally to distinguish between a correct application of an idiolectic understanding and a merely seemingly-correct application deprives the distinction of unilateral content is quite unsupported unless an appeal is made to (something like) the principle that there is content to a comparison only if it can be justified. And it is the same with the idea that the community's propensities for consensus judg-

[1] *RFM* II. 71; v. 14, 18. Cf. *OC*, 110, 204.

ment are insufficient to give sense to the 'double-element' conception, essential to investigation-independence, of on the one hand how we ought to judge if we are to remain faithful to our understanding of the expressions involved, and on the other hand how in fact we judge that we ought to judge. But no way is apparent to support the relevant principle in turn save by urging the question: what, if the principle is not accepted, is it to *understand* the claim that a comparison is correct? And no grounds are apparent for thinking that a satisfactory answer to that question cannot be given save the general anti-realist arguments against the intelligibility of attributing grasp of concepts of which there is no distinctive manifestation. If those arguments are rejected, then there is, so far as I can see, no obstacle to embracing the investigation-independence of decidable statements. If, and only if, one admits the need to describe how an understanding could be *revealed* of what it is for our consensus verdict on a decidable statement to fit the alleged investigation-independent fact of the matter, or of the contrast between cases where there is such a fit and cases where there is not, will one feel pressured to doubt the admissibility of the 'double-element' conception. And it is by his readiness to admit this type of need in general that Dummett's anti-realist distinguishes himself.

If the anti-realist ought to reject investigation-independence, then he has, of course, a new and very direct argument against the admissibility of the idea that statements may be determinate in truth-value in a manner transcending our capacities of recognition. One who endorses the rule-following considerations is, we knew, bound to object to the realist's view of our understanding of statements whose truth-values we cannot guarantee to be able to determine. But now it apppears that one who objects to the realist's view on the sort of grounds which Dummett has developed may well be committed to the radical stance which Wittgenstein seems to have adopted. For the answer to our question, whether there is defensible middle ground between an endorsement of realism and an endorsement of Wittgenstein's ideas, seems to be negative. The rule-following considerations originate at bottom from an application to the idea of our continuing to use an expression in the same way as before of a corollary of the fundamental anti-realist thesis that we have understanding only of concepts of which we can distinctively manifest our understanding. So the realist's world picture, however truncated, appears at root incoherent with anti-realist doctrine when the latter is given its widest application. Whatever difficulty there may or may not be in giving operational content to such a picture from a generalised anti-realist point of view, it appears certain that we cannot give it motivational respectability. Rather, it seems that the only notion of objectivity which the anti-realist can allow himself is the ordinary contrast between areas where disagreement is taken to betoken error or misunderstanding, and areas where it need evince only differences in attitude.

Is there, then, any world-picture especially appropriate to a general-ised anti-realist metaphysic? Perhaps it would be as well to forego all pictures; after all, the body of argument will not be any less clear for the lack of one. Nevertheless we have arrived at a position from which we can give a certain sense, so a certain superiority, to Dummett's initially bewildering hybrid: for while its dynamic 'becoming' aspect serves to highlight the anti-realist's implicit repudiation of investigation-independence, it remains a fundamental fact about us that we do not in general *choose* how to describe a new situation, but admit the correct-ness of particular descriptions with a sense of no option; it is in that way that reality is 'not of our making'.

XII

Anti-realism and the Rule-following Considerations

Sources

RFM: v. 33
PI: i. 241
LFM: lectures x, p. 101; xix, pp. 183–4; xxxi, p. 291

1.   Our problem has been: how, if at all, can we reconcile the various prima facie revisionary aspects of *RFM* with Wittgenstein's explicit non-revisionism? We started out with a 'nap-hand' of such potentially revisionary trains of thought, but by the end of Chapter VIII the field had been reduced to two; viz. Wittgenstein's points of sympathy with the intuitionists' complaints about the classical explanations of the quantifiers, and the rule-following considerations. And now it appears, if the reflections of the preceding chapter are sound, that these in turn are one in origin. For no one who rejects the investigation-independence of decidable mathematical questions can both accept the classical assimilation of the quantifiers to infinitary truth-functions and rest content with the classical belief that '$(x)Fx$', 'x' of infinite range and 'F' decidable of every object in that range, has to be determinately true or false; while the ideas which Dummett has presented as exegesis of the intuitionists' thought on the nature of mathematical truth generate doubt not merely about the sort of mathematical objectivity in which the platonist characteristically believes but about the investigation-independence of all decidable mathematical statements.

The doubt about each of these kinds of objectivity originates, I claimed, as a consequence of the basic anti-realist thesis that it is legitimate to attribute to ourselves grasp only of those concepts of which we can distinctively manifest our grasp. It will do no harm to develop the parallel in a little more detail.

(*a*) Traditionally there is a tendency to rely upon an uncritically accepted notion of 'knowing what it is for *P* to be true' as describing the essential character of our understanding of any statement, *P*. How is such knowledge to be manifested in cases where the realisation of P's truth-conditions is conceived as a state of affairs which must elude our capacities even of 'working' verification? Such knowledge can have no features save as are reflected in what it enables a possessor of it to do; and what it enables a possessor to do can be manifest only in his responses to circumstances of a non-transcendent kind. When, in order for P to be true, a state of affairs will have to obtain of a kind—as we should ordinarily suppose—which no human being could intelligently claim to have verified to obtain, then the best anyone can do by way of displaying his putative knowledge of what it is for P to be true will be by responding appropriately to *other* kinds of states of affairs: states of affairs which warrant the assertion, or denial, of P, that is, which we conceive of as constituting good evidence for or against P; states of affairs which warrant the ascription to someone of the belief that P; states of affairs in general which bear on the warranted assertability of some compound statement of which P is a constituent; and states of affairs in which an opportunity is presented to manifest one's grasp of the inferential liaisons into which P enters with other statements. In short, all that can be manifested is a knowledge of the *use* of the statement P. The question therefore immediately arises: what is the legitimate content of the play traditionally made with the notion of *truth*? What is the role of 'knowing what it is for P to be true' when all that can be displayed is a knowledge of the use of P in response to non-transcendent states of affairs?

There seem to be three possible answers. The first is a re-affirmation of realism. Knowledge of transcendent truth-conditions is in the nature of the case, it will be claimed, knowledge which cannot be fully manifested in the sorts of practical ability described. That is, a knowledge of what it is for a transcendent truth-condition to be realised does not *reduce* to a knowledge of the use of the appropriate statement, but is rather something—the possession of a 'conception' or whatever[1]— which endows and informs knowledge of its use. From our grasp of what it is for it to be true that all men are mortal, for example, *issues* our knowledge of what is good evidence for, or against, that statement, our knowledge of what is good evidence for someone's believing the statement, and our knowledge that either that statement or its negation must be true.

The second answer, usually commended by Dummett, is to reject 'knowledge of what it is for P to be true' as a satisfactory description of what an understanding of a verification-transcendent statement, P, consists in. Once we insist that knowledge of meaning be capable of being fully manifested—a large part of the point of the thesis that meaning is use—we see that it cannot in general amount to knowledge of truth-conditions since it is only in the case of a restricted class of statements that there is any manifestable ability for knowledge of truth-conditions to *be*. What we traditionally took to be knowledge of truth-conditions thus turns out often to be knowledge of a quite differ-

[1] Cf. Strawson 73, p. 16.

ent sort; and the orthodox assumption that truth should play the central rôle in the philosophical theory of meaning is an error.

A third approach would be the view that there is no serious objection to thinking of our understanding of verification-transcendent statements as consisting in a knowledge of their truth-conditions provided this knowledge is then reductively construed in terms of the practical knowledge of use which, it is agreed on all sides, is all that can display that understanding.[1] On this view, to talk of 'knowledge of what it is for *P* to be true' is no analysis, or explanation, of what it is to understand *P* but is simply an idiomatic way of talking about understanding *P*; to put it another way, the equation of statement-understanding with knowledge of truth-conditions is not a substantial philosophical claim. We therefore reject the essential realist notion that grasp of truth-conditions is the *core* of statement-understanding, of which knowledge of the various aspects of the use of statements—their assertion and rejection conditions, their inferential connections, etc.—is a bestowed consequence. Rather, to 'know what it is for *P* to be true' just is to possess practical knowledge of the use of *P*. It follows, in particular, that it is an error to suppose that the validity of classical logic, and especially of the law of excluded middle, for all statements of sufficiently exact sense is *imposed* by the general character of our understanding of them. Save where *P* is effectively decidable, there is nothing in our 'knowledge of what it is for *P* to be true' to *justify* the supposition that one of *P* and its negation has to be true. There is, indeed, no cogent defence of that belief.

(*b*) Traditionally we have also assumed uncritically a notion of 'knowing what it is to use *P* in the way in which it was originally explained to one to use it'; or if *P* is a hitherto unencountered sentence, of 'knowing what it is to use *P* in a way that continues the patterns of employment bestowed on its constituents by previous training'. This assumption is, indeed, even more deeply rooted in our ordinary thought about the character of language-mastery than the assumption that to understand a statement is to grasp what has to be the case in order for it to be true. For we think of it as essential to any mode of learned linguistic competence that it involves the deployment of knowledge of this kind.

Consider the example of any particular effectively decidable predicate, ϕ. What will manifest a knowledge of what it is to use ϕ in the same way as was explained and illustrated during ones 'apprenticeship'? If we grant that there can be nothing to such knowledge save features which are manifest in the performance of someone who possesses it, then we have fully described its character as soon as we have fully described

---

[1] In proposing to construe knowledge of the truth-conditions of *P* in a reductive way, we should not, of course, enter into any general commitment to a reductive construal of what it is for those truth-conditions to obtain; to understand the truth-conditions of a verification-transcendent *P* will involve grasp of its assertability-conditions, but it will also involve grasping that the assertion of *P* under those conditions is always in principle defeasible. Cf. Wright 80.

what someone who possesses it is distinctively able to do. But what a possessor of such knowledge is distinctively able to do is simply to use ϕ in a manner acceptable to his teachers and others who have undergone a similar training. To have grasped what it is to use ϕ in the manner which explanations and examples were meant to convey is just to have derived from one's training the ability to participate in a community of assent in the use of ϕ: a community embracing ones teachers and inside which disagreements in the use of ϕ, if ever irresoluble, take place within a framework of agreement about what would decide who was right.

If knowledge of the meaning of ϕ is a matter of knowing what it is to continue a certain explained pattern of use, the pattern which speakers of one's language all aim to continue when intending truly to ascribe or withold ϕ, the fact is that all that can be done to display this knowledge is to participate in an on-going community of assent in the use of ϕ; ones use of ϕ must be such, ideally, as will elicit the agreement of the community at large if put to the test—or at least the agreement of an 'expert' subsection—and such persistent disagreement as may arise must do so only because information whose relevance is acknowledged on all sides is not available. The question therefore arises: what is the legitimate content of the play traditionally made with the idea of grasping a *pattern* of use, of knowing what it is to use ϕ in the *same* way as that in which one was trained to use it? What is the rôle of this knowledge when all that can be displayed is the ability to participate in a community of assent in the use of ϕ?

Again, there seem to be three possible responses. The first would be to accept that knowledge of what it is to use an expression in a way which continues the pattern which one was taught to follow is knowledge which cannot be fully manifest in ones practical linguistic skills. Knowledge of what it is to use ϕ in the same way as that which one was taught, the way the community has always used it when aiming at the truth, is therefore knowledge which does not *reduce* to the ability to participate in an on-going community of assent with respect to ϕ. Rather it is something in virtue of which alone an on-going community of assent is possible; it is our possession of knowledge of this kind which *explains* our ability to agree in our use of language, and without it we should not agree. Our agreement in our use of ϕ is a product of our individual possession of the relevant piece of knowledge; and the fact, of continuation or breach of the pattern, of which in any particular case it is knowledge is therefore something to which our agreement *corresponds*. So the 'double-element' conception is inescapable.

The second response would be to reject the idea that any essential part of what it is to understand ϕ is happily conceived as knowledge of what it is to use ϕ in the same way as that in which it has been used on prior occasions. Once it is insisted that our understanding of language can have no facets save those manifest in our use of it, we have to acknowledge that there is nothing for our alleged knowledge, that to ascribe ϕ here would continue, or breach, its prior manner of employment, distinctively to consist in. Our understanding of ϕ, which we took to be knowledge of that sort, turns out to be of a quite different character, and needs recourse to no such comparison in order to be satisfactorily described. The traditional assumption, that understanding an expression is a matter of grasping a distinction between uses which do and uses which do not conform to an antecedent pattern, is therefore mistaken.

The third possible response—that, I believe, of Wittgenstein—would be to allow that an understanding of ϕ may be harmlessly, if unilluminatingly, described as grasp of a certain general pattern of use, *provided* that this grasp is then reductively construed as the ability to participate in an on-going community of assent in the use of ϕ. To grasp the pattern of use for ϕ which ones training aimed to illustrate and explain is just to come to understand ϕ; and the knowledge of use in which that understanding consists is manifested as, and so goes no further than, the ability to participate in the linguistic community as far as the use of ϕ is concerned. It is an error to conceive, as did the first response, of understanding ϕ as explanatory of the latter ability, to think of shared grasp of a pattern as underlying and issuing in our propensity for consensus in the use of ϕ. The relation is not that of *explanans* to *explanandum* but of *analysandum* to *analysans*. And the single most striking corollary of this adjustment is that there is now nothing to be found in the idea of grasp of what it is to use ϕ in a particular, antecedently explained way to *justify* the view that if we attribute that grasp to ourselves, we must conceive of any 'ϕa', where the identity of a is known, as determinate in truth-value irrespective of whether we ever investigate the matter or what our findings are if we do. In order to defend that view, a far more robust conception is required of what is known by someone who knows what it is to use ϕ in the way which has been explained to him than any account can warrant which pays due heed to the manner in which that knowledge is manifested; there is, indeed, no satisfactory defence of the view.

2.   The problem, then, is to explain how someone who accepts these broad anti-realist ideas—which I shall continue to refer to as 'the rule-following considerations'—can avoid revisionism. How can a philosophy of language which involves a wholesale rejection of investigation-independence, for both a priori and contingent statements, avoid becoming a revisionary philosophy of mathematics? Platonism thinks of the content of statements quantified over a denumerably infinite domain as specified by determining that their truth-conditions are to be those of an appropriate conjunction, or disjunction. The belief that, where 'F' is effectively decidable or otherwise

sharply defined, one of any such contradictory pair of statements,
'(x)-Fx' and '$(\exists x)\ Fx$' for example, *has* to be true, whether or not we can
find out which, would therefore appear to require that each conjunct,
'$-Fa$' or disjunct, 'Fa', be thought of as determinate in truth-value independently of any investigation we might undertake—a requirement for
which there is now no apparent justification. Of no statement can it any
longer defensibly be supposed that we will be bound to count certain
aspects of reality as consistent, or inconsistent, with its truth if we are to
remain faithful to our understanding of it. An omniscient being could
know which way the consensus will go in any particular case; but even
for such a being there is no sense in the idea of a fit between the verdict
of consensus and what the understanding possessed by those who
partake of the consensus dictates for that particular case.

To be sure, the previous chapter contained no absolutely *conclusive*
demonstration that a philosopher who globally rejects investigation-
independence abrogates all means for a defence of the validity of
excluded middle for undecided effectively decidable statements. But
what does appear certain is that such a philosopher can provide no
defence of the validity of the sort of application of excluded middle with
which we have mainly been concerned—the simplest intuitionistically
controversial type of application, where effective decidability is absent
only because a denumerably infinite range of quantification is involved.
For what could give rise to the principle's validity in this kind of case?
Surely it has to originate in the way we conceive the truth-conditions of
the individual disjuncts; and this in turn has to be explained in terms of
our concepts of the senses of the quantifiers and of the atomic statements through which the quantifications are made. But there simply
seems to be no other course than that, which we have all along taken to
be characteristic of platonism, of conceiving of the atomic statements as
determinate in truth-value, and construing that as settling the truth-
values of the quantified statements; and for this course there now
appears to be no warrant. So why is Wittgenstein not more explicitly
revisionary? Why does he not come into the open and say that there is *no*
justification for regarding excluded middle as always validly applicable
to such alternatives, and that proofs which appeal to its validity in such
cases are therefore defective on philosophical grounds?

One answer to this question, implicit in Dummett's discussion,[1]
would be simply to call attention to the conflict of Wittgenstein's
conventionalism with conceding revisionary force to these general anti-
realist ideas. Indeed, Wittgenstein's situation here is accepted by
Hacker and Baker in their review of the *Grammar*[2] as involving a
fundamental tension. We have to accept that two foundational aspects
of Wittgenstein's thought about mathematics are in flat collision: the
'constructivist' idea that it is by reference to the notions of proof and
disproof, rather than truth and falsity, that the sense of mathematical

[1] 23 (2, pp. 502–3). [2] 44, p. 282.

statements is to be thought of as grasped, and the radical conventional-ist thesis that there are no constraints on what we accept as a proof of a particular statement—in particular, therefore, none imposed by the general character of our understanding of such statements. Clearly, if someone holds that we are in some sense free to accept or reject any proof, and free to accept as holding necessarily any particular state-ment, then he signs away the means of significant criticism both of particular methods of proof and of our acceptance of particular state-ments as necessary. Radical conventionalism is inconsistent with mathematical revisionism not, obviously, in the sense that the conven-tionalist may not coherently reject certain traditionally accepted forms of inference; but in the sense that he cannot consistently present as *cogent* certain reasons for doing so. So the suggested explanation of why Wittgenstein does not face up to the revisionary consequences of his ideas is because to do so would be inconsistent with something he prices very high: a radical conventionalism about necessity.

In fact it is not particularly interesting to speculate why Wittgenstein did not draw out conflicting aspects in his views. The interesting question is rather whether there is a reconciliation of these various ideas. Can Wittgenstein's general anti-realism, expressed in the rule-following considerations and in his points of sympathy with the intuitionist's critique of classical mathematics, be shown to cohere with his conventionalist view of necessity and general non-revisionary approach? Or is there indeed a fundamental 'disharmony of main themes' in Wittgenstein's later philosophy of mathematics, as suggested by Hacker and Baker? The answer, I believe, turns on a fundamental question in the philosophy of language, which we must now approach.

3.   A natural worry at this point is whether one constituent in the possible 'fundamental disharmony' may not be spurious. Surely, it might be suggested, Wittgenstein as radical conventionalist is a mere caricature, derivable only from a concentration on his slogans about proof and necessity and a wilful ignorance of their motivation. Natur-ally, if someone holds that we are free *arbitrarily* to accept or reject any proof, he has to disallow that there can be pointful criticism of particu-lar proof methods. But those elements in Wittgenstein's thought which bear a conventionalist interpretation in no way require that the freedom in question is one which we can intelligibly *exercise*—at least to ourselves. The point is rather that the ideas of a kind of contract in under-standing and of the constraint of fidelity to a certain pattern of use, with which we tend to bolster the opposite view that certain combina-tions of sense just are productive of necessary truth, apprehensible in principle by any rational being, turn out to be without substance. There can be no obstacle in the nature of 'the concepts themselves' for some-one who proposes to reject the necessity of a statement generally

regarded as such. It remains true that his doing so will be a criterion for
his possession of an understanding of the statement, and concepts
involved, divergent from that of the community at large; but this
divergent understanding is not something apart from his deviant move
of which that is to be seen as a mere *application*. The 'concepts themselves' have no finally stable nature. Thus it is wrong to take it as a
consequence of Wittgenstein's view that there cannot be *any* worthwhile criticism of someone's accepting a statement as necessarily true;
the consequence is rather that there cannot, ultimately, be decisive
criticism, that he cannot be convicted of 'breach of contract' in relation
to correct understanding of the involved expressions. (He cannot for
example, as argued at the end of Chapter V, be convicted of using a *false*
criterion of miscounting, if he accepts certain deviant arithmetical
equalities.)

But then, what sort of criticism *can* there legitimately be? In particular, can the intuitionists' critique of classical concepts and explanations
still coherently be developed? It is perfectly true that the intuitionists'
objection to the relevant aspects of classical mathematics is not happily
expressed as that the classical mathematician is led to regard statements
as necessarily true which are actually, in an objective sense, no such
thing. If this were the objection, it would obviously be inaccessible to
one who accepted the rule-following considerations. The objection is
better expressed as that we do not have the sort of concept of the content
of certain statements which is needed to sanction the means whereby
apprehension of their truth, and so necessity, is classically achieved.
But how does this refinement help to make the intuitionists' criticisms
consistent with the kind of conventionalism which Wittgenstein undoubtedly held? It remains unclear how a Wittgensteinian could significantly advance any such criticism. What is it for proof-methods to
'accord' with the general character of our understanding of the intended
conclusion? If we come across a tribe who seem to be employing what
are apparently classical—or even wilder—mathematical techniques, do
we not, from a Wittgensteinian point of view, just have to accept that as
expressive of—or, better, *constitutive* of—the character of their understanding of certain relevant concepts? If we do, there is no room for an
intuitionist to say that the classical mathematician *cannot* understand,
for example, existential generalisations over an infinite domain in such
a way as to permit their proof by *reductio* of their universally quantified
negations; for there is the pattern of *use* unfolding before his eyes.

In order to get clearer about the difficulty, consider an analogy with a
board-game in which, for example, there are a variety of pieces associated with different powers. In order to be able to play the game,
someone has to learn three things: what moves are legally admissible for
him in any situation; when someone has won; and that the aim is to
win—or at least to prevent anyone else doing so. Of course, if someone
is familiar with games at all, he already knows the third item—the force

of describing a particular situation as, say, a win for Italy. So an explanation of the game will normally confine itself to a statement of information in the first two categories. Rules will be given entailing answers to questions of the form: Is such-and-such a move admissible here? Has so-and-so won here? Now, we tend to think of correct understanding of the rules as a guarantee of the ability to play properly—as a state of information which a proper explanation will endow and which then will ensure that no intractable dispute will arise. For Wittgenstein, as we know, this is a misleading picture. In the first place the rules have no character save as is determined by our understanding of them; and our understanding of them has no character save as is manifest in the open set of applications of them which we may be called upon to make. But secondly, even this is misleading; for there is no objective *kind* of application of any particular rule grasp of which it is the aim of our explanations to bestow, and of which we already have a grasp, so that we can regard ourselves, when we agree of a particular move that it accords with the rule or not, as recognising a ratification-independent fact. We say, for example, 'No, that is not the kind of thing we meant'; but we need not have anticipated the possibility of this particular kind of blunder, and then we too, as it were, find out what we meant by our reaction to the move.

That is one thing which Wittgenstein, distinctively, has to say about rules; and it immediately gives rise to a corollary about the status of derived rules. The status of a rule as a consequence of those in the rule-book is not something objectively settled for us by the way in which they, and our ordinary rules of derivation, are understood; our understanding has no such autonomy—we are not *committed* by it to accepting, or rejecting, any particular so far unconsidered derivation. Confronted with a particular derivation, we shall probably have a firm and concerted reaction to it. But we differ from a radical interpreter of our language *only* in our possession of trained and, to us, mandatory-seeming responses within it; it is not that we *know*, whereas he can only make an educated guess at, what response in a particular case conforms to our understanding.

Wittgenstein's conception of necessity involves thinking of necessary statements as comparable to the rules of such a game.[1] They are, so to speak, among the rules for the game of language-use; they supply criteria for the description of circumstances and the correct implementation of procedures. ('I must have miscalculated.') In terms of our game analogy, allied to the rule-following considerations, we can now model three respects in which necessity may be described—perhaps not altogether happily—as on Wittgenstein's view conventional. It is, to begin with, conventional what rules go into the rule-book; we are laying down an activity, and are subject to no external constraints. Correspondingly we saw in Chapter V how we could not in the end make good

[1] *RFM* v. 28.

the objection that principles of inference are answerable to the truth-values of the statements which they enable us to link as premisses and conclusion. (We did not, however, uphold Wittgenstein's further suggestion that there could be practices fundamentally different from our own but still pointfully describable as 'calculations' or 'inference', and incommensurable with ours in point of reasonable acceptability.) Secondly, it is an open question what is a proper application of the rules in any particular case; it is a matter for our successive judgments. We do not bind ourselves to conform to something objective in accepting a statement as necessary; we no more thereby determine an objective pattern of use for the concepts involved than counting a particular statement as necessary is originally answerable to an antecedent such pattern. And finally, as a consequence of the second point, it is an open question to what other conventions we are committed by the acceptance of certain rules, of certain statements as necessary.

If this analogy does indeed capture the essential features of Wittgenstein's conventionalism, then its effect is to reinforce the suspicion that if Wittgenstein's view is accepted, there can be no telling revisionary criticism of classical techniques or, indeed, of any others. For do we not now have to accept classical mathematics, alongside intuitionist mathematics, simply as different 'games'? How can it be claimed that we cannot profess to understand the logical constants in the way which classical mathematics requires? For there is the game, being played. The practice and application of classical mathematics can be learned; there is a (highly profitable) game whose rules are classical. No doubt it is an inept account of Wittgenstein's position to suggest that it involves that we are free, on impulse as it were, to accept or reject proofs. But a less inept sketch of his conventionalism seems to do nothing to make it any clearer how Wittgenstein can coherently sympathise with the intuitionists' complaints to the extent that he does, or challenge the force of Cantor's diagonal argument as classically interpreted.[1]

The foregoing brings out that the interpreter of Wittgenstein in fact confronts two related difficulties in this area. There is, first, the problem already highlighted of explaining how Wittgenstein's general anti-realism can avoid the sort of revisionary consequences which seem to follow so naturally from anti-realist ideas; only if a satisfactory explanation of this can be given is there harmony between the anti-realism, the conventionalism, and the more general non-revisionism. The situation is striking in the respect that the rule-following considerations are both central to the interpretation of Wittgenstein's conventionalism just outlined and prima facie revisionary of classical mathematics at least. That reflection might prompt the thought that some sort of harmony must underlie the anti-realism/conventionalism tension if only its constituents can be understood aright—though there would be no guarantee that it would not turn out to be a harmony in which the

[1] As, it appears, in *RFM*, App. II, 2.

conventionalism emerged as consistent with revisionary criticism of certain sorts of mathematics, so that a conflict with the non-revisionism would remain. However that may be, there looks to be no prospect of a solution to this first problem unless a firm wedge can be driven between the undoubted destructive effect which the rule-following considerations have on the thinking which tends to underpin satisfaction with classical logic and mathematics on the one hand and the capacity of those considerations to enjoin revisionary criticism on the other. But a second, distinct problem is to explain how it is possible for Wittgenstein, from his conventionalist standpoint, coherently to find fault even with the explanatory background and underlying motivation for a particular type of mathematics. He wants to query the classical interpretation of the quantifiers and the significance of the Diagonal Argument; in both cases tenuous analogies with the mathematics of the finite are, he thinks, 'puffed up' to the point where we tend to mistake the content of mathematical results.[1] And that is just what the intuitionists want to say. But what, for a conventionalist, can be wrong with any analogy, however slight, which issues in communicable, practicable mathematics? What objection can there be, for example, to the notion of a completed, randomly selected, infinite subset of integers if mathematicians are able to interest themselves in the pure mathematical treatment of the notion and to agree about the relevant proofs?

We shall defer consideration of the second problem, concentrating for the remainder of this chapter on whether we can make some initial headway with the first.

4.   We have considered two strategies—quasi-platonism and the argument of §§ 4 and 5 of the last chapter—for denying, or at least restricting the scope of, the revisionary threat of the rule-following considerations. Neither was satisfactory. But we have not considered perhaps the most direct strategy. It is this. Integral to the traditional semantics for classical logic is the idea that each significant declarative sentence may be regarded as determinately possessing one of two truth-values. And no truth-functional compound statement can be thought of as determinate in truth-value unless the same can be said of at least some of its constituents. But, if the rule-following considerations are accepted, determinacy in truth-value is bestowed only by explicit ratification. Thus it appears, keeping to the example of excluded middle, that no disjunction can be recognised to be valid by means which leave unratified the truth-values of its individual disjuncts. A disjunction is true if and only if one of its disjuncts is true; so if our concern is with a class of statements for which truth is consequent on ratification—which, it seems, may now be a comprehensive class—then there cannot be a true disjunction, a fortiori not a valid one,

[1] App. *II*, 3.

of such unratified statements. Clearly, however, this reasoning is compelling *only* if it may be assumed that truth remains distributive over disjunction; whereas just this is called into question if we are concerned with a class of statements for which truth, or falsity, is consequent upon decision—or something akin to it. For with such statements, it is plausible, we may precisely decide that a disjunction is to be true *without* deciding which of its disjuncts is.

Consider again Wittgenstein's fictional example. If the author of the poem has decided nothing about the hero's family structure, it may well be appropriate for him to respond to the question, 'Does the hero have a sister or not?' in the way which Wittgenstein suggests. But the question is ambiguous: it may mean, 'Is "x has a sister" true of the hero, or is "x does not have a sister" true of him?'; or it may mean, 'Is "x has a sister or not" true of the hero?' To the first question, if the criterion of correctness of an answer is to be supplied by the author's concept of his hero as so far decided, it may well be appropriate for him to say simply that he has not yet decided. But not, surely, to the second; for he is presumably inviting us to imagine the hero of the poem as a normal man—let us suppose so, anyway—and that is already a constraint upon us to think of him as having any property which any real man must have; for example, that of having a determinate family structure. The ambiguity in the question is not, of course, generated by the possibility of truth depending upon decision, or explicit ratification. It is the ambiguity of 'Is the disjunction true?' versus 'Which disjunct is true?' (If I want tea, and am asked whether I want tea or coffee, I can, taking the question in the first way, reply simply, 'yes'.) The rôle of truth as depending upon decision is to generate the possibility of the correctness of an affirmative answer to the first question without there being any determinately correct answer to the second.

We have to concede that both the platonist and quasi-platonist accounts of the validity of 'Either "550" occurs in π or it does not' are undermined by the rule-following considerations. Nevertheless, the argument goes, this is consistent with the validity of the disjunction in terms of Wittgenstein's general conception of what validity is: the schematic rule, $P \lor -P$, may be enshrined in the 'grammar' of the language. Such a schema, or an equivalent, for example, Double Negation Elimination (whose unassailability Wittgenstein questions elsewhere)[1] may just *be* part of the machinery which we count as constitutive of admissible proof and argument. And this machinery, according to Wittgenstein, does not need a defence. The platonist conception of excluded middle as 'solid' and 'unshakably certain' is untenable; the account given of its certainty involves errors about the concepts of truth and meaning. But to point the errors out is not to establish that the principle is *suspect*; the argument that it is so itself involves assumptions, notably that the validity of any disjunction

[1] See above, Chapter VII, p. 118, note 2.

requires the determinate truth of at least one of its disjuncts, and this too can be called into question. Indeed, as the analogy with fictional statements suggests, the idea that truth is consequent upon (something akin to) decision itself calls it into question. But if this teaches us that we cannot explicate the validity of the principle in terms of a disjunction-distributive notion of truth, the proper conclusion is not that its validity is therefore suspect—for we *use* it as valid—but that disjunction is hence not to be so interpreted in this context.

How good an argument is this? On the face of it, it is absurd to build on an analogy with fiction, just because in fictional contexts we should expect to be released in all sorts of ways from a responsibility to the notions of truth and falsity as ordinarily understood. It is also unclear that it is correct even in fictional contexts to question the distributivity of truth over disjunction (whatever notion of truth is now relevant). Suppose, for example, we are told that the hero was the youngest child of a large family, and later in the poem that he had no brothers. We can imagine critics debating the skill with which the poet had depicted the psychological effects on a youngest son in a family of sisters. Such a debate would seem at least logically appropriate. But its being in point at all depends upon the inference that the hero had many older sisters—a modus tollendo ponens step, which would seem to depend upon the distributivity of truth over disjunction. For if a disjunction can somehow be true without either of the disjuncts being true, then it seems that nothing can follow about the truth of one of the disjuncts from the falsity of the other.

A natural response would be that if we have decided that a disjunction is to be true without deciding on the truth of any particular disjunct, and then plump for the falsity of all but one of the disjuncts, then that *is* to decide that the remaining disjunct is true. The principle underlying this response would seem to be that if an agent has severally decided that a number of propositions, $P_1, \ldots, P_n$, are to be true, it is ceteris paribus legitimate to take it that he has decided that every logical consequence of the set, $\{P_1, \ldots, P_n\}$, is to be true. Such a principle may be well entrenched in our habits of literary criticism. But it is unclear that it can survive if we are concerned not with fictional statements but with the implications of the rule-following considerations for ordinary ones; for, to repeat, *explicit* ratification is then what counts. And it is unclear, in any case, whether it would still be legitimate to think of $\{P_1, \ldots, P_n\}$ as having exactly the same consequences as those it sustains when any occurrences of the logical constants are interpreted in the classical truth-distributive manner.

Modus tollendo ponens is not the only classically valid principle about which a doubt may seem to arise. Let us allow that excluded middle may be validly applied to the constituents of a (quasi-) platonist infinite conjunction for '"550" does not occur in π' without there having to be in any particular case a determinately correct answer to the question, which disjunct is true. Then, it appears, we can allow the validity of '$(x)(Fx \lor -Fx)$', 'x' ranging over finite initial segments of π and 'F' expressing the property of containing the sequence '550'. But this is not all that classically is wanted; there is the matter of the transition to '$(x)-Fx \lor (\exists x)Fx$'. If the validity of '$(x)(Fx \lor -Fx)$' does not require that one of 'Fa' and '$-Fa$' be determinately true for any particular a, how can the validity of the transition now be explained? Clearly we cannot validly infer that one of '$(x)-$Fx' and '$(\exists x)Fx$' is determinately true and the other determinately false; but that is not decisive against the transition since it appeals to the old concept of disjunction for which distributivity holds.

It is not wholly clear how to make the matter perspicuous. A natural ploy would be to introduce an operator, D, $=$ 'it has been decided that', in terms of which to capture the entertained breakdown in distributivity; thus we reject the transition, $D[P \lor Q] \rightarrow D[P] \lor D[Q]$; (cf. drawing up a short list of candidates for a post). Now, plainly, this transition does not hold:

$$D[(x)(Fx \lor Gx)] \rightarrow D[(x)Fx] \lor D[(\exists x)Gx];$$

the question is, does this one:

$$D[(x)(Fx \lor Gx)] \rightarrow D[(x)Fx \lor (\exists x)Gx)]?$$

In finite cases at least it might seem so. Suppose I am to write out a finite series of coloured strokes and decide that I shall certainly use a black crayon and maybe a red one also, but no others; then I have decided of any particular stroke that it will either be black or red, though we can suppose that I have made no specific decisions about the colour of any particular strokes. But now it seems that I have thereby decided either that all the strokes will be black or that at least one will be red, without having made either particular decision; for if I keep to my original decision, either all the strokes will be black or there will be some red ones.

What is appealed to here is the principle that $D[P] \rightarrow D[Q]$ holds good just in case implementing P entails implementing Q. But the model fails us in the infinite case; for now there is no question of completing the series of strokes. Nothing amounts to implementing the original decision. We can still say that, for any initial segment of the series, I have implicitly decided either that it will be all black or contain at least one red stroke; but there seems no way of getting a purchase on whether such a decision may be regarded as implicitly taken for the whole series.

A Wittgensteinian could be expected to reply that it is not at all desirable that we should be able to articulate a *commitment* to '$(x)Fx \lor (\exists x)-Fx$' in terms of our 'decision' that '$(x) (Fx \lor -Fx)$' holds good. For if the rule-following considerations are to be non-revisionary, we do not

want a commitment to a principle which an intuitionist would reject
However, if it is allowed at all that the validity of a disjunction may not
require determinacy in truth-value on the part of its disjuncts, then we
can say the same here. Thus the validity of '$(x)Fx$ V $(\exists x)-Fx$', or of
'$((x)(Fx$ V $-Fx)) \rightarrow ((x)Fx$ V $(\exists x)-Fx)$', can be seen as a consequence of a further decision, one which it is open to us to take but in no
sense required of us.

More generally, he can be expected to argue, the feasibility of the
present non-revisionary approach does not depend upon the uneasy
tactic of taking fictional statements as a paradigm; that idea need only
be taken as suggestive. Nor are we obliged to be able to provide a
vindication of any 'intuitions' of validity, for example that of the principle of modus tollendo ponens, in terms of a notion of disjunction which
is not truth-distributive. The general effect of the rule-following considerations is that the justifiable assertability of any statement, inside or
outside mathematics, always awaits our explicit verdict on the matter.
If we think we must tailor our conception of validity accordingly, we
shall naturally look for guidance to areas, like fiction, where some such
idea—not exactly this idea—is already our prejudice; and what we find
is that while, in some sense, distributivity seems to fail, inferences
which seem to depend upon it continue to be upheld. But the whole idea
that the rule-following considerations may be revisionary begs a substantial question, the fundamental question in the philosophy of
language to which I alluded earlier. The question is, with what right do
we see the standing of any particular principle of inference as answerable to *explanation* in terms of a general semantics of truth—or justifiable assertability, if that is preferred—rather than as a constituent in the
apparatus whereby we arrive at explicit verdicts of truth, or justifiable
assertability? We find, if Wittgenstein's ideas are allowed, that we
cannot *explain* our apprehension of the validity of excluded middle in
terms of a recognition, for example, that one of the alternatives about π
must be determinately true and the other determinately false; then, if
we attempt to hang on to the law while conceding this, it appears that
the distributivity of truth over disjunction is going to be compromised;
and it then becomes unclear how to explain the validity of other
disjunctive principles, for example modus tollendo ponens, or reasoning by disjunction elimination. But it does not matter whether such
explanations can, with a little ingenuity, be given. The point is rather,
why are we insisting that such an explanation ought to be *possible*? Why
are we not content just to accept our principles of inference as we find
them—as 'rules of the language-game'? It is, for example, just a fact
about our concept of disjunction that inferences involving it are
counted as valid which it is natural to explain in terms of a distributive
notion, but that when truth is thought of as consequent upon explicit
ratification, not all traditionally accepted principles are obviously so
explicable. Well then, that is how it is with disjunction.

Our massive intuitive resistance to this line derives from a certain conception of language-mastery, issuing in the recent concept of a programme for a 'theory of meaning'.[1] We think it should in general be possible to explain our understanding of the truth-, or assertability-, conditions of *compound* sentences in terms of those of their sentential constituents. This is a special case of the more general belief that our understanding of our language is such as to require the possibility of a theory in which the sense of every declarative sentence can be revealed as determined by its essential semantic structure and pre-assigned semantic characteristics of its constituents. This idea encourages the hope that a systematic theory of meaning for a language should, at least in some cases, be capable of explaining the validity of inferential schemata in the language.[2] The explanation would consist in showing how the unconditional truth, or justifiable assertability, of the schema followed from the semantic analysis offered by the theory for a suitably corresponding statement; for example, in the case of a principle of derivation, for a corresponding conditional. But it may be a major operational implication of Wittgenstein's conception of logic as antecedent to truth that the compound schemata of logic should not be seen as subject to such an account at all. The validity of principles of inference would then no longer be seen as a consequence of their essential semantic structure (or that of suitably corresponding statements). Rather logic would stand aloof, susceptible neither to reproach nor endorsement in terms of considerations within the theory of meaning; and this would hold whether we are concerned with a formal theory for a specific object-language and the logic in play in that language, or whether—as, for example, in Dummett's writings on intuitionism, and previously in this book—we are bringing to bear on the statements of logic and mathematics very general, informal considerations to do with understanding. For Wittgenstein, I suspect, either kind of assessment would be out of place. For the rôle of logic and mathematics is precisely to *fix* meanings, by entering into the determination of conditions of truth, or justified assertability, of contingent statements. Thus, rather than seeing the statements of logic and mathematics as answerable to a theory of, or general considerations about meaning, we should somehow assign to them a constitutive rôle in such theories. I shall return to this idea in the next chapter.

---

[1] The concept for whose development Davidson has been so largely responsible, and which forms the centre of attention in the essays in Evans' and McDowell's 32.
[2] See, for example, Davidson 20, pp. 184–5, 16, p. 318; and 19, p.144.

XIII

Revisionism

Sources

On 'winning' and 'truth'; expressions of the 'redundancy' view:

RFM: 1. 5; App. *I*, 1–6
  PI: 1. 136
  PB: First Appendix, p. 311
LFM: lectures VII, p. 68; XIX, p. 188; XXIX, p. 276

On the possible indirectly revisionary effect of philosophy:
RFM: App. *II*, 18
  PG: II. 25 (pp. 381–2); 26, (p. 385)
LFM: lectures I, pp. 16–17; XIV, p. 141

1.   Wittgenstein's 'constructivist' strand—his sympathy with aspects of the intuitionists' critique of classical mathematical concepts and proof methods, and his endorsement of a proof-conditional conception of the meaning of mathematical statements—issues, according to the interpretation which we have proposed, from a general anti-realism of which the rule-following considerations are the other most distinctive expression. This constructivism seems to war, as we noted in the previous chapter, on two different fronts with Wittgenstein's conventionalism. First, one would expect a general conventionalism about necessity to involve a laissez-faire view of mathematics, whereas the constructivism looks to be revisionary of classical mathematics at least. (From a conventionalist point of view, the intuitionists—it is natural to think—would be comparable to people who claimed that for generations Chess had been played all wrong, and that we should substitute for the Queen a piece of quite different powers of movement and capture.) And, secondly, it is unclear that Wittgenstein's conventionalism leaves him room to criticise even the underlying motivation for classical mathematics. For a conventionalist, it cannot be a *mistake* to accept classical mathematics; so what *else* can be wrong with the explanations, however fanciful, by which its practice is communicated? How can there be, as the intuitionists suppose, an *illusion* of understanding where there is an agreed practice?

Now, as far as the first point of apparent conflict is concerned, an obvious suggestion to explore would be that Wittgenstein's intuitionistic sympathies precisely cease to be revisionary when taken in the context of his general philosophy of logic and mathematics. This suggestion would require that some assumption is made by the intuitionists, when they translate their criticisms of classical mathematics into a programme of reconstruction, which is called into question by Wittgenstein. A short path to this conclusion would be to move straight to the radical position sketched at the end of the preceding chapter. The queried assumption would then be that the validity of statements of logic and mathematics is correctly seen as liable to answer to the sorts of consideration falling within the province of the philosophical theory of meaning (in the most general sense). But there is, on the face of it, a less radical path leading to a similar position.

The argument would be that the anti-realism of the intuitionists is revisionary if and only if the concept of truth, although no longer interpreted platonistically, is still covertly viewed as fundamental to our understanding of mathematical statements. Of course the founders of intuitionism typically express themselves in terms of constructivist rather than realist imagery; and we have seen, besides, how difficult it is likely to prove to harmonise any world-picture of a realist flavour with the generalised anti-realism which best supports the intuitionists' objection to the rôle assigned by platonism to a verification-transcendent conception of mathematical truth in our understanding of pure mathematics. Nevertheless, it is likely that the intuitionists' acceptance of classical logic for effectively decidable statements was motivated by, in our terms, a belief in their investigation-independence. And if we attribute to them a belief in the investigation-independence of all demonstrable or refutable mathematical statements, then the appropriate picture to use to describe the intuitionists' thought about mathematical truth—whether or not they had any right to it—is one in which the imagery of exploration and discovery predominates. The mathematician, then, is still concerned with charting a determinate conceptual realm; he does not invent it as he goes along. It is just that the 'realm' has no properties save as may be determined by proof; *sense* is given to statements concerning the mathematical world only by reference to conditions whose satisfaction we can determine by proof. It is only of the obtaining of such conditions that we can significantly aim at discovery.

Now, Wittgenstein crucially differs from such a position precisely in his rejection of the notion of mathematical truth; proof in mathematics is exactly *not* a means whereby something is found out. The recurrent discussion in *RFM* of the contrast between proof and experiment has as its principal target to break the hold exerted over us by the idea of proof as an instrument of discovery. Proof in mathematics, whatever it does, does not lead us towards the recognition of certain super-physical facts.

And the conclusion of a proof is thus not something for which an account of the truth-conditions, whether these are interpreted platonistically or not, can be regarded as fixing the sense.

To regard both Wittgenstein and the intuitionists as endorsing a proof-conditions conception of the meaning of mathematical sentences masks, therefore, a deep-lying difference in their respective conceptions of the nature of proof. But how exactly in the present context does this difference make a difference? Suppose a classically valid proof is devised of Goldbach's conjecture, in the course of which it is assumed that either the conjecture or its negation holds good. Why will an intuitionist reject such a proof? Because, he will argue, its cogency requires our possession of a notion of truth for the conjecture in terms of which we can apprehend that either it or its negation *must* be true. But we have no such notion. If it is in terms of proof and disproof that truth and falsity in number-theory are to be understood, there is no longer any reason to suppose that either the conjecture or its negation must be true.[1] Here a proof is still conceived as a way of 'winkling out', so to speak, a predeterminately necessary truth; but it will fail in its object if it has essential recourse to principles whose validity can only be explained by reference to a notion of truth not essentially associated with the possibility of proof. The point of the proof is to tell us something about the natural numbers; if in the course of it a demand is made on the notion of 'true of all even numbers' which dissociates it from 'essentially recognisably true of all even numbers', then the proof breaks faith with the only notion of mathematical truth which we can understand; and thereby fails in its object.

The situation, the argument continues, is quite different if the proof is regarded in the manner of Wittgenstein. For now proof in mathematics is not to be thought of as a means of arriving at independently true descriptions of aspects of reality; so it is arguably no objection to the Goldbach proof that, *were* it so conceived, the idea of truth needed to validate it at some stage would have to be interpreted as a verification-transcendent one. Because there is no longer any question of a correct description of anything, no determinate mathematical reality to *be* described, pure mathematics and logic are freed from the need to ensure that only principles of inference are used whose validity can be explained by reference to an idea of truth acceptable to an anti-realist. The proof is seen now not as the passage from certain correct descriptions (intuitively acceptable postulates and previously derived theorems) to a new correct description, previously unrecognised as such; rather, as is now familiar, we are to regard the construction as

[1] Which is not the same thing as saying that we have to recognise the possibility that neither the conjecture not its negation is provable. Intuitionistically, that is an absurd 'possibility' since to recognise that it was realised would be to have a proof of the negation of the conjecture—given the intuitionists' reading of negation. The intuitionists' doubts about excluded middle are not coherently interpretable as expressing recognition that statements which are not effectively decidable may not be decidable at all—a point which Dummett's early exposition in 22 (p. 107 sq.) rather muddles.

effecting a new paradigm of description—we shall have, for example, a new criterion for correct enumeration of the prime predecessors of a particular even number, or for the existence of prime numbers in a certain range. Since the idea of the *correctness* of the new criterion is not germane, neither are the intuitionists' impositions on the proof-methods whereby, as they conceive of it, its correctness may be recognised.

None of this, a proponent of the argument will claim, prevents Wittgenstein from endorsing the intuitionists' objections to platonism. An application of excluded middle made in a proof like that just envisaged will very probably be motivated by a platonist philosphy of mathematics; in that case, Wittgenstein will oppose the *motivation* upon grounds, among others, of the unintelligibility of the idea of truth therein appealed to. But that, in Wittgenstein, is not the only complaint. Going far beyond anything involved in intuitionism, as we have been understanding the character of that position, is Wittgenstein's rejection of the whole idea of mathematics as descriptive of conceptual *facts*. And this rejection undercuts the revisionary force of rejecting a platonist concept of truth in mathematics. The motivation for classical mathematics does not, indeed, matter in the end—except as a source, and indication, of philosophical muddle. For the *practice* of classical mathematics is intelligible enough from an anti-realist point of view: proof in classical mathematics is still an effectively decidable notion, and to master the use of statements in classical mathematics is to master their conditions of classical proof. The essential point, however, is that to endorse a proof-conditions conception of the meaning of mathematical statements is not, for a Wittgensteinian, to endorse an 'hygienic' sub-species, anti-realistically purified, of explanation of meanings in terms of truth-conditions. There is thus no well-taken anti-realist objection to the rôle, for example, of excluded middle in classical proofs. This has to be understood, rather, simply as part of mathematical procedure, part of the 'motley' of techniques whereby the mathematician invents new norms of description, new rules of language.

According to this argument, then, it is the vestigial realism with which the intuitionist continues to regard pure mathematics that leads him to a revisionary standpoint. In effect, if not in preferred imagery, he continues to think of mathematical truth as constituted by how things are in the mathematical realm; and, thinking that way, it is bound to seem to him that the classical explanations of the logical constants must be refined to accommodate the non-transcendental character of the mathematical realm. Making the perfectly orthodox supposition that, in order for a disjunction to be true, it is necessary and sufficient that one of its disjuncts be true, he therefore concludes that the truth of a disjuction can only consist in the *recognisable* truth of at least one of its disjuncts; so where a guarantee of decidability is missing, we have no right to assert excluded middle. But pure mathematics, as viewed by Wittgenstein, is not an area in which correct assertibility is a matter of secured conformity with objective facts in this way. Here, correct assertibility is, as it were, a game-internal notion; we are concerned not to describe but to *pronounce*. And there is thus no reason why inferences via excluded middle should not enter into the conditions of legitimate pronouncement.

In order to make plain, then, how Wittgenstein can endorse the intuitionists' complaints about platonism while avoiding the apparent revisionary consequences, there is no need, according to this suggestion, to call into question the entire assumption that logic and mathematics are answerable to the philosophical theory of meaning. But there is a major difficulty with the suggestion. For it makes heavy play upon Wittgenstein's specific conception of logic and mathematics. It is because he regards logic and mathematics as not being fact-stating regions of discourse that Wittgenstein's anti-realism is excused revisionary consequences for them; the error of the intuitionists is exactly to carry over into these areas aspects of the idea of truth appropriate to contingent statements. But then let us take a contingent example whose situation mirrors as closely as possible that, according to our preconceptions, of Goldbach's conjecture; a statement of which, as we ordinarily think, a verification may be possible if it is true, but whose truth does not guarantee its verifiability—say 'X is not the father of Y's child'. Wittgenstein's philosophical non-revisionism is a global conception; but the anti-realist case for a revision of classical logic where this kind of example is concerned is quite unanswered by the suggestion which we have been considering. If we claim, for example, that one of such a statement and its negation must be true, we commit ourselves, from an anti-realist point of view, to the existence of an apprehensible state of affairs constituting a verification of one or the other; and no reason is apparent why there has to be such a state of affairs. The preconception—of which this reasoning issuing in a doubt about excluded middle is an illustration—remains unchallenged that the validity of a schema of logic, as applied to contingent statements at least, is something which has to issue from its essential syntactic structure and the general sematics of the logical constants. The concern of logic, applied to contingent contexts, is still to supply us with the means of making truth-preserving inferences. If truth has to be interpreted as recognisable truth, logic is constrained accordingly: valid inferences must now guarantee that if their premises are recognisably true, so are their conclusions; and in general only those schemata will be acceptable as valid of which we can recognise that they cannot but be true, even if the classical truth-tables are taken to deal in decidable truth and falsity and no assumption is made that these exhaust the possibilities.

The proposed account, then, of how Wittgenstein's anti-realism can avoid being revisionary is insufficiently general; it relies on

Wittgenstein's rejection of the statement-making character of logic and
mathematics, and does nothing to explain how an anti-realist can rest
content with classical logic where genuine, but non- effectively-
decidable, contingent statements are concerned. In fact it is clear that
so long as we continue to suppose that the logical validity of a schema
has to be, in the kind of way adumbrated, consequential upon syntactic
and general sematic features, philosophically inspired criticism of the
classical accounts of the logical constants is going to be potentially
revisionary of the classically accepted body of logical laws. The proposal pointed in effect to a reason why Wittgenstein might have rejected
this supposition for reasoning *within* logic and pure mathematics; but
what we need to understand is why he might have rejected it quite
generally. Again, the thought occurs that his conception of logic and
mathematics as 'antecedent to truth' may provide such a reason—a
thought to which we shall shortly return.

2.   The second locus of apparent conflict distinguished above was
the question whether Wittgenstein is even in a position coherently to
sympathise with the intuitionists' complaints about the informal conceptual 'foundations' of classical mathematics. Once we abandon the
idea that pure mathematics describes an objective abstract realm, what
can be wrong with the way the classical mathematician postulates
operational analogies between finite and infinite domains of quantification, or between finite and infinite sets in general?

In order to try to clarify the matter, let us revert to the analogy with a
game. Suppose the game has a certain peculiarity: the distinction
between configurations on the board which constitute a win for Italy
and those which do not is not effectively decidable. The concept of a
win for Italy is explained in such a way that there is no necessity, if Italy
achieves a winning position, that it be possible to recognise as much. It
may be possible, by ingenuity, to bring the fact to light; but it does not
have to be possible. We can suppose that for other countries the notion
of winning is effectively decidable.

On the face of it, this might seem to spoil the game. The contestants
might play past a win for Italy without realising it. Just for that reason,
however, the possibility is idle. Italy has to try to win but he will not be
counted as having done so unless he can show that he has. So what
someone learning the game has to grasp is how to *demonstrate* that
such-and-such is a win for Italy; and what he has to try to do when
playing as Italy to achieve a position in which he can *demonstrate* that he
has won. Correspondingly, the other players have to try, in the course
of trying to win themselves, to preserve a situation in which such a
demonstration cannot be given. There is thus a sense in which the
stated rules of the game, or at least the section on an Italian win, are a
sham; the residue in the idea of an Italian win, which bids us admit the

possibility that Italy may win although no one can recognise the fact, is otiose. If we want to describe the practice of the game, then what Italy is aiming at, and what the other players are trying to avoid, is a demonstrable Italian win. For Italy, the demonstration is part of winning; it would make no difference to the practice of the game if the section of the rules on winning was emended accordingly.

It is essential that this is so; we cannot intend, or try to bring about, a situation which no one will be able to recognise if we succeed; there is nothing to distinguish one such intention from any other. All that Italy can *try* to do is to achieve a situation in which it can be shown that he has won.

The purpose of the game analogy was to provide a working model of certain essential features of Wittgenstein's conventionalism about logic and mathematics. First, the rules of the game are subject to no external constraints; we invent them in the course of defining an activity—we cannot pick *wrong* rules. Secondly, like all rules, what is to count as their correct application, is in a sense, open. Thirdly, the status of further rules as consequences of the original rules is not a predetermined fact, settled as soon as the latter are chosen, but something that requires our verdict. And finally, the rules serve as norms of description for the situations that arise in the course of the game; they determine the conditions of legitimate assertability of, for example, 'I must have missed my turn'.

Now, none of these points, obviously, affects the fact that the practice of the game has to be taught; and that this is likely to be done by explaining the rules. Wittgenstein, we may allow, has undermined certain pictures which we might have of what such an explanation achieves; but he has done nothing which ought to be interpreted as aiming to show that giving the explanation is somehow vacuous, or that we could just as well give any old rules. The game has to be explained *properly:* we have to give an intelligible account of the situations which the various rules cover and what they say about them. And we have to present victory and the avoidance of defeat as intelligible objectives. If we do not, then people will not learn to play the game.

Arguably, however, the original rules concerning an Italian win are *not* fully intelligible. For nothing counts as following them rather than an emended version making it part of the concept of an Italian win that Italy can prove it. We cannot aim to follow rules save those pertaining to situations which we can recognise; and this is perfectly consistent with Wittgenstein's leading idea that our grasp of the kind of situation involved does not, as it were, leap ahead of us and classify cases at which we have not yet arrived.

The analogy suggests that an anti-realist critique of the usual semantics for classical logic and mathematics need not after all conflict with Wittgenstein's conventionalism; that the view that the existence of classical techniques as a communicable practice settles all questions about intelligibility is too crude. The point of the analogy is the now familiar comparison between the notions of winning and truth.[1] As in playing the game we are to aim at winning, so in 'playing' mathematics we are to aim at making 'true' mathematical statements. But the notion of truth appealed to in platonist semantics corresponds to the original notion of a win for Italy in the game; and there are corresponding limitations on the rôle which it can play in describing the practice of mathematics and the intentions of the players. The mathematician is in the position of Italy: he is trying to articulate 'true' mathematical statements but he will not be credited with having done so unless he can supply proofs. Italy, comparably, is trying to engineer a situation in which he can rightly claim to have won. So it is idle to describe the mathematician as aiming at mathematical truth if this is understood so as to have no essential connections with demonstrability; what he is aiming at is *proof*. The notion of truth has a legitimate part to play in the practice of the mathematical game only in so far as it is tied to situations which the players can recognise—can bring about by the construction of proof.

If this analogy is successful, then to go so far as to allow that the rules of either 'game' were even completely *arbitrary* would still be consistent with the desirability of expurgating from their formulation any use of the concepts of truth, or winning, dissociated from what the 'players' can recognise. For the rules have a double rôle. They are not merely that which would-be players must grasp; their statement also forms the essential core of a *theory* of the game—a systematic description of its correct practice such as an observer might with patience discover. When they take this latter rôle, it seems particularly evident that reference to unrecognisable objectives can play no essential part; for it will only be by reference to situations which *he* can recognise, and can reasonably suppose that the players can recognise, that the observer will be able to interpret the players' goals at all. Thus, while the practice of classical mathematics may well be intelligible enough, the standardly associated theory of that practice, proceding by reference to a verification-transcendent concept of truth for infinitary quantification, is defective. Grasp of that notion plays no part in the practice of the game; rather what has to be grasped is the notion of proof which it is spuriously invoked to explain.

The suggestion, then, is that a conventionalist view of logic and mathematics is quite consistent with endorsement of anti-realist reservations about the traditional platonist *theory* of the classical practice of those subjects. For the platonist's 'intended interpretation' of any branch of classical mathematics is, he is bound to conceive, the core of the theory appropriate to his practice of that branch: it is a conceptual structure grasp of which, in his view, would have to be attributed to him

[1] *RFM*, App. *I*. 1-6; cf. Dummett 22 (Pitcher, p. 95), and Wright 77, p. 233.

by any observer of his mathematical practices whose aim was to arrive at a correct theoretical description of those aspects of his behaviour; for it is by reference to the 'intended interpretation' that the species of *truth* has to be elucidated which, in his own view, governs the platonist's pratice of the relevant branch of mathematics. The anti-realist objection is then that no satisfactory account seems to be possible of what grasp of the distinctively platonist aspects of the intended interpretation of, say, classical number-theory, or Cantorian set-theory, consists in: that there is nothing to distinguish the performance of someone who grasps the possibly verification-transcendent truth-conditions of Goldbach's conjecture from that of someone who grasps merely the deductive liaisons and proof-conditions which the medium of classical logic generates for that statement; and that there is nothing to distinguish the performance of someone who grasps the classical notion of a non-effectively-enumerable infinite subset of the natural numbers, and the contribution that notion makes to the content of Cantor's Theorem, from that of someone who grasps merely that the classical theory of infinite cardinals is going to have it that the familiar, performable operation of arbitrarily selecting a proper subset of a finite set is to have a determinate but unperformable infinitary analogue. The platonist view is that transcendent conceptions inform the practice of classical mathematics; that they are essential to a proper grasp of the truth-conditions of statements in all non-trivial branches of classical mathematics, and that the rôle of pure mathematics is, indeed, their exploration. A conventionalist will reject the idea that mathematics *explores* anything; but he is not thereby prevented from mounting an anti-realist challenge against the conceptions which the platonist thinks it explores. For the essence of the platonist position is that classical mathematics deploys axioms and deductive machinery which are *appropriate* to the transcendent conceptions—that we recognise the validity of excluded middle for Goldbach's conjecture as a *consequence* of our conception of its truth-conditions. He has therefore to face the demand for an explanation of how a grasp of such conceptions can be displayed *except* in the mathematical practices which they are supposed to validate; an equivalent demand would be for a satisfactory introductory explanation for them, since such an explanation could hardly fail to make it clear what someone would have to be able to do in order to show that he had understood it—while, conversely, to have an account of what someone who understands a particular concept is distinctively able to do is to have an account of the information which a satisfactory explanation of it must get across. It seems to me a coherent view to hold both that this demand cannot be met in the way the platonist needs, and that his more or less fumbling attempts to meet it—the desperate recourse to the idea of infinitely doubling ones pace, etc.—serve well enough—like the original rules about an Italian win—to introduce the 'game' he is proposing to play.

3.   Can we, now, take a further step and argue that Wittgenstein's conventionalism would actually be consistent with endorsement of something akin to the intuitionistic reconstruction of classical mathematics? It depends on what we understand by 'expurgating' all play upon possibly transcendent objectives. What would this come to in the case of the board game? It might be sufficient to introduce refer-ences to 'demonstrability' in the section on an Italian win, so that no changes in the actual practice of the game are going to be enjoined by the change in the rules. In that case we shall merely be concerned to give a purer description of how the game is actually carried on, such as an observer might arrive at; for all that he can observe is that Italy is never counted as having won unless his position can be demonstrated to have a certain feature. But the nub of the matter is the question whether, if the rules are being refined in this way, all the formerly admissible means available to Italy in an attempt to demonstrate that he has won should continue to be so. In particular, should he be allowed to assume that any particular board configuration either constitutes a win for him or does not?

It may be that this was permitted and will continue to be so. Still it is *arguably* a bad rule because the conception of a win needed to validate it is not the same conception as that in terms of which the players' objectives are now to be explained. For Italy, a win is now essentially a win which he can demonstrate. This alone might seem to leave open the supposition that winning is understood in such a way that the disjunc-tion is valid: any particular position is either such that an Italian win can be demonstrated for it, or it is not. But here we are in danger of overlooking that it is an intention of the other players to *prevent* an Italian win; that is, to attempt to ensure that no position is arrived at in which an Italian win can be demonstrated. If only such states of affairs are to be appealed to in the rules whose realisation, or avoidance, the players can hold as intelligible objectives—that is, could recognise, should they obtain—then a configuration which is not an Italian win ought to be interpreted as one of which it can be recognised that no demonstration of an Italian win in it is possible. As a practical strategy, the other countries might have to settle for less, concentrating on preserving a situation in which it could at least be reasonably believed that no such demonstration was possible. But there seems no reason why the game should have to be such that in any particular situation Italy could either demonstrate a win, or it could be demonstrated—or at least reasonably believed—that he could not. Obviously, it depends on the exact character of the example; and we have not troubled to give it a sufficiently exact character. But, at least, there *could* be a game with this characteristic, for which the disjunction in question would thus be debatable.

In the context of a mere game it would be strained to conclude that it was *wrong* to allow the players use of such a disjunction in their reasoning in the course of the play. But there would at least be no explaining why it was right. The status of the disjunction would be purely formal; it would just be a piece of machinery which the players could use if it seemed expedient. There would be nothing in the concept of an Italian win as displayed in the actual conduct of the game, nothing which an observer could detect, which would explain *why* such a rule was accepted—why, indeed, reasoning involving essential appeal to it should be thought of as a *demonstration*.

At any rate, it seems that a dispute about the appropriateness of using such a disjunction in the course of the game is not preempted by its being a conventional practice in the various respects detailed. There is thus a tentative case for saying the same about mathematics and logic. To accept Wittgenstein's conventionalism, and associated rejection of the concept of mathematical truth, would not preclude sympathy with an anti-realist critique of classical methods. On such a view, mathematical 'truth' is a fiction; there are no objective liaisons between concepts which it is the task of mathematics to describe. But, just for that reason, it is only in the character of actual mathematical practice that an account of the constraints on correct assertability in mathematics can be sought; and the simple fact is that we are trained to assert mathematical statements as the conclusions of proofs. There is thus arguably a kind of hypocrisy if among admissible proof techniques are to be found principles of inference whose validity we propose to explain by a semantics involving a proof-transcendent conception of correct assertion.

Lest there be any misunderstanding, my suggestion is merely, tentatively, that a Wittgensteinian conventionalist who agreed with the intuitionists' criticisms of the platonist interpretation of classical mathematics could also intelligibly sympathise with the intuitionistic revisions; there is no claim at this point that he would be constrained, or that he ought, to do so. I shall return to this suggestion in the next chapter. However there are three immediate consequences if it is correct. First, we have to qualify the diagnosis mooted in §1 of why the intuitionists are led to revision: it may still be that a vestigial realism makes it natural to allow their doubts about platonism's transcendent conceptions to issue in a new style of pure mathematics, but such realism cannot be required or implied by a readiness to do so. Secondly, the particular 'apparent disharmony of main themes' found by Hacker and Baker in Wittgenstein's later philosophy of mathematics is only apparent; Wittgenstein's 'constructivism', in the sense glossed at the start of the chapter, has already been argued to be consistent with his conventionalism, and, if the present suggestion is correct, would have *remained* so even had he allowed it revisionary consequences. Thirdly, the situation suggests that the conception of logic and mathematics as 'antecedent to truth' cannot *require* a non-revisionary philosophy of logic and mathematics; or, if it does so, the explanation will have to be that the game analogy fails to capture some crucial element of Wittgenstein's conventionalism which the 'antecedence' doctrine incorporates.

4.　If Wittgenstein could consistently have endorsed not merely the motivation for the intuitionists' revisionary programme but the programme itself, nevertheless, as we know, he refrained from doing so. Rather he embraces an explicit non-revisionism. He writes (IV. 52):

> The philosopher must twist and turn about so as to pass the mathematical problems by, and not run up against one which would have to be solved before he could go further. His labour in philosophy is, as it were, an idleness in mathematics. It is not that a *new* building has to be erected, or a new bridge built, but that the geography *as it now is* has to be judged.

Compare *Investigations* I. 124:

> Philosophy may in no way interfere with the actual use of language; it can in the end only describe it. It cannot give it any foundation either; it leaves everything as it is. It also leaves mathematics as it is, and no mathematical discovery can advance it.

In his classes on the philosophy of mathematics in the spring of 1939, Wittgenstein emphasised at the outset that one must at all costs avoid 'interfering with the mathematicians'. And in the *Grammar* (II. 39, p. 458) he rebukes the intuitionists for their revisionism:

> The whole approach that if a proposition is valid for one region of mathematics it need not necessarily be valid for a second region as well, is quite out of place in mathematics, completely contrary to its essence. Although many authors hold just this approach to be particularly subtle and to combat prejudice.

How are we to interpret this attitude? Consider again the argument for withholding from Italy the right to assume, in the course of the reasoning involved in the game, that in any particular configuration he has either won or not. The argument was that there is no reason to accept such a disjunction as valid if the concept of winning is understood in the way relevant to the objectives of the players; when winning is so understood, a situation may be legitimately described as an Italian win only if it can be recognised as such, and as *not* an Italian win only if it can be reasonably claimed that an Italian win cannot be demonstrated. Now, this calls the disjunction into question only if, as we are supposing, the game is such that there is no apparent necessity that all board configurations be of one of these two kinds. But the argument also assumes that legitimate assertibility is to be conceived as distributive over disjunction so that in order to be entitled to assert the disjunction generally, we have to be able to guarantee the legitimate assertability of one of the disjuncts individually in any possible board configuration.

This is the assumption on which, in the previous chapter, the revisionary threat of the rule-following considerations at first seemed to depend. But, as noted, a deeper assumption is involved, of which the distributivity of truth, or legitimate assertability, over disjunction is merely a special application. The assumption is that the truth, or legitimate assertability, of any compound statement is a function of the truth-, or assertability-conditions, of its constituents; more, that when the manner of compounding involves only the vocabulary traditionally conceived as 'logical', the truth, or legitimate assertability, of the compound will be a function of the actual truth-, or assertability- *status* of the constituents. The validity of such a compound is thus always to be conceived as a *consequence* of this fact: that, however the world happens to be, the status of the constituents in point of truth, or legitimate assertability, is bound to be such that, in view of the specific truth, or assertability, function expressed by the main connective, the compound statement in question is itself true, or legitimately assertable. A paradigm of such explanation is, of course, a truth-table exhibition of a classical two-valued tautology. In general, the assumption is that the acceptability of any stated principle of inference depends upon its receiving a sanction in terms of the general semantics of the language in which it is stated. I cannot see but that we must interpret Wittgenstein as calling this assumption into question.

Imagine a piece of reasoning within a physical theory which at some stage assumes of a particular unrestrictedly general hypothesis that it must either hold good or not. There is no notion of verification for such a hypothesis; so any anti-realist is bound to construe its meaning in terms of its conditions of legitimate assertion. Now, if legitimate assertability is taken to be disjunction-distributive, the legitimate assertion of the disjunction requires that one be in a position to claim the existence of standard scientific grounds for the assertion of one of its disjuncts. As a consequence, there is no reason why such a disjunction *must*, ultimately, be legitimately assertable; it becomes itself a hypothesis, for which we may have no ground whatever. But our ordinary conception of such reasoning is that no such *hypothesis* enters into it. So it seems we have a choice: to abandon the ordinary conception, or to provide some other account of the semantics of the disjunction in terms of which we can explain its validity. Wittgenstein, if I am right, implicitly proposes that we should adopt neither course. There is no reason to attempt to *justify* our conception of what is hypothetical in such reasoning and what is not, nor therefore to change it if justification is not forthcoming. If reasoning of such a pattern forms part of a productive, successful practice, that is justification enough. And if we conceive of it as valid, rather than hypothetical, at the relevant stage, then that is how it is—an observer attempting to characterise our notion of validity would have to record as much.

Wittgenstein's later philosophy of language generalises the anti-realism of the intuitionists. And in at least one respect his philosophy of mathematics involves an important advance over the intuitionists' historical position: the case for repudiating a platonist conception of mathematical truth is no longer muddled with a psychologistic conception of mathematical understanding. The cardinal difference so far stressed is that Wittgenstein's anti-realism embraces the rule-following considerations and the radical reappraisal of lay-philosophical preconceptions which they occasion. But the present contrast is equally fundamental: Wittgenstein and the intuitionists embrace radically opposing conceptions of the character of the interplay between philosophy and linguistic procedure. For a generalised intuitionism, philosophy would practise an unceasing critical examination of our linguistic habits, the concepts which we employ and the inferential transitions which we allow. But for Wittgenstein, the rôle of the philosopher can only be that of apologist. Philosophy is to explain and justify where explanation and justification are possible; but where they are not, its work is finished when it has recorded as much. That is not to say that philosophy cannot lead to changes in linguistic or, more specifically, mathematical habits. A particular kind of mathematics may have a deep-seeming but spurious motivation; when philosophy has laid bare the poverty of this motivation, it is natural to expect that people will lose interest in that sort of mathematics. That philosophy may well prove revisionary in this indirect way is anticipated in several places in Wittgenstein's later writings. But, in contrast with the intuitionists, Wittgenstein will not have it that philosophy can discover that mathematicians are going about their business wrong; it can only discover that they are talking nonsense *about* it.

It is difficult to see how to decide which side has the truth of the matter; it might almost seem to be a question of taste. But the dispute is not without practical implications. As I have been suggesting, one fundamental question on which it directly bears is the susceptibility of accepted logical principles to vindication, or revision, in terms of general philosophical semantics. And this in turn leads to a question concerning the status which object-language principles of inference ought to receive in a fully articulated theory of meaning for that language. It is on the latter question that the remainder of this chapter will concentrate.

5.   The goal of constructing such a theory, as it ought to be conceived, is that of giving a perspicuous representation of the nature of mastery of the language in question. Such a representation is naturally taken to involve providing an account of how recognition of the sense of any of the infinitely many significant sentences in his language which a speaker has not previously encountered is possible. Since such a recog-

nitional ability is learned,[1] it seems it ought to be possible finitely to specify the essential syntax and semantics of the language in such a way that a speaker's understanding of a new sentence can be seen as derived from a recognition of its essential structure discerned by the syntax, and knowledge of the various semantic rôles assigned to its constituents within the framework of the resulting various syntactic types.

The modern advocacy of the interest of this conception, of course, is largely due to Davidson. For Davidson, the desired theory takes the form of a syntactic analysis and associated, recursively yielded specification of the truth-conditions of an arbitrary well-formed declarative sentence of the language; we need a finitely axiomatisable theory which yields a syntactic analysis of the sentence in such a way as to issue in an explanation of its truth-conditions on the basis of an account of the semantic rôles of its essential syntactic constituents. On Davidson's conception of the matter, as is familiar, the game is won when a theory is constructed which yields *in this way* the T-theorem corresponding to an arbitrary object-language sentence, P: 'P' is true $\leftrightarrow P$.

We are not at present directly concerned with whether such a theory would indeed be in any interesting sense a theory of meaning for the language in question; or whether a correct such theory would indeed codify the character of a speaker's understanding of (the declarative part of?) his language; or whether knowledge of the language can be conceived as, in some sense, *implicit* knowledge of such a theory. Our concern is rather with the bearing of Wittgenstein's ideas upon Davidson's claim that such a theory will be a theory of *logical form*, which will consequently discern formally valid patterns of inference in the object-language and thereby be explanatory, and so potentially revisionary, of object-language inferential practices. In Wittgenstein's view legitimate philosophy can have no such bearing on actual linguistic practice. Does it follow that, for a Wittgensteinian, Davidson's project is wholly out of order?

It seems inescapable that the appropriate form of explanation for such a theory to give of the meaning of compound sentences will be an account of their truth-conditions in terms of those of their sentential constituents which their syntactic analysis bids us regard as playing a sense-determining role. Of course, it is enormously difficult to see how to do this with, for example, attributions of belief. The case most immediately receptive to this sort of explanation is that of sentences arrived at by straightforward truth-functional combination of sentential constituents. This much of the theory is easily adapted from the traditional semantics for classical propositional logic.

Suppose we have such a theory, stated in a given language, for a finite

---

[1] Or so one would suppose. Chomskians would prefer to regard it as environmentally 'triggered'—like puberty; it is unclear whether this view would be supposed by them to undercut the need for finite axiomatisability.

number of atomic statements of the same language. Consider the task of enlarging the theory so as to have it cope with any compound statement formed out of these statements by (reiterated) application of any of a finite number of unary and binary truth-functional connectives. It would hardly occur to one to do anything but formulate the theory recursively, so that, to take a simple example, the derivation of the truth-conditions of 'P or Q' would proceed:

'P' is true$\leftrightarrow P$.   (Theorem.)

'Q' is true$\leftrightarrow Q$.   (Theorem.)

'P or Q' is true$\leftrightarrow$'P' is true or

  'Q' is true   (Recursion for 'or'.)

'P or Q' is true$\leftrightarrow P$ or Q $\left\{ \text{Meta-rule}: \dfrac{A \leftrightarrow B,\ C \leftrightarrow D}{A \text{ or } C \leftrightarrow B \text{ or } D} \right\}$

The derivation for 'P or it is not the case that P' would be exactly the same save that the second clause would be replaced by:

'It is not the case that P' is true$\leftrightarrow$it is not the case that P (Theorem),

and the rest of the derivation would be substituted through accordingly.

In this approach it is presupposed unthinkingly that the theory of meaning is to apply alike to all sentences of the relevant type of syntactic structure, irrespective of their modal status. The theory is, therefore, at least to give an account of the truth-conditions of necessary statements in this category, whether or not it explains in addition how it is that those truth-conditions necessarily obtain. In fact, though, it is apt to seem desirable that it should do both somehow, and that the latter explanation should build on the former. For example, if we are satisfied about the validity of all sentences in the relevant fragment of the object-language of the form, 'A or it is not the case that A', the theory should provide means for proving in any particular case the theorem:

'P or it is not the case that P' is true,

purely on the basis of the analysis it provides of the syntactic structure of that type of sentence and the semantic rôles of its constituents. This seems desirable because we tend to conceive of the validity of sentences of that type, if they are indeed valid, as recognised purely in virtue of the *very understanding* (of the fragment) which the theory seeks to describe.

An anti-realist, of course, will reject the presupposition of such a theory of meaning, that knowledge of the sense of a declarative sentence is always to be equated with knowledge of its truth-conditions. What, if anything, we should be trying to explain is knowledge of the appropriate *use* of a new declarative sentence; and this knowledge cannot be thought of as knowledge of truth-conditions in the multitude of types of case for which verification is not a possibility. The connection between truth and correct use can only be that one is to aim at making true statements; if there is no way of recognising the truth of a particular statement, then it is quite unclear how to interpret knowledge of its truth-conditions—whatever such knowledge might now be held to consist in—in such a way as to make it clear how this knowledge would entail knowing under what practical circumstances the assertion of the statement would be considered justified. But it is the latter knowledge which is criterial of understanding.

To sustain this objection, however, does not immediately require that we have to abandon the whole idea of a theory of meaning of this type. The immediate requirement is only that the concept of legitimate assertability be taken as fundamental from the outset. If we had a complete theory of the former type for a particular language, no reason is apparent why its anti-realist replacement could not retain the syntax of that theory; but the semantic rôle of the essential syntactic constitutents in a sentence would now have to be analysed as a contribution towards determining the conditions of legitimate assertion of the sentence, rather than its truth-conditions. In particular, therefore, the paradigm of explanation of the sense of a compound sentence would become an explanation of its assertability-conditions in terms of those of its sentential constituents. Such are the explanations given by the intuitionists of the logical constants. Once again, we should naturally require the resulting theory to apply to *all* object-language declarative sentences, irrespective of their modal status. An explanation would be requisite at least of the conditions of legitimate assertability of necessary statements; without it there would be no prospect of fulfilling the further desideratum of an explanation of how it is that these conditions obtain in any possible informational situation.

The suggested reason for holding the latter sort of explanation desirable was that the capacity to recognise necessity is naturally conceived as bestowed by language-mastery; and the thought is not merely that this capacity can only be displayed by a language master—it is that it is the *very same ability* as that exercised whenever the sense of a new declarative sentence is recognised, that recognition of necessity wholly proceeds via recognition of structure and semantic rôle. To be sure, it is evident in any case that a satisfactory theory of meaning cannot just keep silent about inferential practices in the object-language community: for speaker's conceptions of the truth- or assertability-conditions of their statements will be conditioned through and through by the statements' inferential connections. So the theory must incorporate some sort of description of the inferential practices of object-language speakers. This, however, does not yet commit us to the idea that the theory should in the relevant sense explain those practices; it is when it is supposed that the validity of object-language principles of inference, and logical laws in general, is consequential upon the general semantics

of the language that it comes to seem that a satisfactory theory ought to
so represent validity, ought to articulate *how* the validity of a principle
flows from the clauses of the theory germane to its sentential expression.

It is background thinking along these lines, I believe, which encourages the hope that a theory of meaning may furnish *deep* explanations of
necessity and validity, systematic articulations of how they 'flow' from
meaning. But this hope is surely badly confused. How, for example,
unless we have '*A* or it is the case that *A*' as a theorem in the metalinguistic proof-theory, could we provide a derivation, for the sort of fragment
which we considered, of: '*P* or it is not the case that *P*' is true? In order
to 'explain' in this way the validity of the most basic object-language
logical schemata, we shall need recourse in general to a metalinguistic
analogue of the very schema concerned. Of course, there is no objection
to 'explanations' of this trivial sort, if our goal is simply to characterise
object-language practice; but the target was deep explanations of validity, tracing it back to its source in meaning and structure. Not but what
deep explanations could not have been forthcoming in *every* case anyway. The theory will have to assume a determinate proof-theory if it is
to be capable of application at all; and the explanations of the standing
of object-language analogues of theses in this proof-theory are thus
bound to have this kind of triviality. What is unclear is whether, in
terms of the present conception of a theory of meaning, non-trivial
explanations of our intuitive idea of the 'flow' of necessity from meaning
could be provided in *any* case.

I do not raise this point in order to cast doubt upon the interest of a
systematic theory of meaning of this general sort, but to bring out an
unclarity in the idea that such a theory *could* offend against Wittgensteinian non-revisionism. For, as noted above, the capacity of a theory of
meaning to prove revisionary of logical, or mathematical, practice is one
with its capacity to explain them. When we are devising a theory for our
own language, the general manner in which a revision might issue is if
our impression of the validity of a particular schema should receive
no support from a correct account of the semantics of a sentential
expression of it. But if every such schema receives this support just
in case an analogue of it features in the metalinguistic proof-theory,
and if none does if no such analogue features, then the notion of 'support' seems to have been completely trivialised. If a theory of meaning
cannot explain inferential practices in the studied language in a more
substantial sense than this, how can it possibly suggest that they are
wrong?

The answer is that questions about the rights and wrongs of inferential practices in the object-language will resurge in the form of issues
concerning the desirable content for the proof-theoretic part of our
theory. It is here that we can expect the sorts of informal consideration
with which we have been repeatedly occupied—considerations about

the nature of understanding, language-learning and the permissible rôle of a transcendent interpretation of the central notion in the theory—to come into play. Suppose, for example, that a community for whose language we are constructing a formal theory of assertability, accepts inferences via excluded middle for sentences which, we are satisfied, are correctly translated into certain non- effectively-decidable statements in our language. If we have persuaded ourselves that excluded middle has no legitimate place in a logic for such statements, we shall be reluctant that our theory should supply a theorem stating of the sentence which expresses excluded middle in the object-language that it is unrestrictedly legitimately assertable. We shall want, however, to be able to prove the analogue of a T-theorem for every sentence, P, in the object-language; that is, presumably, a theorem prescribing coincidence in *assertability-conditions* between each such P and the meta-linguistic A which we are confident correctly translates it. But the effect of such a theorem will be to associate with P conditions of justified assertion which do not everywhere coincide with those assigned to it by speakers of the object-language. The discrepant case will be precisely when the community asserts P on the basis of an inference via excluded middle; only if 'A or it is not the case that A' figured in our proof-theory could the development of an apparent counter-example to the relevant theorem be prevented. And, given the usual machinery, we shall have that principle in our proof-theory if and only if we can demonstrate that the sentence which expresses excluded middle in the object-language is unrestrictedly legitimately assertable. Our theory, then, will be implicitly revisionary in just this sense: we shall discount as malpractice the object-language community's persistent acceptance that sentences are legitimately assertable in a particular type of situation, rather than regard it as evidence against our meaning-delivering theorems.

If—and it is a very substantial 'if'—Wittgenstein would not have thought the idea of a systematic theory of meaning for a language to be, as a philosophical enterprise, totally misconceived, it is clear that all his sympathy would have been with the second, anti-realist type of project. But he would have found the suggestion quite unacceptable that the devisors of such a theory might wilfully but justifiably accept a systematically incorrect description of actual practice in the object-language community—as though, to borrow a well-worn example, it was up to the planets to conform to Galileo's laws. After all, we are supposed to be seeking an *empirical* theory of the use of the object-language. 'But what if the object-language community are using an *incorrect* logic?' Wittgenstein will not allow any sense to the idea. The logic used by the community enters into the determination of the conditions of legitimate assertion of the declarative sentences in the language; a theory which is to describe the use of the language *must*, therefore, provide means for incorporating a theorem to the effect that excluded middle, or any other schema which object-language speakers regard as unconditionally legitimately assertable, is so.

But how is this to be achieved? If we simply add an axiom-schema to the effect, say, that '*P* o no es verdad que *P*' is unrestrictedly legitimately assertable, then, where *A* is the metalinguistic statement associated by the theory with a particular object-language *P*, it seems there would be no way of blocking the inference to the meta-theorem:

(i) *A* or it is not the case that *A*;

for we shall presumably have a theorem to the effect:

(ii) '*P*, on no es verdad que *P*' is legitimately assertable ↔ *A* or it is not the case that *A*.

But we, as anti-realists, may very possibly accept no such theorem as (i). How, then, in the attempt systematically to describe object-language practice, can we avoid spoiling *our* proof-theory? The only apparent strategy is to avoid *having* any theorem to the effect of (ii). And this implies that the original unthinking assumption that the clauses of the theory are to apply alike to all sentences, irrespective of the modal status which they are granted in the object-language, must now be abandoned. There would have to be no *T*-theorems, or assertability-theoretic analogues of *T*-theorems, for instances of schemata accepted by the object-language community as valid; sentential instantiations of such schemata are to be exempted from analysis in terms of the recursions of the theory.

Only if our theory embraces a full and accurate description of principles of inference accepted as valid by object-language speakers can it furnish a satisfactory account of their linguistic practice. The question is therefore: what form should such a 'full and accurate' description take? What we have seen is that the simplest course: that of registering principles which the object-language community accepts as valid by the adoption of axiom schemata to the effect that any instance of such-and-such a sentence type is true, or unrestrictedly legitimately assertable, is always in potential tension with the desideratum that the theory be embedded in a satisfactory logic. (The point applies equally to the case where object- and metalanguage are the same.) The only policy guaranteed to avoid trouble is to exempt object-language principles of inference from the scope of the base- and recursive clauses of the theory. And this policy amounts to a total surrender of the goal of deep explanations of the validity of object-language logical laws. If such is the policy, then, whatever philosophical fruit a recursive theory of meaning may be hoped to bear, insights of that sort will not be among them.

It is natural to feel that the 'simplest course' is not the appropriate way to register object-language principles of inference; that we ought merely to record that principles are accepted, rather than dignify them as true, or unrestrictedly legitimately assertable. But the trouble is, of course, that there is no distinction between accepted principles and principles which are truly, or unrestrictedly legitimately assertable unless object-language speakers' acceptance of a principle can rightly be regarded as *mistaken*. And Wittgenstein's non-revisionism prohibits any such assessment. It therefore appears that a theory of meaning of, in the broadest sense, a Davidsonian kind could be reconciled with Wittgenstein's views only by foreswearing all its ambition to cast philosophical light on the character of valid argument in the object-language; though, for the reason outlined earlier, the propriety of that ambition needs further supporting argument in any case.

A theory the scope of whose base- and recursive clauses was qualified in this way would still be liable to misdescribe object-language practice if object-language principles of inference did not exactly coincide with those of speakers of the metalanguage; a difference between the use of any particular object-language sentence, P, and that of the metalinguistic A which the theory uses on the right-hand side of the meaning-delivering theorem for P would always be liable to break through. But provided the theorists accepted the divergence as a limitation on the accuracy of the theory, rather than as something which it was proper to disregard, there would be no implicit revisionism.

In Chapter XV we shall consider more generally what bearing Wittgenstein's later philosophy of language, especially the rule-following considerations, has on Davidson's philosophically somewhat inchoate proposal. But first we must consider the motivation and possible justification, of Wittgenstein's non-revisionism, and especially its connection, if any, with the conception of logic and mathematics as 'antecedent to truth'.

XIV

Non-revisionism and 'Antecedence to Truth'

Sources

RFM: v. 45
PI : I. 496–7
PG : I. 68; 133–4
LFM: lectures xxiv and xxv

'Rules of inference are involved in the determination of meaning':
RFM: I. 10–16; v. 23
BGM: App. *I*, 11–16
PI : footnote p. 147
PG : I. 14–18; 24
LFM: lectures viii. pp. 80–2; xviii, pp. 179–81; xx, pp. 190–4

Non-revisionism:
PI : I. 133
PG : II. 29 (concluding remark); 40 (concluding para)
BlB : pp. 18, 28
LFM: lectures v, pp. 49–50; xxiii, p. 223

The *Investigations* on the nature of philosophy:
    I. 38, 90, 109, 118–33, 254–5, 464, 599

1.   What connection is there between Wittgenstein's conception of logic and mathematics as 'antecedent to truth' and his non-revisionism concerning those disciplines? Each of these aspects of his thought implies that logic and mathematics do not have to meet a certain constraint which it is our intuitive inclination to impose on them; the constraints, respectively, of being faithful to contingent fact and of admitting validation in terms of a philosophically sound underlying semantics. So we are asking in particular: does the idea that the first of these constraints is empty absolve logic and mathematics from meeting the second as well?

2.   Wittgenstein's antipathy towards philosophically inspired revision of logic and mathematics is explicit throughout the writings of his middle and later periods. In conversation, I have heard the suggestion that this is mere disingenuousness on Wittgenstein's part—a kind of philosopher's mock modesty. There is no finally defeating such an interpretation, of course, but it seems to me highly unattractive. Rather, it appears to have been a fundamental element in Wittgenstein's conception of the necessity of logical and mathematical statements that their acceptability in any particular case cannot prove to be an illusion based on philosophical error. This needs interpretation, not dismissal.

In fact, as we have noted before, the view is not specifically confined to logic and mathematics. *Investigations* 124 is plausibly read as an expression of non-revisionism about the totality of our linguistic practices. That philosophers have no business 'interfering with the mathematicians' is, it seems we must suppose, to be viewed within the scheme of Wittgenstein's later philosophy as a consequence of the thesis that philosophers have no business interfering with *any* aspect of our linguistic practice.

It is unclear why exactly Wittgenstein took this general line. It seems plain that he conceived it as a consequence not merely of his particular way of doing philosophy but of the very nature of the subject, when properly viewed. The conception is illustrated by passages like these: *Investigations* 126:

> Philosophy simply puts everything before us, and neither explains nor deduces anything. Since everything lies open to view, there is nothing to explain . . .

127:

> The work of the philosopher consists in assembling reminders for a particular purpose. If one tried to advance *theses* in philosophy, it would never be possible to debate them, because everyone would agree to them.

In general, the picture of the nature and place of philosophy which comes over in the *Investigations* is somewhat as follows. We are to think of language use in general as divided into two broad levels: the first consists of functional 'language-games' where language is used as a tool for practical purposes—where it is, as it were, in gear and drives us along the road. But when we go in for linguistic activity of the second level, we unwittingly throw the gear lever, and language becomes like an engine idling (132); it is then that we say the things which lead to philosophical puzzlement, giving way, for example, to the temptation to generalise about different regions of discourse and falling victim thereby to surface grammatical similarities between them. We have an urge (109) to misunderstand the workings of language-games of the first kind; we are easily seduced into talking a great deal of nonsense about

them; our thinking about them tends to be coloured by all sorts of
superstitions (49), for example, that to every significant symbol
corresponds something signified. (In Wittgenstein's view, platonism
about mathematics primarily originates in an application of that 'super-
stition'.)

What is philosophy to do about this? 122:

> A main source of our failure to understand is that we do not command *a
> clear view* of the use of our words.—Our grammar is lacking in this sort of
> perspicuity.

The task of the philosopher is to try to introduce such perspicuity, to
give us that clear view. It is with this object that he will 'assemble
reminders' and call our attention to well-known facts. If this is done
well—if the right facts are arranged in the right way—the temptation to
misinterpret so much of what we say will simply disappear; we shall
discover that we were talking 'plain nonsense' (119).

This is meant only as the baldest sketch of the conception of
philosophical error and its correction which emerges in the *Investigations*. My point is simply that it would only be via a thorough examination of this conception that Wittgenstein's general non-revisionism
could be understood.

To carry through a satisfactory such examination would be no easy
matter. For it is difficult to reconcile Wittgenstein's pronouncements
about the kind of thing which he thinks he ought to be doing with what
he actually seems to do. Not that his actual treatment of the particular
issues seems flatly inconsistent with his general methodological ideas.
Rather, we can put the would-be interpreter's difficulty like this: it is
doubtful how anyone who read only a bowdlerised edition of the
Investigations, from which all reference to philosophical method and the
nature and place of philosophy had been removed, would be able to
arrive at the conclusion that the author viewed those matters in just
the way in which Wittgenstein professes to do. At the time I write this,
the complaint is justified that the great volume of commentary on the
Investigations has so far done very little to clarify either how we should
interpret the general remarks on philosophy so as to have our understanding enhanced of Wittgenstein's treatment of specific questions, or
conversely. (What are the 'well-known facts' arranged in the course of
the Private Language discussion?) Wittgenstein's later views on philosophy constitute one of the so far least well understood aspects of his
thought.

No doubt part of the explanation of this omission has to do with the
discouraging difficulties which his conception, at any rate as sketched
above, seems to involve. How exactly is the intended contrast to be
made out between practical, unexceptionable language use and its
therapy-needing, degenerate counterpart? Little direct assistance is
provided by Wittgenstein himself save by metaphor—engines idling,

language going on holiday, etc.—and example; the contrast, for example, between propositions of pure mathematics proper and (254)

> what a mathematician is inclined to say about the objectivity and reality of mathematical facts.[1]

What we need, however, is a sharper *theoretical* account of the distinction. Only thereby will we be enabled to assess Wittgenstein's view of the rôle of grammatical analogy in motivating philosophical puzzlement. And in the absence of such an account, it seems no more than article of faith that where philosophical puzzlement does indeed prove to be generated in that way, a sufficiently careful rescrutiny of the 'facts' will reveal the analogies in question to be unfounded.

Theoretical clarification of the distinction is also necessary if we are to understand the scope of the class of propositions which philosophy is to leave alone. It is clear that when it is claimed that language is more or less all right as it is,[2] it is only the first type of discourse which is being considered. Part of the philosopher's task will be precisely to expose as nonsense much of what happens in discourse of the second type. IV. 6:

> Someone makes an addition to mathematics, gives new definitions and discovers new theorems—and in a *certain* respect he can be said not to know what he is doing.—He has a vague imagination of having *discovered* something like a space (at which point he thinks of a room), of having opened up a kingdom, and when asked about it he would talk a great deal of nonsense.

This is the kind of thing that needs therapy; and the test of the cure will presumably be that the patient no longer wants to say this kind of thing. Wittgenstein's philosophy, like any other, will be revisionary of any linguistic habits which it exposes as expressive of philosophical error, or confusion.

It seems probable to me that Wittgenstein had no clear general account of his distinction between language in harness and language on holiday. Consider the example of transfinite set theory. In point of *practicality* there is all the difference in the world between pure mathematical statements of this sort and the laws of inference codified in the multiplication tables which we teach our children. What are we to say about a branch of pure mathematics for which, under its intended interpretation, there is *no* natural application—no application which gives work to just the aspects which seem to give it its mathematical interest: the hierarchy of successively more 'massive' infinite sets, and so on. Is this not an example of *mathematics* going on holiday? Wittgenstein does not want to say so. He wants to say (IV. 16) that there is a 'solid

[1] Cf. *RFM* IV. 5–7.
[2] *PI* I. 98. Cf. *PB*, 158, and *BlB*, p. 28.

core' to it all, though it does not emerge at all clearly what he thinks this is. He wants to say that it is our interpretation which makes it all seem exciting and mysterious. But in this kind of mathematics one feels that interpretation and practice are so closely interwoven that if the interpretation is rejected, nothing may be left.[1] To reject the platonist interpretation of Cantorian arithmetic leaves one at a loss for a sense of its propositions' meaning—though no doubt some sort of constructive interpretation could be extemporised. A rejection of the platonist interpretation of the multiplication table, on the other hand, leaves one's intuitive understanding of its propositions—how to verify and apply them—untouched.

Wittgenstein's faith in the existence of a 'solid core' within classical mathematics of however exotic a kind is presumably a product of his reluctance to place any proposition from within mathematics, or science, on the second level of discourse, thereby ranking it with the sort of nonsense that mathematicians, physicists and cosmologists are inclined to talk *about* their subjects. However, one suspects that as soon as we had a reasonably sharp theoretical account of the distinction which Wittgenstein wants, it would be clear that much of what went on on each side of it would have to be viewed by way of reaction to what happened on the other—that interpretation habitually fed back into practice, which in turn changed in ways which prompted further development of the interpretation, and so on. One source of examples for the point might be language concerning one's own and others' mental states; in particular, if a dualist conception of the mind originates in certain disanalogies in grammatical behaviour of the first- and other-person pronouns,[2] it seems certain that the effect of that conception in turn is to affect profoundly the sort of thing which, concerning the mental life of others and oneself, it is thought to make sense to say. If this general suspicion is correct, it is likely that the effect of the grammatical *übersicht*, which the official Wittgensteinian objective is to construct, will be to suggest revision not merely in our 'interpretational' discourse—however it is to be distinguished—but, by bringing into review aspects of their motivation and supposed point, in elements of our 'working' language as well.

I do not propose to pursue this extremely difficult general question. I simply want to stress that in order fully to understand Wittgenstein's non-revisionism about logic and mathematics, we need to understand his whole conception of philosophy; in particular, we need to grasp the point of his repeated emphasis that the philosopher merely describes and arranges the familiar and uncontroversial. And in order to do this, we need to be able more clearly to see Wittgenstein himself doing just that in the *Investigations* and *RFM*.

For this reason I do not consider that an affirmative answer to our

[1] Wittgenstein is sensitive to the point. The remark about inducing people to leave 'Cantor's Paradise' 'of their own accord'—*LFM* lecture xi, p. 103—rather gives him away; and *RFM* iv. 7 is an explicit acknowledgment.
[2] See *BlB*, pp. 66–9; also *PB*, 57–8.

question, whether Wittgenstein in effect committed himself to a non-revisionary conception of the relation of philosophy with logic and mathematics by his conventionalist conception of the latter disciplines as 'antecedent to truth', would provide a complete account of the motivation for his non-revisionism. Indeed, nowhere to my knowledge does Wittgenstein explicitly recognise such a commitment. Nor is the interest of the question merely whether we can use one aspect of Wittgenstein's thought to bolster another. Rather, if the entailment exists, then it is right to allow doubts about the legitimacy of the standard interpretation of classical mathematics to ramify into revisionary criticism only if the antecedence doctrine is wrong. And if the latter is wrong, then the implications for our philosophical understanding of the necessity of statements of logic and mathematics will be deep-reaching; in particular, it is unclear that any sort of conventionalist theory of necessity will then be viable, so that mathematical intuitionism, if it is to be a programme of reconstruction, will have to presuppose some sort of realist conception of logical and mathematical necessity.[1]

3.   In order to try to clarify their connections, let me attempt to précis, and also to develop somewhat, our understanding up to this point of the antecedence doctrine and of the implications of a general non-revisionism about logic and mathematics. I shall take the latter first; and the former in §4.

Classically, the meaning of compound statements is construed as a function of the (possibly transcendent) truth-conditions of their constituent clauses; and in the special case where the compound is achieved through linking clauses by means of the logical constants, its actual truth-value is taken to be determined by those of its constituents. In particular, the quantifiers are conceived as possibly infinitary such truth-functions. Suppose, however, that we reject the classical truth-conditions conception of declarative sentence meaning in favour of a conception that views conditions of warranted assertion as the fundamental notion; how are the logical constants then to be viewed? It seems inevitable that we should take them as precisely *assertability-functions*; that is, as determining the assertability-conditions of compound statements in which they are the principle connectives as functions of those of the constituent clauses which they therein connect. The explanations of the logical constants given by the intuitionists are simply specialisations of this idea to the case where the only sort of assertability-conditions which interest us are constituted by the having of *proofs*.

With this shift in interpretation comes a simple systematic change in

---

[1] If that were to prove to be the situation, then a revisionary intuitionism based upon the sort of general anti-realism of which, I have argued, the rule-following considerations are themselves an application would presuppose, in order to be justified, that those considerations could be reconciled with a realist view of necessity.

the conception of logical validity: whereas, classically, a valid schema, $\phi(P_1, \ldots, P_n)$, was conceived as one which was true irrespective of the truth-values of $P_1, \ldots, P_n$ now—it is natural to suppose—validity will consist in the circumstances that $\phi(P_1, \ldots, P_n)$ will be warrantedly assertable irrespective of the status of each of $P_1, \ldots, P_n$ in point of warranted assertability. Certainly, bearing in mind the informal explanations of the logical constants which the intuitionists offer, this looks like the conception of validity which, by their practice, they have implicitly adopted.

A valid logical schema, then, is now a schema any instance of which will be warrantedly assertable in any state of information, irrespective of the assertability-situation in that state of information of its constituent clauses. What, though, *bestows* this status on a schema? What would be expected, following the classical account as model, is that when a schema had this status, that it did so would be a consequence purely of its structure and the character of the assertability-functions associated with its essential structural constituents. That is to say, the feature of the classical account, that validity is everywhere consequential on the general semantics of the logical constants, would be preserved.

It is preservation of this feature which makes it seem that an assertability-conditions logic, if it is to be applicable to the full range of statements which we make, is bound to discard certain classically valid principles. Consider, for example, the rule of Double Negation Elimination, taken as a conditional: $--P \rightarrow P$. Classically, a conditional is valid just in case its consequent clause is true under all logically possible circumstances in which its antecedent clause is true. Assume then that the double negation of P is true. We also have (i) that the classical negation of a proposition is true if and only if that proposition is false. So, assuming $--P$ is true, it follows that $-P$ is false; whence, negating both sides of (i) it follows that P is something other than false. That would be consistent with a truth-table for negation which assigned $-P$ the value, *False*, when P received some other assignment than *True* or *False*. But if we now appeal to the classical conception of meaning, that we determine the meaning of P by stipulating that it is to be true under such and such circumstances, and false under *any others*, then there can be no such assignment; and we can conclude that P is true whenever its double negation is.

Contrast the situation under an assertability-conditions conception of statement meaning. Now, generalising the intuitionistic account of the conditional in the obvious way,[1] a conditional is valid if it can be recognised that in any possible state of information in which its antecedent was warrantedly assertable, so would be its consequent. But the general conception of statement meaning now no longer has anything corresponding to the 'and false under any others' tag in the classical conception; the general form of explanation of statement meaning is simply a determination of what states of information justify the asser-

[1] Cf. Chapter XI, p. 211.

tion of the statement. Nor is it any longer appropriate to explain the meaning of negation, as above, in terms of truth-conditions; rather, we have to explain the assertability-conditions of the negation of P as a function of the assertability-conditions of P. Generalising the intuitionistic account, we have, for example:

the negation of P is legitimately assertable in a state of information, Σ, just in case Σ recognisably warrants the assertion that a state of information recognisably warranting the assertion of P cannot be achieved.

On this account, we are justified in affirming the double negation of a statement P just in case our overall information entitles us to assert that a state of information cannot be achieved in which we should be entitled to assert that a state of information cannot be achieved in which we should be entitled to assert P. With other than effectively-decidable statements there is, therefore, scope for a lacuna between realisation of the assertability-conditions of $--P$ and realisation of those of P: to be entitled to rule out the possibility of being entitled to rule out the possibility of a justified assertion of P is not, in general, the same thing as being entitled to assert P. For example, let P be the statement, 'Somewhere in the universe outside this solar system there are intelligent life forms'. It seems fair to say that we are now in a position to rule out the possibility of our achieving a state of information in which we should be entitled to rule out the possibility that we should ever get sufficient grounds for making that claim; which, manifestly, is not at all the same thing as *having* such grounds.

What the foregoing is meant to illustrate is that we are likely to feel obliged, on coming to sympathise with the sort of outlook of which Dummett has urged us to see intuitionist mathematics as a special application, to *revise* classical logic only because we retain the assumption, implicit in the traditional semantics for classical logic, that a schema is valid, if at all, only in virtue of its structure and the pre-assigned semantic properties of the essential logical vocabulary. The onus, as it were, is to be on each classically valid principle to show that it survives on the new conception of statement meaning and the associated re-explanations of the logical constants. It seems to me, however, that there is nothing in an anti-realist outlook as such which prohibits questioning this assumption; all that is at risk is a certain amount of complication. If, as seems likely, the logic appropriate for a generalised anti-realism will have much in common with intuitionist logic, the great majority of classically valid schemata will survive in any case; it is just that some principles—those whose classical validity rests on excluded middle—can no longer be *seen* to be valid in terms of anti-realist semantics. What, then, is there to prevent an anti-realist from regarding classically valid principles in this category precisely as *stipulations*, comparable to any meaning-determining explicit conventions? So viewed, excluded middle would be a *partial determination* of the concepts of disjunction and negation, supplementary to—by way of being a special exception to—the general intuitionist-type explanations of those notions.

Everyone is content to allow that the necessity of *some* statements is purely conventional; that we adopt them as necessary precisely by way of fixing, or further determining, the sense of expressions which they contain (for example, 3 feet = 1 yard). It is just that this seems an enormously unattractive account for the majority of cases. Our attitude to the majority of cases is rather that necessity is to be seen as a *product* of pre-established semantic features; these are exactly the cases where it is our inclination to say that necessity is recognised rather than conventionally conferred. Wittgenstein, of course, challenges the substance of this distinction. But the non-revisionist's question here need not be Wittgensteinian. It is simply this: why were we hell-bent on the idea that the validity of all valid schemata in propositional and predicate logic should be in the relevant way non-trivially *consequential*?—that it should always have to be in virtue of independently fixed properties of the logical constants that their validity holds? As an anti-realist, he grants, indeed insists, that a repudiation of the realist conception of meaning represents an advance in our understanding of these matters; but rather than concluding that a substantial part of classical logic has been exposed as ill-founded, he concludes merely that our inferential practices are not after all susceptible to the high degree of systematic explanation which the classical way of looking at things teaches us to expect. Rather than revise classical logic, he is ready to allow that there are no uniform explanations of the logical constants which are both philosophically respectable *and* adequate to explain the validity of all the schemata involving 'all', 'some', 'not', 'if . . . then', 'or' and 'and' which we habitually employ. We have entertained as a general anti-realist explanation of disjunction:

> a state of information justifies the assertion of 'A V B' just in case it *either* recognisably justifies the assertion of A *or* recognisably justifies the assertion of B *or* can be recognised to be capable of effective enlargement into a state of one of those two sorts.

An anti-realist of the present non-revisionary temper would hold that this account requires supplementation with, for example: '. . . save that "A V $-A$" is justifiably assertable in any state of information whatever'.

This view[1] at least deserves a reply; the more so, if we are unhappy with the idea that a 'correct' theory of meaning has any business venturing revision of what goes on in the studied 'language-game'. But the view need involve no enquiry into the character of necessity in

[1] Adumbrated also in Wright 77, concluding paragraph.

general; in particular, it need involve no examination of the idea that meanings can so be finally established as to predetermine validity in particular cases. The view need do no more than take over the ordinary distinction between those necessary statements whose necessity is conceived of as flowing from their structure and the pre-determined semantic properties of their constituents, and those in dignifying which as necessary it is precisely our intention further to determine the semantics of their constituents. The challenge to the would-be revisionary anti-realist is to explain why the validity of excluded middle, in particular, must consist in the circumstance that any of its instances are necessary in the former way.

We must therefore distinguish Wittgenstein's non-revisionism from this position for two reasons: first, Wittgenstein would not accept the above 'ordinary distinction' among necessary statements; and secondly, his own non-revisionism ultimately originates in his whole conception of the nature of philosophy. Our question is therefore rather whether the antecedence doctrine provides a motive for putting to an intuitionist essentially the same challenge but without drawing the distinction between two kinds of necessity in that, for a Wittgensteinian, dubious way; for example, like this: why, having rightly rejected as spurious a certain sort of explanation of the validity of classical logical principles, do you continue to suppose that logical validity must, or ought, always to be capable of explanation in that broad kind of manner? More strongly, does the antecedence doctrine actually require us to reject the supposition that such explanation should universally be possible?

4.   As we have interpreted it, the central point of the antecedence doctrine is to repudiate the orthodox idea that principles of inference have a responsibility to avoid imposing incorrect constraints on the status, in point of truth, or assertability, of the contingent propositions among which they enable us to make inferences; that is to say, the idea that, at least sometimes, contingent statements may be seen as determinate in truth- or assertability-status independently of the inferential liaisons which we construct among them by accepting certain statements as necessary which involve them or the concepts which they contain. According to Wittgenstein's opposing view, there is no way in which accepting, or rejecting, the necessity of a statement can betray, in the manner which the orthodox idea would suggest to be a possibility, our prior understanding of the concepts involved or of the statements on whose behaviour in inferential contexts that action will bear.

One way of formulating the point would be that whatever principles of inference we accept will contribute towards determining the meaning of the sorts of statement they enable us to link by inference. But there is another chain of argument in Wittgenstein's writings of whose apparent intention that formulation is a very natural description, and which, as I see it, is essentially prefatory to the main point about 'antecedence to truth'. Consider, for example, the discussion of the rule, '*(x)Fx* → *Fa*', in i. 10-13. Wittgenstein notes that we have the idea that the rule is imposed on us by the meanings of '*(x)Fx*' and '*Fa*' respectively; and proceeds to attempt to rob the idea of its point—the suggestion that some sort of deep semantic fact underlies the rule—by the simple reflection that no explanation of those meanings could be considered adequate which did not *explicitly* stipulate that the relation formulated in the rule obtained. Thus our instinctive rejoinder to the suggestion that the rule might coherently be rejected, viz. 'If someone doesn't allow the rule, he cannot mean the same by "all" ', so far from pointing to a deep source of the validity of the rule, is actually quite vacuous. The idea of our having a responsibility to the meanings of the inferentially linked statements comes to little if there is no explaining those meanings save by reference to the very principle of inference in accepting which we are held to discharge that alleged responsibility.

Now, there is an inclination to suppose that Wittgenstein has here picked on a particularly favourable example for himself, and that the situation is different in other cases where there is some 'epistemic distance', so to speak, between the principle of inference in question and what we should ordinarily take to be an adequate explanation of the meanings of the (types of) statements from and to which it allowed us to infer. For example, it is plausible to suppose that ostensive training in the use of colour vocabulary already suffices to make plain the virtues of 'Nothing is simultaneously red and green all over', and that no explicit reference to that principle need feature in an adequate explanation of the colour concepts involved. Here there is a very powerful temptation to think of the meanings of 'red' and 'green' as a factor *behind* the statement of incompatibility in virtue of which it holds necessarily; there is a temptation to give a special stress to locutions like 'If he understands "red" and "green" as we do—has got out of the ostensive training that we intended he should[1]—he must grasp the cogency of the principle'; as though the antecedent depicted some ulterior state which imposed assent to the statement of incompatibility. But Wittgenstein's point about 'all' carries over to this kind of case too. What account are we to give of what it *is* to understand 'red' and 'green' as we do, or to have 'got out of the ostensive training what we intended he should'? If our account mentions acceptance of the principle, what we say will have the triviality of the 'all' case; if, on the other hand, we concentrate on other aspects of understanding 'red' and 'green', it is at least behaviourally conceivable that someone might display total normality in those aspects yet refuse to grant the validity of the principle.

The dilemma generalises to any case where there seems to be an appropriate epistemic distance between paradigmatic forms of explanation of certain concepts and the acceptability as necessary of a par-

[1] Cf. Dummett 23 (2, p. 494).

ticular statement involving them. Whichever horn we adopt, its intention is to drive us to admit that our acceptance of the necessity of the statement is a factor which determines the involved concepts, rather than something which their predeterminate character imposes on us. An adequate explanation of those concepts either makes explicit the acceptability of the statement in question as necessary or it does not. If it does, then that the statement has that status is part of the explanation. If it does not, then it is at most a striking fact about *us* that the explanation secures our assent to the statement; for it is conceivable that others, on receipt of the same explanation, might behave in all respects as we do save that the necessity of the statement doesn't strike them. But then no sense remains in which we can say that the explanation imposed the necessity of the statement; to grant its necessity is a further step—something over and above the minimum required of one who has understood the explanation.

I am not concerned to evaluate this argument now. Suffice it to say that the second part of the dilemma, if conventionalist capital is to be made out of it, will require fairly rapid appeal to the rule-following considerations; for an argument will be needed that the 'further step' does not consist merely in an acknowledgment of an objective feature of the patterns of use which were explained—a feature which, though not strictly explicit in the explanations, was nevertheless part of their content. Rather, I want to bring out the sense in which the argument prefaces the antecedence doctrine. We have the fiction of somebody who, on receipt of a normal training in certain concepts, proceeds to apply them normally in all respects save that he is not inclined to accept as necessary a statement whose necessity it was our inclination to suppose to have been predetermined by the character of the training which he has received. But one reason why this fiction seems intelligible is just that not being inclined to accept the necessity of a statement is quite consistent with doing nothing which *conflicts* with the statement. That is to say, whether or not the argument succeeds in pointing to a sense in which the necessity of the statement is not predetermined by the character of the training received in its constituent concepts, it does nothing to explain how any *other* principle, incompatible with the one which we allow, could be reconciled with that training. The man whose ostensive education in colours does not incline him to accept the necessity of 'Nothing is simultaneously red and green all over' can behave normally in all other respects precisely because he need not, on that sole account, do anything which violates the statement. But could he behave normally in other respects, if he was inclined to accept as necessary some principle which was actually incompatible with that statement? Surely, one feels, rigorous adherence to such a principle would be bound to lead to behaviour in contingent contexts which would be taken as demonstrative of misunderstanding of the original ostensive training; for it would require misdescription of contingent fact. It is one

thing to hold the ostensive training in colour vocabulary does not predetermine the necessity of the statement of incompatibility. But it is quite
another, stronger thesis to maintain that standard such training leaves it
open to us to adopt as necessary any principle which is at variance with
that statement.

It is this stronger thesis which is the essence of the antecedence
doctrine. The central point is exactly the impossibility of anything
properly described as a *conflict* between any principles of inference on
the one hand and the pre-assigned truth-conditions, or assertability-
conditions, of the contingent statements whose use, if accepted as valid,
those principles would partially govern. There is no possibility
of our being required to deny, as the result of an unfortunate 'choice' of
principles of inference, that the assertability- or truth-conditions of
certain contingent statements obtain when in fact they do obtain, or to
affirm that they do obtain when in fact they do not.

We saw (Chapter V) a general and a more specific reason for saying
so. The general reason is an application of the rule-following considerations; it is a direct appeal to the non-objectivity of continuing to use an
expression in the same way as before, or, if what is involved is a new
sentence, of using it in a way consonant with its syntax and the way in
which we have used its parts before. There is no question of our
inferential principles requiring us to affirm what is incorrect or to deny
what is correct, because it will be our verdict which is ultimately
constitutive of correctness or incorrectness. Even if P and Q are both
statements of which previous use has been made, it is a confusion to
think of patterns in their assertability-conditions which extend of themselves unratified to new cases, so that we have a responsibility to ensure
that we admit as valid no principle of inference bearing on P and Q
which is in potential conflict with those patterns. Rather, whatever
principles of inference we have, and whatever judgments about the
truth, or assertability, of P and Q which they enable us to make in
particular contexts, will constitute fidelity to the sense of those statements. (Whatever we do, an observer will have to tailor his interpretation of P and Q accordingly.)

This general point, however, does not yet make it explicit how we
could in practice avoid apparent inconsistency, or display misunderstanding of certain contexts, if we used, for example, some quite
different arithmetic while retaining the *operational* meaning—as given
by counting, aggregation, etc.—of arithmetical signs. Would such an
arithmetic not simply lead us to false predictions? Plainly, one sort of
clash is possible. Our having given operational meaning to arithmetical
expressions consists fundamentally in the fact that we are in general
able to agree about what ought to be asserted on the basis of criteria
which have nothing to do with calculation; paramountly, counting. So
it may be that we shall get situations where all that is stopping us
affirming a pair of propositions is their apparent inferential discord;

that is, were it not for the fact that they are by our arithmetical rules inconsistent, we should have been content in terms of the operational techniques—counting, aggregation, careful observation to detect omissions, etc—to assert them both. But it would be tendentious to regard our refusal to assert them both as involving ignorance of contingent fact, or as evincing a misunderstanding of their meaning. For the non-calculational techniques for assessing number were always regarded as capable of misapplications of various sorts; and we were always ready to allow that other sorts of development were capable of frustrating our efforts to get correct information using those techniques, even when they were not misapplied. The clash takes the form not that we are constrained to deny what, by standards we had previously accepted, could be conclusively determined to be the case; but that we are constrained not to affirm what we should *otherwise* have supposed ourselves to have determined to be the case. So also someone who had adopted as necessary some principle about colour which somehow constrained him, say, to regard a particular object as simultaneously red and green all over could save himself from an outright expression of a misunderstanding of his ostensive training by conceding, for example, that the object did not *look* green.

As noted in Chapter XI, the claim 'I have conclusively verified P' is always liable to correction, is always defeasible. *One* way in which such a claim may always be defeated is if it proves via accepted principles of inference to be in discord with other accepted data; that is, if it leads to a conflict with the result which we 'ought' to get by arithmetic, geometry or logic.[1] What is at stake in our 'choice' of a set of principles of inference is the range of circumstances in which we shall judge ourselves not to have negotiated successfully the divide between the apparent acceptability, on the basis of non-inferential criteria, of statements in the class among which those principles enable us to make inferences and their *really* being acceptable; the divide between appearance and reality. It is part of the idea of objectivity—for an anti-realist, perhaps the whole part—that we have use for the idea of *error*, that we can give content to the distinction between a judgment's seeming to be acceptable and its really being so. So any conceptual scheme rich enough to incorporate that idea must have in currency practical criteria for the occurrence of error. Necessary statements, by linking and co-ordinating the acceptability of contingent statements in a manner dictated by the inferential transitions which they sanction, provide one central source of such criteria.

That necessary statements play this rôle is not, of course, a point inaccessible to a realist about necessity. But the realist, accepting the objectivity of error, will continue to believe that there is in any particular case an antecedent fact about whether or not a statement which seems correct, or acceptable, really is so; and he will continue to hold

[1] Cf. *RFM* ii. 55.

that principles of inference, if they are to be *sound*, must square with such antecedent facts. At this point, however, the Wittgensteinian will bring the rule-following considerations to bear on the notion of objectivity appealed to.

To accept the antecedence doctrine is not to be committed to holding that necessity is subject to no constraints. The workability of any system of concepts requires, in particular, that the gulf between appearance and reality does not become too wide; whereas it seems certain, as things are, that the employment of a radically alternative system of arithmetic or of a radically alternative 'grammar' of colour would create too pervasive a division. Only statements can manageably be accepted as necessary which can play a part in this way in a workable conceptual scheme. That is not at all the same thing as saying that we have to see to it that they soundly reflect antecedent contingent fact. (We shall return to this topic in Part Three. See especially Chapter XXII.)

5. Now, can we answer our question? What relation is there between the ideas just sketched and a general non-revisionism about logic and mathematics?

There is at least a threefold connection. To begin with, the non-revisionist is committed to holding that—bar, perhaps, inconsistency[1]—there is never any philosophically compelling reason for modifying a branch of logic or mathematics. But the kind of *unsoundness* which, if the antecedence doctrine were false, would be possible—viz. the entailment of false, or non-warrantedly assertable, contingent statements by true, or warrantedly assertable ones—would be something of which a remedy would unquestionably be needed. So the non-revisionist is committed to a defence of the antecedence doctrine.

Secondly, someone who holds the antecedence doctrine thereby absolves his logic and mathematics from meeting any conditions whose *sole* point is to insure against that kind of unsoundness. And there is no doubt that one factor in our feeling that validity in logic should everywhere issue from a philosophically satisfactory underlying semantics is that we conceive of that constraint precisely as a safeguard against unsoundness. The antecedence doctrine thus undercuts one likely motive for a willingness to revise systems whose consistency is not in doubt. (It would be a mistake, however, to suppose that any anti-realist who opposed revision of classical logic for non- effectively-decidable statements would be committed to disowning the desirability of this 'safeguard'. For, as noted, he may hold simply that the underlying, assertability-conditions semantics necessary to validate classical logic is more complex an affair than the platonist and intuitionist paradigms accustom us to look for.)

---

[1] On which Wittgenstein has, of course, distinctive views; see below, Chapter XVI.

There is, thirdly, a deeper way in which the antecedence doctrine wars with the frame of mind which makes it seem that the anti-realism of the intuitionists ought to be revisionary. The antecedence doctrine rests on a repudiation of the objectivity of truth for contingent statements; where 'objectivity' is to be understood in the sense of *investigation-independence*. Obviously, once we allowed investigation-independent determinacy in truth-value for so far unassessed (decidable) contingent statements, we should implicitly admit an objective question whether a verdict reached by inference corresponded with the result which would be yielded by a more direct (correctly implemented) non-inferential investigation, so far not carried out, into the truth of that verdict; which would be fatal to any global conventionalism. Now, there is, I think, a presumption of the same notion of objectivity buried in the cluster of ideas which motivate us to revise classical logic once it is seen that principles like excluded middle cannot apparently be non-trivially explicated as valid when the spurious philosophy of understanding implicit in classical semantics is abandoned. Our idea was, surely, one with Davidson's idea outlined in the last chapter; it was that laws of logic ought to be, in a certain sense, deep truths; that their validity ought to be a reflection of certain very general semantic and structural features of our language, features of which we, as mere masters of the language, need have no self-conscious awareness. (A systematic theory of meaning would have to *bring them out*.) The status of a law of logic, so conceived, would be quite indifferent to whether or not it was ever thought about, or written down; that it was valid would be conceived as implicitly settled by its structure and the semantic rôles assigned in contingent contexts to its essential constituents. So viewed, the expression of any logical law is a distilled transcription of a certain most general , structure-originating aspect of correct linguistic practice; it would be immaterial whether speakers of the language ever articulated the principle, or indeed ever did any logic in explicit, propositional form.

Once in the clutches of this conception, it will seem to us that the validity of any of the most general logical laws ought to be capable of some sort of explication; that it should be possible to trace its validity back to the essential structure of any context which amounts to an application of it and to the general semantics of the expressions which will feature in any such context, viz. the logical constants. We come to the view that what goes for most necessary statements must go for the truths of logic without exception: namely, that their validity originates in pre-established semantic features.

To think of logical validity as everywhere originating in this way is to think of it as predetermined by the ulterior, general semantics of the language, the latter in turn conceived as a stable, meaning-determining totality of material and combinatorial rules which settle the content of every well-formed sentence in the language. But what is it to think of the content of a so-far unconstructed, unconsidered sentence as already settled—if it is not a mere piece of embroidery on the mundane fact that we shall probably agree about its use? It is just to suppose that facts about meaning in a particular language run ahead of its speakers' ratification of them, that their use of the language conforms to certain determinate, objective patterns. So the roots of the general conception of logical validity which leads the anti-realist to want to revise classical logic involve an assumption which it is the essential point of the 'antecedence to truth' doctrine to reject. The antecedence doctrine requires that we can rightly abandon the notion that there is a ratification-independent fact about the correct use of any particular statement in any particular situation with which any principle of inference bearing on the use of that statement had better accord; whereas the outlook on logic which makes it seem that anti-realism must be revisionary of classical logic is a reflection of a general conception of language which views logical laws as a particularly deep sort of necessary truth issuing from the underlying framework of semantic and combinatorial rules which determine the meanings of all the sentences in the language—in short, exactly the conception that there *is* a ratification-independent fact about the correct use of any particular statement in any particular situation.

The second and third considerations suffice, I think, to explain Wittgenstein's occasional outright hostility to the revisionism of the intuitionists; for he could reasonably have suspected their stance to have been based upon prejudices—a felt need for safeguards; and an inchoate 'organicism' about language as a whole, issuing in the conception of logical validity described—which he rejected. But nothing has emerged to indicate that the antecedence doctrine contains a *requirement* that the intuitionists' ideas, or other similar considerations, should not be allowed a revisionary edge. Rather, the antecedence doctrine puts a challenge to the would-be revisionist: provide a reason for changing logical, or mathematical practice in the ways you propose which does not invoke the need for safeguards against the sort of unsoundness which the antecedence doctrine dismisses as mythical, or rest upon imputation to the semantics of our language of an objectionable objectivity.

6.   It is not clear to me that this challenge cannot be met. It seems a fairly fundamental fact about the philosophical consciousness that it seeks to be reflectively content in its habits of thought and language; in particular, it needs to be able to think of its inferential practices as well motivated, that is, as either susceptible to interpretation following a generally applicable pattern, or at least as serving some practical point. But it seems certain that some principles of inference, classically accepted as valid, will prove inexplicable in terms of any plausible

anti-realist general semantics. They will be 'explained' only by being catered for explicitly as special cases. Nor does it seem likely that it will be possible to point to practical objectives which would be frustated by dropping such principles; quite the contrary, their effect will tend to be merely to mask certain distinctions—like that, for example, which the intuitionist tries to mark by querying the unrestricted validity of Double Negation Elimination.

It is one thing to hold that philosophy is incapable of exposing *error* in our ordinary linguistic practices; quite another to hold a dogmatic non-revisionism. To be sure, some passages in Wittgenstein suggest that he intended only the former, milder view. But others, as we have seen, suggest something stronger. Why is the philosopher abjured from 'interfering' with ordinary linguistic (mathematical) practice? If it was a defective conception of language and reality which motivated the acceptance of distinctively classical principles, then, now that we know better, it seems hardly a good reason for refusing change that certain beliefs which might have made changes seem mandatory are equally defective. One of the most traditional of philosophical benefits is exactly a sense of harmony between our practices and the interpretation of them which philosophy is able to supply. A whole cluster of long-established philosophical questions—personal identity, induction, knowledge of other minds, etc.—arise because our habitual use of certain concepts seems to prove to be in tension with an account of their justification and point which, for whatever reason, has some appeal for us. That we should educate ourselves to back away from the urge to seek *justification* in philosophy is, of course, a recurrent theme in Wittgenstein's later work. But I should have thought that the kind of intellectual satisfaction, the quest for which prompts us to philosophise at all, would be more likely achieved if, when a clash arises between practice and its natural interpretation, we attempt, rather than simply noting the fact, to bring our linguistic-conceptual habits into line with the sort of interpretation which we wish to be able to give them. It is our capacity to give explanations, to so interpret our procedures that they emerge as well-founded, which —to put it crudely—allows us to feel that we know what we are doing.

I have no further argument for this view of the question, nor any wish to be dogmatic about it. But for anyone who looks to philosophy for this sense of harmony, and who is sensitive to the difficulties in the traditional philosophical realism which Wittgenstein's own thought exposes, there is every motive to interest himself in the development of a logic and mathematics appropriate to the opposing anti-realist outlook, whatever it may prove exactly to be, and in the exploration of its other practical implications. For such a person the other course, seemingly endorsed in at least some passages in Wittgenstein, of holding a rigid line between philosophy and ordinary linguistic activity and, if perplexities do indeed arise from the snares which our 'grammar' sets for us, of always refusing to allow the fruits of philosophical insight to filter back into a 'straightened out' grammar, will seem—at any rate, so long as we lack a more satisfactory clarification of the nature and motives of Wittgenstein's later general conception of philosophy—to have little to commend it.

XV

The Idea of a 'Theory of Meaning'

Sources

PI: i. 23–4; 81–2
BlB: pp. 17–18; 25; 67–8; 73–4
BrB: ii. 2–4; 6
LFM: lectures ii, pp. 23–4; xi, pp. 108–9

1.   Some philosophers, of whom Davidson has been perhaps the most committed representative, have advocated work towards the construction of formal theories of meaning for natural languages as a fruitful way of approaching traditional problems in the philosophy of language. Others have agreed to the extent of accepting that extensive philosophical gains can be expected to accrue from consideration of the question: what exact form should such a theory ideally take.[1] What does Wittgenstein's later philosophy of language have to say to proponents of these views?

It is, in one way, astonishing that it is possible pointfully to ask this question at the present time; astonishing, that is, that less than thirty years after the publication of the *Investigations* it is possible to search almost entirely fruitlessly in the writings of advocates of the above approach for a clear indication of where they consider that their programme complements, supersedes, advances or undermines the ideas of the most original philosophical thinker of the twentieth century. In any other field of enquiry a situation of this kind would be considered scandalous; philosophy, however, has long been accustomed to the vagaries of its fashions.

The question is, of course, of very wide scope, raising a host of exegetical problems about Wittgenstein's writings and an army of questions concerning the claims which might legitimately be made on behalf of the sort of formal theory proposed. I do not propose to try to answer it here. The aim of this chapter is no more than to indicate some of the members of the army: questions which, I think, will fairly rapidly occur to anyone familiar with Wittgenstein's later writings who devotes some thought to the sort of programme being put forward.

We have a formal theory of meaning for a natural language when we

[1] See, for example, Foster 34 (32, p. 4); McDowell 57 (32, p. 42); and Dummett 29a, p. 97.

have an axiomatic system, deployment of which enables us in principle to resolve any well-formed declarative sentence of the language into its essential semantically contributive constituents and to determine its meaning on the basis of the structure thereby revealed and the assigned semantic values of those constituents. The theory may, as Davidson has proposed, incorporate an axiomatic theory of truth; or it may utilise as its central semantic concept something other than truth. In what follows we shall be concerned with the constraints which Davidson imposes on any axiomatic theory of truth which is to be suitable for the purposes of a theory of meaning; but these constraints would be no less intuitively cogent if the central theory was to be, say, a theory of assertability. And the points to be made will apply equally, if at all, to all theories of meaning fitting the above general characterisation.

The crucial issue is: to what questions is such a theory necessarily able to give answers? Clearly, there are plenty of well-established questions in the philosophy of language which a theory of this kind need not be capable of addressing at all. To begin with, our characterisation does not require that the theory say anything about commanding, asking whether, wishing; it needs supplementation with, in the Frege/Dummett terminology,[1] a theory of *force*—a theory which makes explicit what difference it makes whether a sentence to which the original theory assigns a particular truth- or assertability-condition, is understood as a command, or question, or wish, or assertion, or any other kind of linguistic act which we deem to constitute a further kind of force with which an utterance can be made; and goes on to describe the grammatical indicators, if any, which the studied language uses to mark these various kinds of force. Moreover it is clear, secondly, that a whole range of limitations are consequent upon the fact that, in the case where the language in which the theory is stated is the same as, or an extension of, its object-language, the theory need not, in order to comply with our characterisation, say any more about the meaning of any particular semantically atomic expression than can be conveyed by incorporation of an appropriate 'homophonic' axiom; that is, an axiom in which that very expression is *used*, in whatever way the theory considers appropriate, to state its own meaning. For suppose that such an axiom is all the theory has to offer by way of elucidation of the semantic rôle of a particular predicate ϕ. Then the construction of the theory will neither deliver nor presuppose any account of the *vagueness* of ϕ, if it is vague; the *theoreticity*, or *observationality*, of ϕ, if it is theoretical, or observational; the *reducibility* of ϕ in terms of other predicates, if it is so reducible. In these ways, and in general, the theory will fail to speak to the question: in what does an understanding of ϕ consist—what are the criteria for mastery of the use of ϕ? And it will keep silence on perhaps the most fundamental question of all: what *is* a language—what distinguishes language-use from any other rule-governed, goal-directed

[1] 25, pp. 302–3.

activity, and what makes mouthings and inscriptions into uses of language?

Where, then, will light be cast? One answer would be that a theory of the kind characterised can still be a theory of speakers' understanding of the object-language. For we ought to distinguish between giving an account of what an understanding of any particular expression consists in—what distinguishes anyone who possesses that understanding from someone who lacks it—and giving an account of what is known by anyone who understands that expression. It is one thing to state a particular item of knowledge; another further to explain what having that knowledge essentially is. So, prima facie, a theory of the 'modest' [1] kind may still be presented as an account of understanding; it is just that it purports to characterise only what is understood by any particular expression, that is, what is known by anyone who understands it, and does not venture to explain what having that understanding amounts to.

It appears to be open, then, to the proponents of such a theory to present it as a theory of understanding, a theory of speakers' knowledge. That Davidson originally so intended his particular brand is strongly suggested by his insistence that an adequate such theory must supply the means of deriving for each declarative sentence in the language a meaning-delivering theorem in (i) a *structure-reflecting way* on the basis of (ii) a *finite axiomatisation*. [2] Surely, the intention of this pair of constraints is to ensure the capacity of the theory to explain the 'striking' fact that mastery of a typical natural language will involve the ability, on the basis of finite training, to understand indefinitely many sentences which its speakers have never heard before. The target is, apparently, a model of this epistemic capacity; it is desired to understand how recognition of the meaning of any given new significant sentence is *possible*. The answer is to be: by deploying the information encapsulated in the axioms germane to the parts of the sentence in the manner determined by the structure of the sentence, that is, the manner reflected by the mode of derivation within the theory of the relevant meaning-delivering theorem.

It would be unwise to assert that *only* this underlying conception of the significance of the theory can explain the organisation which Davidson proposed it should have. It could make a difference, to begin with, whether, for example, the theory is being constructed as the end of a programme of radical translation, in one language for another, or whether the goal is to use a particular language to state its own semantics. It might very well be that in the former case it would only be by observing Davidson's constraints that we could actually construct a theory yielding a meaning-delivering theorem for each of the object-language's declarative sentences. But in the latter case, unless the

[1] Dummett 29a, p. 102.
[2] See, for example, 21, p. 81.

relevant conception is presupposed, it is, to say the least, a very nice question why,—assuming that we have available an effective criterion for determining whether an arbitrary sequence of symbols is a significant declarative sentence in the language in question—we should not simply arrogate the general form of the T-theorem, or whatever form the meaning-delivering theorems take, as an infinitary *axiom schema,* thereby saving a great deal of work. If the goal is not to provide a model of how the meaning of an arbitrary declarative sentence can be recognised, but merely to state what its meaning is, then—putting on one side presently irrelevant complications about the appropriate form for such a statement to take—it is hard to see what objection there could be to such a course.

How, though, are we to receive the suggestion that a theory which both correctly characterised the meaning of all significant expressions in the object-language and satisfied Davidson's two constraints, would *explain* the capacity of speakers of the object-language to recognise the sense of new sentences? It is part of mastery of a language neither to know a theory of this kind explicitly, nor to be able to formulate one, nor even to be able to recognise a satisfactory formulation if one is presented. (For aught we know, there may *be* no such satisfactory theory for English.) But it seems that if such a theory really is to explain the performance of speakers of the object-language, then there must be some sense in which knowledge of its content can be attributed to them. Accordingly, there has been a tendency to invoke some sort of notion of *implicit* knowledge to supply the needed contact of the theory with the studied speakers' actual performance.[1]

The notion can seem a natural one. Suppose, for example, we were to teach certain dumb illiterates to play Chess; that is, not merely to mimic chess play, but to play correctly and strive to win. Then it would seem unexceptionable to describe such people as *knowing*, albeit inarticulately, exactly that which the rules of Chess state. Taken in one way, there is, indeed, nothing objectionable about such a description. The description need convey no more than that the people in question habitually comport themselves in such a way when playing Chess as to satisfy any criteria which we might wish to impose on someone's having understood an *explicit* statement of the rules of the game. But in that case, to attribute implicit knowledge of the rules to them is to do no more than obliquely to describe their behaviour: it is to say that they behave in just the way in which someone would behave who successfully tried to suit his behaviour to an explicit statement of the rules of the game.[2]

It is doubtful, however, whether it can be *this* notion of implicit knowledge which is relevant to the concerns of a theorist of meaning who conceives his goal to be an explanation of how a master of the

[1] See, for example, Dummett 29 (32, pp. 69–72).
[2] Cf. Foster 34 (32, pp. 1–4).

object-language is able to recognise the senses of novel sentences, and accepts Davidson's two constraints as necessary if that goal is to be achieved. For to have, in the described sense, implicit knowledge of the contents of a theory of meaning amounts to no more than being disposed to employ the object-language in the manner which would be followed by someone who, being sufficiently agile intellectually and knowing sufficiently much about the world, successfully tried to suit his practice with the object-language to an explicit statement of the theory. So construed, talk of speakers' implicit *knowledge* of the theory is, indeed, misleading; there is no real suggestion of an internalised 'programme'—all that such talk involves is that speakers' practice fits a certain compendious description. Now, however, the question resurges: what reason would the theorist have to accept Davidson's two constraints, that the theory be finitely axiomatisable and that the mode of derivation of each meaning-delivering theorem reflect the structure of the relevant sentence, if his objective were only the achievement of such a description? For the fact is that it is only the lowest-level output of the theory—the T-theorems, or what corresponds to them in a non- truth-theoretic version—which purport to describe isolable aspects of speakers' practice. So if, as when the theory is stated in a language for that very language, it is possible to peel off, as it were, the descriptive part of the theory, for example by taking the general form of T-theorem as an infinitary axiom schema, there is apparent, once the objective is restricted merely to describing speakers' practice, no reason for having the full-blown theory. To suit one's behaviour to what is stated by the theory is just to suit one's behaviour to what is stated by its practice-describing part. So if the theorist's aim was merely to formulate a theory knowledge of which would suffice for competence in the object-language, there would not necessarily be any point in observing Davidson's constraints; no advance reason why it would not help his purpose to axiomatise more simply, even if infinitistically, the descriptive part of a theory which met those constraints. A theory of that truncated form would, of course, seem *hugely* unexplanatory. It would contribute not a jot to the purpose of devising a framework by reference to which traditional philosophical questions about language-mastery could be formulated and answered. But if the aim is no more than deliverance of an account of the meaning of each declarative sentence in the object-language, and if we were content that each instance of the relevant axiom schema was of an appropriate form for such an account to take, then what more could we want?

Davidson's program has excited interest, it seems to me, for two main reasons. First, it aspires to cast philosophical light on meaning without recourse to any intensional notions; but second, and more important, it seems to promise a framework in which we can *explain* the (potentially) infinitistic character of mastery of the sort of natural language which we all speak. What I am suggesting is that if it is to do the latter, it must make sense to attribute to speakers of the language in question implicit knowledge of the whole of the theory in a richer sense than can be reduced to an imputation to them of a disposition to suit their practice to the theory's requirements. For implicit knowledge, in the latter, weaker sense, of the content of a Davidsonian theory is just the same thing as implicit knowledge of its T-theorems. And it is manifestly no *explanation* of a man's ability to understand novel sentences in his language that he has *that* knowledge; to attribute that knowledge to him is no more than to attribute to him the very ability that interests us.

It appears, then, that a more substantial notion of implicit knowledge of its content is going to be required if a theory of meaning is to explain how speakers of its studied language are able to recognise the meanings of new sentences. The question is whether there is any respectable such 'more substantial' notion.

2.   Wittgenstein's thought contains the germ of a number of challenges to the idea that has emerged: the thesis that the potentially infinitary character of mastery of a typical natural language can be explained by appeal to its speakers' implicit knowledge, appropriately richly interpreted, of the contents of a suitable theory of meaning.

First, and most obvious, the thesis seems to involve thinking of mastery of a language as consisting in (unconscious) equipment with the information which the axioms of the theory codify, information which systematically settles the content of so far unconstructed and unconsidered sentences. Such a conception is far from patently coherent with Wittgenstein's repudiation of the objectivity of sameness of use. The first challenge is to demonstrate *either* that it is right to part company with Wittgenstein here *or* that there is no real tension—that the explanatory claim made by the thesis does not require that we conceive of the meanings of novel sentences as ratification-independent, nor presuppose that people's agreement in their use of language has a foundation of the sort which the rule-following considerations would argue to be mythical.

There immediately arises, secondly, the challenge to show that the thesis is not, at bottom, simply an inflated version of the old muddle about 'universals' and predicate understanding. It is at best harmless to talk of understanding a predicate as 'grasp of a universal'; and it is harmless only if we do not think that we have thereby achieved some sort of *explanation* of people's capacity to agree about the application of the predicate. All that such terminology actually achieves is a measure of embroidery upon the phenomenon of agreement; for there is no criterion of what it is to have 'cottoned on' to the same universal save a disposition to agree in application of the predicate. But if the fiction of universals is no explanation of people's capacity to agree in their basic predicative classifications, it is far from clear that the assumption of shared implicit knowledge of the contents of a theory of meaning can provide any genuine explanation of the capacity of speakers to agree in their use of new sentences.

The third challenge is anti-realist. There is no such thing as a verification that a theory of meaning states precisely what someone implicitly knows. For the theorist, the theory has, like any theory, the status of a hypothesis; while a speaker of the studied language who sincerely avows that the theory is a correct formulation of his under-standing may always, by his subsequent practice, reveal that his avowal was mistaken. But the kind of explanation of the infinitary character of his language mastery which the theory seeks to provide seems to require us to suppose that there is some 'fact of the matter', that a speaker's use of the language is essentially the deployment of some *particular* body of information which some particular theory states. Each axiom of *that* theory will be actually true of the speaker in question. The challenge, then, is to defend this particular application of a verification-transcendent notion of truth; or to show that it is not needed.

Each of these challenges deserves a separate, detailed treatment which I shall not venture to try to give. It may be that in each case the challenge can be met, either by an appropriate refinement of the requisite notion of implicit knowledge or by some more direct defence of it. I claim merely that, until it is shown that the play which the thesis makes with a rich notion of implicit knowledge can be defended against these three challenges at least, there is no reason to think that it is within the power of the sort of theory which we are considering to serve up anything which could rightly be considered an *explanation* of the infinit-ary nature of mastery of its object-language.

3.   The foregoing interpretation of a theory of meaning was con-trasted with a more restricted one, in terms of which we would steer clear of the attribution of rich implicit knowledge of the entire contents of the theory to speakers of the object-language, contenting ourselves with an attempt merely to describe systematically their linguistic prac-tice. But it may seem unclear whether, even if the governing motive is this more restricted objective, Wittgenstein's views can be made consis-tent with thinking of language as amenable to description by the sort of theory envisaged.

There are two reservations which it is natural to feel. The first concerns Wittgenstein's apparent emphasis on the indefinite *variety* of 'language-games' which can be played within a particular language. The goal of a theory of meaning, as now restrictedly interpreted, is no more than deliverance of a statement of the meaning of each declarative sentence in the studied language. But mastery of the language will likely require a capacity to effect, and respond to, a variety of kinds of speech-act; and, indeed, there is no necessity that the speech-act effected by a particular utterance of a particular sentence should be explicitly signposted by its mode of construction. In any case, such a theory, as noted earlier, so far says nothing at all about commanding, wishing, questioning, exhorting—nor, insofar as an analysis of the notion has to be given by contrast with these other operations, about asserting. So such a theory is at best a first step towards a full description of the practice of the language. Wittgenstein's stress on an *indefinite* variety, however, might appear—if it is warranted—to call into question its capacity even to be that.

It is tempting to think that a first step can be taken in this way because it is natural to suppose that, if it is possible to use one and the same sentence assertorically, interrogatively, imperatively, etc., the content of doing so will be in each case the value of a function, appropriate to whichever speech-act is effected, taking as its argument the same speech-act neutrally explained idea of the meaning of the sentence. This seems natural because it appears that in order to understand any utterance of, for convenience, a decidable sentence, it is sufficient to know under what circumstances that sentence could be used to state something true and what speech-act the utterer intends to effect by its use on this occasion. To know, for example, the truth-conditions of 'John's car is green', and to be master of the distinction in general between asserting, commanding, questioning and wishing, is to be in a position to understand any utterance of that sentence if given the information which—if any—of those kinds of utterance it is. The view which emerges is thus that we should regard a theory of meaning as essentially no more than the core of a full description of the use of a language, a core which it still remains to *envelop* in a theory of force: a theory which, given a specification of the (speech-act neutral) meaning-determining conditions of a sentence, will determine on that basis the content of using that sentence, or an appropriate variant, to effect a question, command, assertion, etc.

Now, it is open to doubt whether Wittgenstein would not have rejected the bipartite conception on which this programme depends; that is, whether he would not have rejected the idea that there *is* any central notion, M,—truth, justified assertion, verification, falsification, or whatever—such that, on the basis of a determination of the M-conditions of a sentence, it will be possible to construct a systematic account of every kind of use of the sentence which might mutually intelligibly be made by speakers of the object-language. Not that, so far as I am aware, there is evidence to think that Wittgenstein would have rejected the distinction between meaning and force altogether.[1] Rather, the point concerns the aspirations to comprehensiveness of a theory founded on the bipartite conception. Wittgenstein writes:

[1] Contrast Dummett, 25, pp. 359–62.

But how many kinds of sentence are there? Say assertion, question, and command?—There are *countless* kinds: countless different kinds of use of what we call 'symbols', 'words', 'sentences'. (*PI* I. 23)

If 'countless' were merely rhetorical emphasis, Wittgenstein's idea could be taken merely as a caution that a full-fledged theory of force would have to be an enormously complex, ramified thing. But:

> . . . this multiplicity is not something fixed, given once for all; but new types of language, new language-games, as we may say, come into existence and others become obsolete and get forgotten.' (ibid.)

If one and the same sentence can really feature, as Wittgenstein appears to be suggesting, in a literally indefinite variety of kinds of use, what prospect is there for a systematic circumscription of that variety?

I am not sure how deep this first apparent clash goes. Doubtless, the goal of a *comprehensive* description of all possible uses of any particular natural language is probably unrealistic. But it could still be reasonable to aim to devise a general framework of which an appropriate modification, or specialisation, would enable us to give a systematic description of any new developments within the use of a particular language, of any of the—if there are so many—indefinite variety of language-games which can be played using its materials. So theory of force need not essay definitive formulation; it is enough if we can explain a concept of the kind of development which we should need to make of it in order to cope with a new sort of linguistic activity. To be sure, it is not easy to see how such an explanation might run in detail. My point is that mere complexity only argues the need for a complex theory; and that if, on the other hand, as Wittgenstein seems to think, the variety in the kinds of use of language open to us cannot in principle be circumscribed, it should still be possible to give an account of the general principles to which we should appeal in order to decide whether, and in what respects, a new kind of use was being made of extant linguistic materials. Plausibly, to have such an account would be to be in a position appropriately to extend whatever sort of theory of force had so far been achieved.

In any case, not every 'kind of use' of a sentence—of the sort which Wittgenstein proceeds to offer in the same passage by way of illustration of his claim—would be of proper concern to the theory of force; it is, for example, intuitively irrelevant to the *content* of an utterance whether it is offered by way of a report of an event, or speculation about it, or as a hypothesis; but it is not irrelevant whether it is uttered as a question or as a command. We want, that is, to draw a distinction *somehow* between genuinely semantic variations in the use of a sentence and pragmatic ones; for it seems that there are quite different species of possible misunderstanding of an utterance which such a distinction is needed to mark. If this distinction can be made out, it is more than likely that much of the variety which Wittgenstein was keen to stress is actually in the latter dimension, falling outside the province of a bipartite theory of meaning; to which description of that variety would then be supplementary. Alternatively, perhaps the burden of Wittgenstein's point is, indeed, that such a distinction cannot in the end be made out. In any case a challenge emerges to the advocate of the philosophical value of formal theories of meaning of the kind characterised at the beginning of this chapter; viz., to provide an account of the distinction between uses of one and the same sentence which genuinely differ in force, and so have different semantic content, and uses of it which differ only in pragmatic respects; and then to make good the idea that an axiomatic theory designed merely to produce a meaning-delivering theorem for each declarative sentence of the object-language is indeed a necessary preliminary if we are to give a complete description of the meaning of every type of utterance which that language provides for.

The second reservation derives from a feeling which, if it were substantiated, would be much more directly fatal to the project of any sort of theory of meaning. The idea is exactly that the rule-following considerations constrain us to suppose that the whole project of a systematic description of any particular language-game is somehow vacuous. For, surely, it is required, if such a description is to be possible, that what will constitute an admissible use of each new sentence which can be used to make a 'move' in that game is something settled in advance. The theory is not going to confine itself to recording actual, historic uses; its rôle will be to formulate generalisations furnishing predictions about what will be counted as correct use, or appropriate response to the use, of new sentences, and of old sentences used on new occasions. So it seems to be presupposed that the language-game in question *has* a certain general character, capturable by an appropriate generalisation-venturing theory. Whereas the moral of the rule-following considerations was that we have to regard correct use, or correct response to the use, of any particular sentence on a new occasion as objectively indeterminate; it is what practitioners of the 'game' do on any particular new occasion which will determine the correct use of an expression, rather than the other way about. But surely there can be no such thing as a general, correct, systematic description of any practice which at any particular stage may go in *any* direction without betrayal of its character; there is simply nothing *there* systematically to be described.

If this doubt was sound, it would not merely be a certain post-Wittgensteinian conception of the proper objectives and methods of the philosophy of language which was in jeopardy. Wittgenstein's own presented methodology of assiduous attention to facts about linguistic usage, and especially of attempting to bring out the criteria which we use for certain sorts of assertion, would be susceptible to the same reproach; for it involves a comparable attempt to capture general traits of our linguistic practice. In fact, though, it seems to me that the difficulty is spurious; that there is no reason to suppose that *all* the purposes which we might have in attempting to generalise about aspects of our own, or others' linguistic practices can be served only if it is legitimate to think of those practices as having an investigation-independent general character. Quite how Wittgenstein's own approach is to be interpreted, once it is conceded that its 'results' cannot have the relevant sort of objectivity, is another matter. But if the goal of a particular theory of meaning, appropriately supplemented by a theory of force and, perhaps, further pragmatic considerations is simply to supply a framework in one language for describing what will count as correct linguistic practice in another, no need is apparent for any assumption which the rule-following considerations would call into question. Or rather, if there need by any such assumption, then the difficulty is one not merely for the theory of meaning but for predictive theorising of all kinds. In order for a particular prediction about what will be accepted as correct use of an object-language sentence in particular circumstances to have practical content from the theorists' point of view, it is required neither that they conceive of the relevant hypotheses in the theory as codifying certain investigation-independent general features of object-language practice, nor that they conceive of it as predeterminate what behaviour by object-language speakers *they* will count as meeting the relevant characterisation if they judge correctly. All that is required is that they can secure a consensus among themselves about whether, in any particular situation, the initial conditions of the prediction are fulfilled, and about whether, if they are, the performance of the object-language speakers meets its requirements. In general, in order for any hypothesis to have a determinate meaning for us, it is required neither that we can render philosophically respectable the conception that it depicts some objective general trait in the studied range of phenomena, nor that it be objectively predeterminate what will have to happen if it is to be falsified in any particular situation; our securable consensus about its standing in any particular situation is enough.

The rule-following considerations caution us against allowing our interpretation of a theory of meaning to aspire to a bogus objectivity. But they do not impugn the legitimacy of at least the most basic purpose with which such a theory might be devised: that of securing a description of the use of (part of) the object-language of such a kind that to know that description would be to know how to participate in the use of (that part of) the language. But if that, and no more, is to be the purpose, there are clearly no legitimate constraints on the manner of its achievement save those without whose imposition it cannot be achieved. There would thus be no point in paying attention to semantic *structure* if a complete description of the use of (the relevant part of) the object-language could be given without doing so; and we noted earlier that, where the metalanguage contains the object-language, there seems to be no very clear reason why it should be necessary for the theory to discern structure in order to fulfil that purpose. But of course, almost all the philosophical interest thought to attach to formal theories of meaning depends upon preoccupation with semantic structure. A further challenge to those philosophers of language who would argue for such theories' importance is therefore to answer this question: can any adequate motive be elucidated for the belief that an interesting theory of meaning must be structure-discerning other than the desire that the theory be capable of explaining, when its contents are viewed as richly implicitly known by object-language speakers, the infinitary character of their competence?

4.   It is another question whether the Davidsonian ploy of using expressions to state their own semantics has a worthwhile contribution to make to the project of giving a systematic description of a particular language-game. To be sure, it is pointless to reproach a theory based on this ploy for failing to provide the means to *explain* the meanings of expressions in the object-language to someone unfamiliar with it. The theory will not claim to explain the meaning of object-language expressions; nor to supply analyses of their meaning, nor to give accounts of what anyone who understands them is thereby distinctively enabled to do. Rather, it will claim merely to state their meaning; and no more accurate or economical means exists for achieving this objective than to *use* the very same expression in an appropriate formulation. There is, for example, no more accurate or economical statement of the conditions under which 'brown' may truly be predicated of an individual than: when it is brown.

There is however some question about the empirical content of a theory which ventures no more than homophonic formulations of this sort. For the fact is that a *change* in the meaning of any particular object-language expression will not require any change in the edifice of such a theory; if 'brown' changes its meaning, the above statement will remain just as economical and accurate. Those virtues depend only on using an expression to state its own semantics *whatever* it means. But if the same theory can survive wholesale changes in the meanings of expressions in the object-language, how can it be conceived as characterising their meanings at all?

There is a serious challenge here; though it is not, I think, one to which Wittgenstein's ideas give a special thrust. It is obviously unacceptable that what purports to be a theory of meaning for a particular language is consistent with the possession by expressions in that language of any meaning whatever. So a proponent of such a theory has two choices. Either he must reject the notion of change in meaning, and thereby of *particularity* in meaning, which the question utilises; or *he* must utilise the very same notion to defend the particularity of content of the theory, arguing that while the edifice of the theory will indeed remain acceptable whatever changes in meaning take place, *what is actually stated* by it will shift accordingly. But either course threatens a rough ride. The former seems to be tantamount to a rejection of the idea that a theory of meaning describes properties of signs, or sounds, which are conventionally conferred; for if it did, change in respect of those properties would have to be a possibility. But a theory of meaning which does not describe conventionally conferred properties cannot be a theory of *meaning* as we ordinarily understand that notion; so what is it a theory *of*? The latter course, on the other hand, presupposes—what no one has yet been able to provide—a satisfactory defence of the general distinction between a change in meaning and a change in speakers' factual beliefs; for in order to make out the claim that the content of, say, a particular homophonic T-theorem has changed, it will be necessary to point to changes in the use of certain of its constituent expressions which reflect something other than changes in the non-semantic beliefs of object-language speakers. I am not suggesting that the theorist has no right to the latter course until such a defence has been provided; but while so fundamental a question remains un-illumined, the claim that insight into the notion of meaning will issue from the construction of homophonic theories of meaning had better be fairly tentative.

A principal lesson of this discussion, conjoined with that of §5, Chapter XIII, is, I believe, that three claims familiarly made for a successful homophonic theory of meaning of Davidson's proposed kind, viz. that it would cast light on the notion of a structurally valid inference, that it would explain the infinitary character of language-mastery, and that it would be an empirical theory, still await persuasive defence. The promise of an illumination of deep semantic sources for certain types of valid inference presupposes that, while in no doubt that a certain inference is valid, we can be unclear about what fundamentally *makes* it valid; I do not see how this presupposition could be defended without recourse to a thoroughgoing realist conception of logical valid-ity. The promise of an explanatory model of the infinitary character of mastery of a typical natural language seems to require that it be intellig-ible to attribute rich implicit knowledge of the contents of such theories to object-language speakers. Finally, the claim that a theory is empirical needs an account of the circumstances in which it would be shown to be wrong—an account which, since the facade of a homophonic theory is proof against all that the shifting tide of its object-language can throw at it, looks likely to require a defence of something close to the classical notion of a proposition. Wittgenstein's later philosophy of language poses obstacles on all three fronts. Of course, there is no doubt that his contempt for 'scientism' in philosophy, his mistrust of the 'craving for generality', and his repudiation of any assimilation of language-mastery to grasp of some sort of calculus, would have led him to deplore recent interest in formal theories of meaning. I have tried to indicate some rather more specific respects in which the preconceptions which nourish that interest are challenged by his work.

PART THREE

Necessity

XVI

Consistency

Sources

RFM: I. 134–6; App. I, 11–14, 17–18; II. 77–8;
      III. 55–60; v. 8–13, 21–2, 27–30
PG: II. 11, 14
OC: 392
Z: 685–92
PB: 160
LFM: lectures XIV, p. 138; XVIII, pp. 174–9; XIX, pp.
      184–90; XXI, pp. 205–11; XXII; XXIII
WWK: parts IV–VI, sections on *Widerspruchsfreiheit*, ex-
      cerpted in translation in *PB*, pp. 318–46

1.   Many of the central themes in *RFM* contrast sharply with
received attitudes to proof, necessity, and mathematical truth. In most
cases these attitudes are not merely prevalent but extremely natural;
what are the chances of anyone, previously unfamiliar with Wittgenstein's later work, reading *RFM* and finding himself in substantial
agreement with the author? With the episodic remarks on consistency
the likelihood must be almost nil. Here the impression is not so much
that of ordinary attitudes or assumptions questioned, as of good sense
outraged. Wittgenstein talks of the

> superstitious fear and awe of mathematicians in the face of contradiction.
> (I. 17)

He suggests in several places that mathematicians might have taken a
pride in producing contradictions, that they might have been

> . . . glad to lead their lives in the neighbourhood of a contradiction. (II.
> 81)

In places he seems to propose that we might just accept the paradoxes
which the Theory of Types was designed to solve; we might, for
example, accept Russell's paradox as something

> . . . supra-propositional, something that towers above the propositions
> and looks in both directions

—truth and falsity—

like a Janus head. (III. 59)

He suggests we might even begin logic with the paradox

. . . and, as it were, descend from it to propositions.

Elsewhere it is allowed, (e.g. v. 8) that such a paradox is disquieting; not, however, because of the derivation of a contradiction, but because the background framework of set theory and logic in which it is developed is a

. . . cancerous growth, seeming to have grown out of the normal body aimlessly and senselessly.

And there is a general tendency to talk of contradictions as in themselves harmless (Appendix *I*, 11), or at least as 'places' where we do not *have* to go.[1]

It is easy to sympathise with Ross Anderson's impatient response to all this: the avoidance of contradiction just is a prime requisite of worthwhile foundational studies—we might just as well speak of the 'superstitious fear and awe of chess-players in the face of checkmate'![2] Nevertheless, this is to reply at the wrong level. Wittgenstein is not disputing the right of mathematicians and logicians to propose to work within such a constraint; rather he wants to question the ordinary conception that this constraint is *imposed* on us.

Wittgenstein's stated aim is (II. 82) to alter the *attitude* towards contradiction and consistency proofs. Among the most important attitudes to which he thinks we incline, and which he wishes to alter, are the following:

(*a*) A calculus with a contradiction in it is in some way *essentially* defective.

(*b*) When a contradiction comes to light, some sort of remedial action is rationally demanded of us; we cannot coherently just let the thing be.

(*c*) There is such a thing as the *correct* logic, or set theory; and the paradoxes show that we have not found it. The problem of 'solving' the paradoxes is a determinate one; it is that of finding the mistakes in the assumptions which lead to them.

(*d*) For any particular branch of mathematics, it is desirable that it be set up in such a way that contradictions can be avoided *mechanically*; that is, so that a slavish, unintelligent and totally aimless application of the rules of inference can never lead to any difficulty.

---

[1] *LFM*, lecture XIV, p. 138.
[2] 68 (2, p. 489).

(*e*) Consistency-proofs are needed—or at least desirable. A system for which such a proof is missing, or unobtainable, is somehow insecure. 'Only the proof of consistency shows me that I can rely on the calculus' (II. 84)

(*f*) A hidden contradiction is just as bad as a revealed one. A system containing such a contradiction is totally spoiled by it. The contradiction is, as it were, a pervasive, general sickness of the system.

The aim of this chapter is to explore to what extent we are now in a position to understand the motivation for Wittgenstein's repudiation of these attitudes, to what extent Wittgenstein's views here are of a piece with the more general ideas with which we have been concerned.

2.   There is no doubt that the attitudes sketched are, or at least have been, widespread; though they have for us now varying degrees of attraction. No one, for example, who has thought at all seriously about the paradoxes will feel at ease with the supposition that they must contain one or more specific errors which, if presented to us, we should be readily capable of recognising as such and excising from our conception of admissible argument and definition. We have learned of a variety of strategies which seem to keep us out of trouble; but none of them has the simple intuitive appeal originally possessed by the 'naive' assumptions concerning class existence and predication, and what constitutes an admissible range of quantification, which featured in, for example, Frege's foundational theory. It is difficult to defend a notion of 'error' in this context for which the criterion is not precisely the potential to generate paradox; and this criterion, naturally, fails to discriminate in point of preferability between the alternative, seemingly successful strategies for avoiding paradox.

The demand for consistency proofs has also come to seem less urgent than it did; not, though, because we have abandoned the view that a consistency proof displays the *security* of a system, but because we have learned from Gödel that the sort of consistency proof to which we would have been inclined to attribute that effect is in key cases probably unavailable.

However, others of the attitudes would, I think, still be vigorously endorsed. We do feel, for example, that if a contradiction came to light in an established branch of mathematics, we could not rationally carry on as though nothing had happened, that some sort of rectification would be demanded. What we need to be clear about is why those of the attitudes which we continue to find attractive appeal to us. To know the sources of this appeal will be to be in a position to know whether Wittgenstein succeeds in attacking the attitudes at source—or whether, less interestingly, he calls them into question only to encourage such an investigation.

Why, then, is an inconsistent system essentially defective? The question is, defective for what purpose? Wittgenstein concedes, for example, (v. 13) that the inconsistency of the system of Frege's *Grundgesetze* ruins it as a foundation for mathematics; but his view, sketched in Chapter VII, is that the whole conception of such a foundation is in any case misguided. Now obviously, to the extent that such a project is thought of as unimportant or misguided, the contradictions which arose in its execution will also be thought of as unimportant. But the general concern which we feel about inconsistent systems has nothing to do with whether they are conceived as having a 'foundational' rôle. Rather, it is twofold: we are concerned about their applicability, and about their truth. If a system is inconsistent, then the inferences permitted within it will not in general be truth-preserving when it is applied to contingent contexts; and if we had thought of it as a systematic description of some abstract conceptual structure, then, again, not all its theorems can be regarded as correct descriptions of the intended structure.

The latter notion actually receives little explicit attention in Wittgenstein's remarks on consistency. But it is, of course, unacceptable for a Wittgensteinian. To talk of a conceptual structure is to talk of a projection of our understanding of the concepts involved; but the Wittgensteinian view is that we may not legitimately think of our understanding of any concept as something with a determinate, objective nature, of which it would be in point to seek a systematic *description*. Not that, on the opposite view, inconsistency would be, as Wittgenstein in one place suggests (III. 56), the *only* 'bogey', the only thing that could be wrong with the system. A system may be unfaithful, we conceive, to an intended interpretation (for example, omega-inconsistent) without being actually inconsistent. But if our idea is that the essential business of pure mathematical systems is to describe determinate conceptual structures and that the notion of truth for its theorems corresponds accordingly—that is, it is appropriate to hold the theorem to be true or false just in case the structure in question does, or does not, have the property attributed—then it seems inescapable that an inconsistency *is* a total disaster, and demands remedy if there is to be any pure mathematics for the structure in question. For not only does an inconsistent system not truly describe the intended structure; it does not *truly* describe anything at all.

Such a picture is obviously one source of the attitudes towards contradiction—(*a*), (*b*) and (*f*) in particular—which Wittgenstein is questioning; a picture which is utterly unacceptable to him. For there just is not anything for pure mathematics to describe. To be sure, we have the apparent practice of asserting and denying mathematical statements; but this does not establish a substantial analogy between these statements and those whose rôle is genuinely to record the facts. It is consistent with Wittgenstein's preferred idea that we should regard mathematical statements as intra-linguistic analogues of what meta-linguistically would take the form of rules, explicitly prescribing forms of description. He writes:

> Might we not do arithmetic without having the idea of uttering arithmetical *propositions* . . . should we not shake our heads, though, when someone showed us a multiplication done wrong, as we do when someone tells us it is raining, if it is not? Yes, and here is a point of connection. But we also make gestures to stop our dog, e.g. when he behaves as we do not wish. Appendix *I*, 4).

Now, whether to repudiate unacknowledged internal relations, as the rule following considerations entail, of itself requires jettison of a descriptivist view of necessary statements is something which we shall shortly have to rethink.[1] But there is no question that the analogy with rules, or imperatives, is fundamental to the sort of positive picture of necessity which Wittgenstein wants to commend. So we should expect that, according to Wittgenstein's view of the character of an interpreted branch of pure mathematics, the trouble, if any, generated by the development of an inconsistency would be best illumined by comparison with an inconsistent set of directives. Just this is the comparison which Wittgenstein does take up repeatedly. Discovery of a contradiction in a mathematical system is compared with the discovery that the rules of a game are inconsistent, or defective in some other way; for example, whoever goes first can always win. Thus ii. 77; ii. 85; and v. 10;

> Suppose I've devised rules for a game of Hare and Hounds, fancying it to be a nice, amusing game; later, however, I find that the hounds can always win, once one knows how. Now, let us say, I am dissatisfied with the game; the rules which I gave brought forth a result which I did not foresee and which spoils the game for me.

Again, v. 22:

> I am defining a game and I say: 'If you move like this, then I move like this, and if you do that, then I do this, etc. Now play'. And now he makes a move, or something which I have to accept as a move, and when I want to reply according to my rules, whatever I do proves to conflict with the rules.

In terms of this analogy, Wittgenstein's questioning of some of the ordinary attitudes to contradiction which we listed is extremely easy to understand. There is, for example, no reason why a game with a contradiction, or some other flaw, in the rules *must* be regarded as essentially defective; nor is there any reason to insist that if the defect

---

[1] See Chapters XIX and XX below.

comes to light, some sort of remedial action is demanded of us. Of course, it *may* be that the game is spoiled because we cannot see how to avoid exploiting the inconsistent, or otherwise defective, elements in the rules; perhaps we will no longer be able to agree whether key moves are admissible or not, or we cannot see how intelligibly to try to win without exploiting the strategy which makes winning an effectively achievable end. But all sorts of cases are possible here. We may find ourselves with a sure grasp of what could be called the 'spirit' of the game; so that in practice the issue whether a move is allowable hardly ever becomes urgent, even though the task of formulating the rules consistently so as to codify this spirit seems to be beyond us.

Wittgenstein is right in suggesting that, with games anyway, what determines an inconsistency, or other peculiarity, brought to light in the rules to be a defect, of which repair is demanded if the game is to continue being worth playing, or even playable at all, is nothing other than the attitude which we feel able naturally to take up to it. If, for whatever reason, we go on being able to play and enjoy the game, then it is aimless to say that the game is defective or that we are thereby shown to be irrational. No doubt *we* almost always would lose interest in a game whose rules turned out to be inconsistent; but we can imagine other people in our own society who were somehow happy to go on playing, making, as it seemed to us, quite arbitrary and eclectic applications of the inconsistent elements in the rules, but fortuitously—as it seemed to us—always agreeing on any particular occasion when appeal was made to the rules about which element in them was to be appealed to; so that irresoluble disputes about the admissibility of a rule never arose for them. We can suppose that if the inconsistency was pointed out to them, they would concede it and say, for example, 'Yes, playing this game is a matter of good sense; you have to be careful who you play with.'

There is no point in insisting that if such people continued to find the game playable, they would have somehow to understand the rules differently, so that for them there was no inconsistency. After all, the hypothesis was that we find it impossible to interpret the way they play—it seems quite arbitrary. To insist that they *must* possess a consistent understanding of the rules is just to stipulate that an inconsistency makes the game unplayable. But the fact is that while, when a set of directives is, by ordinary criteria, inconsistent, we are very seldom able to receive them as enjoining an intelligible pattern of conduct, still it is imaginable that things should generally be otherwise. There seems no reason why it should be possible consistently to codify the practice of everything that we are inclined to describe as a game, in such a way that every move can be explained by reference to the rules. Of course, if it is a matter of codifying an established practice, we shall be reluctant to impose an inconsistent description upon it, preferring instead to say that certain moves are arbitrary. But it is intelligible that such a pattern of activity should result *from* an inconsistent explanation, and if it works as a game for the participants—serves the purposes of the game—then that is all there is to say about the matter. The conception that there is something essentially amiss with such a game is just a projection of our inability to receive such rules as an explanation of how to play. The point of having consistent rules is to ensure that irresoluble disputes concerning the legitimacy of a move do not arise; but if certain people are able to receive such rules and then make a game of it, then for them, it seems, such an insurance is not required.

The analogy also gives us some purchase both on Wittgenstein's opposition to our ordinary idea that, in point of defectiveness, a system whose inconsistency has not yet been discovered is just as bad as one whose inconsistency is known, and the perspective which he tends to adopt towards the problem of solving the paradoxes. It is evident that a 'hidden' defect in a game certainly need not be as obstructive as a revealed one; on the contrary, so long as the defeat remains hidden, that presumably just means that the game is in general being successfully played. It may be that situations occasionally arise with which the players cannot cope and which, as we later say, symptomatised an underlying inconsistency; but the real problems come when the contradiction is brought to light. A hidden contradiction in the rules need not hinder the game at all so long as it remains hidden—so long as a situation does not arise in which it becomes patent that we have inconsistent directives about what to allow. It is, I suppose, a defect of 'Noughts and Crosses' that the second player can always force at least a draw; but this does not spoil the game for children who do not know this or, if they do, do not really know *how* to do it. The same may well be true of Chess, at least if we consider a limit of, say, one hundred moves per game. (Indeed, it had *better* be true because—on a realist conception of the structure of the game, at any rate—the only alternative is that White can always force a win.) A perspective is conceivable from which our capacity to interest ourselves in Chess (with such a move limit) would be viewed as we view the ability of children enjoyably to play Noughts and Crosses. It is the same with a hidden contradiction in a game. It need not even matter if we are reliably informed that the contradiction is there. If we are not able to see it for ourselves and exploit it, then the game remains playable. Again, a superior perspective is imaginable which discerns some subtle contradiction in the rules and views us as in the position of the people described above, save that we know nothing about the contradiction. From such a point of view, our ability to play the game might not seem fully intelligible; someone in such a position might not be able to receive the rules as an explanation of how to play—and he might be able to form no plausible hypothesis by means of which he could predict what sort of selective application we were likely to make in certain circumstances of the inconsistent elements in the rules. There would be no going back from a position of understanding, just as we cannot go back to an innocence when Noughts and Crosses was an absorbing pastime. But it is wrong to suppose that such a game is inherently absurd or that it never existed as a 'proper' game; the game survives as long as our innocence survives.

Mathematics *can* be 'played' just as a pastime. And if we so regard the subject, it is clear that, should a contradiction come to light in a system and we wish to rectify matters, the problem need not be a sharply delimited one, that there need be nothing which may rightly be identified as a *mistake* in the old system which is to be held responsible for the contradiction. The nature of the problem confronting us is more likely to be correctly described in something like the following terms:

> I want to know how I must alter the rules in order to get a proper game. 'But you can e.g. alter them entirely and so give a quite different game in place of this one.' But that is not what I want. I want to keep the general outline of the rules and only eliminate a mistake. That however is vague; it is now simply *not clear* what is to be considered as the mistake. It is almost like when one says, what is the mistake in this piece of music? It doesn't sound well on the instruments.—Now, the mistake is not necessarily to be looked for in the instrumentation; it *could* be looked for in the themes. (II. 77).

There are different areas in which the mistake can be sought, and different conceptions of the kind of mistake to look for. And it is in such terms that Wittgenstein wants us to look upon the problem of 'solving' the paradoxes, and to characterise the perplexity into which they cast us:

> . . . but now the new calculus became unusable in certain parts, (at least for the former purposes.) I therefore seek to alter it: that is, to replace it by one which is *to some extent* different. And by one that has the advantages without the disadvantages of the new one. But is that a clearly *defined* task? Is there such a thing . . . as the *right* logical calculus only without the contradictions? (II. 85; cf. also II. 81)

The analogy with an inconsistent set of rules for a game does not, of course, demonstrate anything. Rather, it poses a question. If we incline to accept Wittgenstein's rejection of a descriptivist account of statements of logic and mathematics, and are prepared to accede to the suggestion that they be regarded rather as rules of description—at least as a general rubric of interpretation—then what does the game analogy leave out of account, if we remain dissatisfied with the attitudes to paradox and inconsistency which it suggests? To regard theorems in logic and mathematics as regulations (if only 'game-internal' ones, for the production of further theorems) does make it natural to see the problem of rectifying an inconsistent system as that of preserving the 'essential character' of the regulated activity while eliminating the contradiction; and that may well be a vague brief. Depending on the case, there may be equally attractive ways of carrying it through, or no attractive way of carrying it through. The contradiction might be so deeply situated that to remove it would be to change the activity in question beyond recognition. This is not to say that there can be no room for argument about alternative strategies for resolving, say, the set-theoretic paradoxes. But if we want both to resist the suggestion that the argument is of the character suggested by the game analogy, and to reject the descriptivist account of the content of set-theoretic statements, then what *is* the character of such argument?

3.   Of course, we are reluctant to accept the assimilation of mathematics to a game. It seems a travesty. But if this reluctance is not traceable to a hankering after an objective domain of pure mathematical truth, we will do well to be clear about its source. I cannot see that it can be anything but this. In an inconsistent game, situations are feasible in which the rules issue contradictory verdicts on the allowability of a move, or on whether someone has won. If, for whatever reason, people are still able to agree about what to do in such situations and the game is still enjoyably played, then it is fair to say that all is well only because the *point* of the game is to be enjoyably played. But mathematics is harnessed to other goals besides fun. When an inconsistent system issued conflicting verdicts about the assertability of some thesis, our *agreement* about what to say, we feel, would not be enough. Some constructivistically-inclined mathematicians might be disposed to feel that it does not much matter what we say in far-out non-constructive regions of classical mathematics. But if we are concerned with a system, like number theory or analysis, many of whose theorems under their 'pure' interpretation are conceived as having determinate applications in contingent contexts, then to each such theorem will correspond a rule of *inference* applicable to contingent statements. In such cases we cannot think of the theorems merely as 'moves', or 'positions', within a game; nor even as derived regulations for the game. For they also regulate quite different activities in which the stakes are likely to be higher than in any game: they license inferential transitions from contingent premises to contingent conclusions—and getting these inferences right can literally make the difference between life and death.

This was the second above-mentioned major source of intuitive doubt about Wittgenstein's attitude. We want to say that inconsistent systems are at best *useless*; that they can have no practical application. The rules of inference which they yield will not in general be truth-preserving when applied; and if we just *choose*—even if we find ourselves spontaneously inclined to choose the same way—among the inconsistent theorems which the system yields, we forsake all guarantee that we will choose *right*—that the favoured theorems will be valid as rules. Alternatively, if we had intended to interpret the system in some empirical domain, we can have no reason to think that those theorems which we conceive as derived within the 'spirit' of the (inadequately formalised) system, will be true; or, if false, will bear on the truth-value of the hypotheses from which they are derived.

Wittgenstein's treatment of these doubts is prima facie very unsatisfactory. His favourite example (II. 78; v. 8, 11) is that of an arithmetic licensing division by zero; that is, of people whose arithmetic is as ours is save that in simplifying an equation they regularly divide through by factors of form, $n-n$. But most of the discussion proceeds in the interrogative. Would these peoples' practice have to throw them into confusion? Is such division necessarily a mistake? What makes it one (II. 78)? Are the results at which they arrive necessarily illegitimate (v. 8)? Do we have to regard their calculations as wrong, or wrongly got (v. 12)? And in the last passage,

> 'But a contradiction in mathematics is incompatible with its application. If it is consistently applied, that is applied to produce arbitrary results, it makes the application of mathematics into a farce or some kind of superfluous ceremony. Its effect is e.g. that of non-rigid rulers which permit various results of measuring by being expanded and contracted'. But was measuring by pacing not measuring at all? And if people worked with rulers made of dough, would that of itself have to be called wrong? Couldn't reasons be easily imagined on account of which a certain elasticity in rulers might be desirable?

Wittgenstein never actually helps us easily to imagine such reasons, still less reasons why it might be desirable to have a *total* elasticity in our concept of a valid numerical calculation. A consequence of the reflections of Chapter IV is that we have to suppose, in order to make sense of the example, that the *applications* which these people make of their arithmetic are essentially similar to ours; that they count as we do, that we can agree with them when a particular calculation is germane, that in general we understand the ways in which they apply their arithmetic and the interest which they have in the outcome of a particular calculation. We have to suppose all this because otherwise it will become tendentious to describe them as allowing division by zero. The objection which we shall immediately want to make to all this is, therefore, essentially that developed in Chapter IV with respect to the idea of measurement with soft rulers. It seems to be immaterial whether we describe such people as not calculating, or as calculating wrongly, or neither. The point is that their rules of 'calculation' allow them to get absolutely any outcome as the result of a particular calculation on particular initial data. So where the results of their calculations are capable of an independent check or corroboration, for example, by counting or measurement, or via another 'proved' equality, it seems certain that these people are going repeatedly to be forced into a position of having to postulate miscalculation, or some sort of error in the process of the check—a miscount or misweighing, etc., or an error in the other calculation—or some sort of physical peculiarity, where *their* ordinary criteria for the application of these concepts strongly suggest that nothing of the sort has occurred. How can we be certain that this will be the situation by *their* criteria? Because it is a precondition of our being entitled to interpret them as calculating at all, a fortiori as allowing division by zero, that we can be reasonably certain that we share the criteria for such judgments with them. (Correspondingly, measurement with sufficiently elastic rulers would force us to postulate changes in the dimensions of an object where other criteria—the look of the object, its capacity to fit into a certain space, etc.—suggested that its size was unaltered, or at least unaltered to *that* extent.) It seems that, provided we and the zero-dividers have a common conception of the non-calculational assertion-grounds for numerical statements—to the extent presupposed by the propriety of describing them as allowing division by zero—it will be child's play to set up for them no end of situations in which their arithmetic will come into prima facie collision with that conception. And to imagine the zero-dividers practical purposes to be quite different from ours—to suppose that they do not count or calculate for anything like the reasons which we do—raises the question, with what right are they identified as *counting*, or *calculating*, at all? So it is unsatisfactory to suggest that there is no ready sense in which these people's calculations are *wrong*; the wrongness resides in an easily demonstrated discord with the very practical objectives, and non-calculational criteria of judgment, their possession of which is needed to give sense to the idea that they have an arithmetic which permits division by zero.

Also apparently unsatisfactory is a particular favourite argument of Wittgenstein's in this context. He suggests that the utility of a branch of mathematics cannot depend on its consistency because:

> Imagine the following queer possibility. We have always gone wrong up to now in multiplying 12×12. True, it is unintelligible how this can have happened; but it has happened, so everything worked out this way is wrong. But what does it matter? It does not matter at all. And in that case there must be something wrong in our ideas of the truth and falsity of arithmetical propositions. (i. 134)

Compare v. 28:

> I mean, if a contradiction were now actually found in arithmetic, that would only prove that an arithmetic with such a contradiction in it could render very good service. And it would be better for us to modify our concept of the certainty required than to say that it would really not yet have been a proper arithmetic.

The suggestion would seem to be that the practical utility of arithmetic, its capacity to render 'good service', cannot depend upon its consistency or, in some ulterior sense, its 'truth'; for our arithmetic has rendered good service and yet we have to concede it as a possibility, although a hugely remote one, that a contradiction might yet be found in it. Might we not be, in some more subtle way, in a position comparable to that of the people who allow division by zero without realising that they can get any result at all that way? And since we have to concede that this is a possibility, do we not also have to concede that the utility of arithmetic, its manifold applicability, cannot depend on whether that possibility is realised or not?

Actually, it would be presumptuous to concede anything of the sort. No doubt we should have to make the concession *if* things turned out as hypothesised, if some inconsistency in the rules of whole number computation came to light (and errors had occurred in the proofs of consistency for the quantifier-free part of arithmetic). But that does not mean that, because we think that we have to envisage the possibility, we are constrained to make the concession *now*. For if there is an essential connection between the general utility of arithmetic and its consistency, then it is presumably only because it is consistent that arithmetic has proved so successful in application. Wittgenstein seems to present us with a feasible thought experiment—waiving any doubts about whether the discovery of contradictions in arithmetic really is fully intelligible. But if the conclusion to which it is intended to persuade us—that there is no essential connection between utility and consistency—is false, the appearance of feasibility is false also.

Still, the intention of the argument is clear. In other places, however, Wittgenstein seems to concede that a connection between consistency and applicability obtains. In II. 78 it is allowed that to admit division by zero would render calculations unusable for our ordinary practical purposes:

> . . . multiplication would surely become unusable in practice, because of its ambiguity, for the former normal purposes. Predictions based on multiplications would no longer hit the mark. If I tried to predict the length of a line of soldiers that can be formed from a square, 50×50, I should keep on arriving at wrong results. Is this kind of calculation wrong then?—well, it *is* unusable for *these* purposes, perhaps usable for other ones.

While such an admission is welcome, we might well continue to quarrel with its exact formulation. For, again, what are these 'other purposes' to which an inconsistent arithmetic would be well adapted? They have to be purposes which make it proper to describe a series of manipulations on paper, for example, as a calculation; for what makes such a structure, over and above being a symbolic manipulation, into a calculation *is* the kind of application to which its practitioners conceive it may be put. More generally, what makes a process into one of inference is what is done with the 'conclusion'. So we want Wittgenstein to illustrate purposes which an inconsistent set of computation rules could serve well enough and which would still allow us to describe a series of manipulations sanctioned by such rules as calculation, rather than a mere ceremony or symbolic ballet. But he never does; and it is doubtful if he could have. For one suspects that the constraints which the notion of calculation places upon its range of admissible applications and purposes are such that none are included which would not be frustrated by an inconsistent system of rules of arithmetical computation. This suspicion is not a veiled stipulation about the sense of the word 'calculate'. It is rooted in our concept of when an activity of a tribe, say, might legitimately be interpreted as a kind of arithmetical calculation, of what it is correctly to interpret as *arithmetical* symbols the expressions which are manipulated in their symbolic transformations.

Nonetheless, Wittgenstein appears, at least sometimes, to have been ready to grant what in this context we really want to maintain: an inconsistent arithmetic could not serve the purposes to which arithmetic is actually put. But on closer inspection, it becomes doubtful whether this admission is made in the form which we desire. Our intuitive view is that such an arithmetic is *essentially* maladapted to our purposes; that it must issue in spurious rules of inference when put to these purposes; that it is bound to come into conflict with, as it were, the arithmetical structure of worldly facts. Nothing in the foregoing considerations suggests this has to be so. It was proposed merely that, once the overlap of intended applications, prerequisite for meaningfully describing a tribe of zero-dividers as such, was assured, it would, plausibly, be extremely easy to set up situations with which they would not be able to cope. But Wittgenstein does not allow that it is in this way in the *nature* of an inconsistent arithmetic to frustrate our practical goals. In the same passage (ii. 78) we find:

> . . . is the calculation with $a - a$ not a proper game? Is it impossible *not* to trip up . . .?

and shortly afterwards:

> It is, I should like to say, for practical, not for theoretical purposes, that the disorder is avoided. A kind of order is introduced because one has fared ill without it—or again, it is introduced like streamlining in prams and lamps, because it has perhaps proved its value somewhere else and in this way has become the style or fashion. (ii. 83)

For Wittgenstein, an inconsistent calculus is not in some deep, theoretical way maladjusted to our ordinary practical purposes; it is rather that such a 'disordered' system just proves not to pay.

4.   What is the truth of the matter? We had better formulate in a little more detail our intuitive reservations about Wittgenstein's view.

Corresponding to any calculation is a rule of inference; in the simple cases with which we are concerned, a rule licensing the co-applicability of the concepts flanking the arithmetical equality which the calculation proves. But these concepts will be linked with other criteria of application than that incorporated in the resulting rule. The effect of the calculation is precisely that we come to regard these criteria as determining the same thing. There are other criteria, for example, for supposing that the number of ϕ's is 23×23 than its being 529. It is the number of tiles, for example, in a square of square tiles with sides 23 tiles long; and an ability to apply the concept, m^2, on the basis of the criterion illustrated here is clearly independent of knowledge of the product in a particular case and even of the ability in practice to calculate it. We should describe someone who did not have the latter two abilities as knowing that the number of ϕ's is m^2 without knowing how many ϕ's there are. There are thus two ways to find out how many: to calculate the product, $m \times m$ or to count up. Now, if we are using an inconsistent system of computation, the result which we get in computing the product can literally be anything at all; there will be a way of getting any result which we cannot reproach on grounds of correctness, (though probably on grounds of over-elaborateness). And in that case we have lost all essential connection between correct calculation and procedures which enable us to discover how many ϕ's there are. More generally, if a system is inconsistent, there is no reason why the inferences which it licenses in practical contexts should reflect the facts; and it is essential to our practical purposes that they should do so.

If this is a satisfactory expression of our intuitive view, it promises to be a task of some difficulty to defeat Wittgenstein's opposing pragmatism about the matter. Why, after all, *must* it prove disadvantageous to infer via inconsistent rules? Certainly there seems no reason why it should do so if we concentrate on any particular single application. Suppose a contractor is paving a yard 23 feet by 23 feet with foot-square flagstones. He sets about the calculation 23×23, but his arithmetic is haywire and he gets the answer 625. Accordingly, he orders 625 foot-square flagstones and when they arrive he finds he is able to complete the job to his satisfaction. How is this possible? Simply because the supplier has mis-cut the flagstones 11 inches by 11 inches each. It is true that the contractor is left with an inch of slack in each row of flagstones, but this reduces to a barely noticeable 1/24 inch between each pair. Perhaps he does not even notice this slight tolerance, or if he does, perhaps he attributes it to a marginal mis-measurement of the dimensions of the courtyard, say. At any rate, his practical purpose—getting the courtyard paved to professional standards—has been well enough served by his defective arithmetic.

This example will not pass unprotested. To begin with, the dimensions involved were obviously fixed so as to suit the convenience of the particular error. It happens to be true that we cannot readily distinguish by unaided eye whether a flagstone is more nearly 11 inches square than 1 foot square. But suppose the contractor's calculation had yielded 999. If when the tiles arrive, they turn out to be suitable for the job, it can only be because they are roughly 8.3 inches square; and now it must be clear to the man just by looking that he has not got what he ordered—999 foot-square tiles. Alternatively, if the tiles look to be about the right size, he will find he has far too many of them to pave the courtyard.

A Wittgensteinian could reply that it is only a contingent fact about us that we are able to tell just by looking at a square flagstone whether it is more nearly 8 inches square than a foot-square in most cases; so no *essential* difficulty is brought to light about applying an inconsistent arithmetic by dwelling on it. But, while any such fact about our discriminations is contingent, is not the game given away if we have to hypothesise increasing degrees of coarseness of visual discrimination to protect the contractor from the patent inadequacy of his arithmetic? Surely if his visual discriminations become *too* crude, the contractor's capacity for the task—indeed, his very ability to measure—will be in jeopardy anyway, and the question of his applying arithmetic to this sort of simple metric situation will not arise at all. If he is to be equal to the job, it must be possible to set reasonable limits on the contractor's hypothesised perceptual abilities; and now it will not be possible to protect him from recognition of the anomaly if his calculation is wild enough and the supplier renders just what is ordered.

This, I think, is wrong. Certainly sufficient degree of crudity of perceptual discrimination would be a great *handicap*; it would, for example, make it very difficult in practice to know when it was in point to re-measure an object, to know when a previous measurement could no longer be relied upon. But the ability to tell of a pair of objects, which one is in a position to compare, whether they are or are not roughly the same in length is logically quite distinct from the ability to tell of a single object with reasonable accuracy within what range of lengths it would fall if measured.

But there are, of course, other points. What if the number at which the contractor arrives by his multiplication is not a square at all? Again, things could work out if it was equal to the product of m and n, m and n lying reasonably close together, so that a flagstone m inches by n inches would not look blatantly unsquare. But there is no reason why things should work out so conveniently. What if the number had no such pair of factors—what, indeed, if it were *prime*? Now he is certain to have tiles over, or too few. And even in a favourable case, for example if 625 were ordered, what does our man say if after completing the work, he happens to count the tiles down one side of the courtyard? If he counts correctly he will find that there are 25—two more than he incorporated into his calculation. What does he say about that?

It is important to see that this sort of consideration does not get to grips with Wittgenstein's attitude. To be sure, nothing is easier than to describe situations in which the attempt to apply the results of a defective calculation would be particularly awkward. But we are not thereby indicating any necessity why such situations should arise. In any particular case it seems we have to concede that things could still work out favourably; the error may be compounded in some way by others, with the effect that they cancel each other out and the job gets done. And now the question arises, why should things not *usually* be so? Is there some metaphysical absurdity about the idea? Imagine a benign demon who protects the contractor's tribe from the consequences of their incoherent arithmetic by swiftly tailoring the facts and/or their apprehension of them to meet their conclusions. If, for example, the contractor orders a prime number of paving stones, the demon might ensure that he does not get what he orders, but some number k, $= m^2$, and that when the stones are delivered the builder miscounts them, concluding that he has received what he sent for.

This fancy brings us to the crux of the matter. It is tempting to say that to have recourse to such pictures is to give away the argument. For it is thereby implicitly conceded that the arithmetic of the tribe is bringing them into conflict with the *facts*—that a change in the facts, or a compounding misassessment of them, is going to be needed to protect the tribe whenever their arithmetic leads them to a bogus result. But what are 'the facts'? From a Wittgensteinian point of view, this whole discussion has involved an illicit supposition. Our objection has been that an inconsistent arithmetic will lead us to adopt spurious rules of inference—rules of inference which will permit the derivation from true premisses of false conclusions. Perhaps we have to concede that in any particular case a false conclusion need not be practically injurious. But it *is*, arguably, methaphysically absurd to suppose that an inferential technique which leads to its being *usually* the case that conclusions were false could generally avoid deleterious consequences. For in order to explain how this might be so, we have no alternative but to open up a general gulf between the true nature of the situations in which the conclusions are drawn and applied, and their character as assessed by those who practice the technique in question; just this is what is achieved by the fantasy of the demon. There have to be compounding errors which the tribe are unable to discern, shifts in the facts which pass unnoticed, etc. If it had been proposed, for example, that *we* might unknowingly be the beneficiary of such a demon, we should have wanted to object along precisely these lines. There is no sense in such a fantasy if—as Wittgenstein especially would wish to insist—the status of the 'true' facts is answerable to our ordinary criteria. But if it is so answerable, we have to recognise that paramount among such criteria are *our* techniques of calculation and inference.

In short, the discussion has been prejudging against Wittgenstein's conception of logic and mathematics as *antecedent* to matters of fact. We have been discussing various ways in which people might get away with false conclusions in practical contexts. But the real point is that there is for a Wittgensteinian no objective distinction, in point of the capacity correctly to assess the world, between our situation and that of the contractor's tribe. We have, that is to say, no right to a point of view from which we may regard our conclusion that the contractor needs 529 flagstones as 'really' true and his conclusion that he needs 625 as false in a corresponding sense. We have the picture that our arithmetic codifies, is faithful to, facts concerning correct counting, the occurrence of errors, mis-measurement—in general, the applied judgments which calculation enables us to make—of an antecedent, objective stature; that we have somehow discharged a responsibility to these facts in adopting the arithmetic which we have—a responsibility in which the tribe has failed—and that we have been able to do so as part and parcel of the ability to recognise the necessity of our arithmetic. But here, as so often, the rule-following considerations are of primary importance in interpreting Wittgenstein's opposing standpoint. It is not just that it is a mystery how we *recognise* that such a responsibility is discharged—there *is* no responsibility to discharge. There are no arithmetic-independent, objective truths concerning correct counting, the disappearance of objects, etc., to which calculation must conform; for there is no objective pattern in our application of these concepts, covering cases to which we have yet to, or did not, apply them.

Suppose the contractor does get into trouble. The stones arrive and there are far too many of them; but when he counts them up, he finds that he has no more than he ordered. Now, it seems, he is bound to conclude that he originally miscalculated, or mis-measured, or that the stones are larger than he asked for. But when he repeats the calculation he makes what we should describe as the same mistake again. And when he checks the dimensions of the courtyard and the flagstones, he finds them respectively 23 feet and 1 foot square. Now he begins to suspect his tape measure, but the calibrations on it seem regular enough. So he double-checks the calculation. Perhaps this time he gets a different result, but again he can find no errors. He may put it down to fatigue; or perhaps he begins to doubt his eyes, or even his sanity . . . We want to say that the explanation of all this is transparent: the man's trouble lies entirely in his arithmetic. Probably we could persuade him of as much. But the criterion for locating the trouble there is supplied by our repeatable, surveyable calculation that $23^2 = 529$. The builder, on the other hand, if we leave him to his own devices, will have to rummage around for some other explanation in the manner just sketched. But, for Wittgenstein, there is no Olympian standpoint from which it may be discerned who is giving the right account of the matter; the concepts of miscalculation, mis-measurement, distortion in the ruler, etc., do not have a life of their own, so that it may be said with objective correctness that here, for example, miscalculation really is the trouble, that it is we who have given the correct explanation. The idea of the 'correct' account here is just that of the account which we give, in accordance with accepted criteria for assessing such matters; criteria which it is the business of arithmetic in particular to supply. By our criteria, the builder is calculating wrong; but there is nothing which we ought to dignify as the apprehension that the rule of inference to which his calculation has led him is in some ulterior sense non- truth-preserving. Statements of the type whose assertion is justified by discordance with accepted arithmetical equalities—'he has miscalculated', 'he has mis-measured', even 'he has hallucinated'—do not have independent, predeterminate truth-values an ongoing concord with which is a pre-condition of any adequate arithmetic. The only superiority which we can claim—and it is no meagre thing—is that we hardly ever get into the kind of turmoil into which, it is practically certain, the contractor's tribe will be repeatedly plunged. But it is no explanation of this to think of ourselves as having apprehended the essential arithmetical structure of the world and systematised it in a uniquely correct arithmetic. We do, *very* occasionally, get into such trouble. When we do, and are reduced to explanations in terms of physical anomalies, hallucinations and the like, it is perhaps worth reflecting that people who 'miscalculated' might in that situation be in no such difficulties.

5.   Let us pass to consideration of the attitudes towards consistency proofs which Wittgenstein thought prevalent and mistaken. Repeatedly he seems to present what would amount to a kind of Cartesian doubt about mathematical certainty. Consider, for example, these passages:

> But then is it impossible for me to have gone wrong in my calculation? What if a devil deceives me so that I go on overlooking something, however often I go over the sum step by step? (i. 135)

> The order convinces me that I have overlooked nothing when I have these six possibilities

—an example to do with a set of possible permutations—

> but does it also convince me that nothing is going to be able to upset my present conception of such possibilities? (ii. 84)

And:

> Can one prove that one has not overlooked anything? Certainly, but must one not perhaps later admit: yes, I did overlook something, but not in the field for which my proof held. (ii. 86).

Part of the point of this is, I think, clear. Wittgenstein is not com-mending a scepticism about mathematical certainty, but attempting to expose what he takes to be an incoherence in the attitude of someone who thinks that a proof of consistency makes things in some way more certain; that it then becomes rational to depend upon a system in a way in which it was not before.

> 'Only the proof of consistency shows me I can rely on the calculus'. What sort of proposition is it that only *then* can you rely on the calculus? What if you just do rely on it without the proof—what sort of mistake have you made? (II. 84)

And earlier:

> . . . *must* a proof of consistency . . . necessarily give me greater certainty than I have without it? (II. 82)

The argument does not seem prepossessing. As directed against a neurotic who is prepared to suspect miscalculation or contradiction anywhere at all, the point is a fair one. Perhaps some workers in Foundations were smitten by such a neurosis when the paradoxes were discovered. How can a consistency proof allay the doubts of someone in such a frame of mind? If he has a general anxiety about overlooked error, it seems he ought to extend it to the machinery utilised in the consistency proof; how he does know that there is not some error in the proof, or that he will not want to revise his conception of the deductive potentialities of the object-system, and later have to admit, 'yes, I did overlook something, but not in the field for which my proof held?' To allay such doubts, the proof of consistency must give reasons for a prediction; not, however

> . . . that no disorder will arise in this way; for that would not be a prediction—it is the mathematical proposition—but that no disorder will arise. (II. 86)

But no mathematical development can refute a doubt on the latter score.

> If I have to fear that something, somehow might sometimes be interpreted as the construction of a contradiction, then no proof can take this indefinite fear from me. The fence that I put round contradiction

—by a consistency proof—

> is not a super-fence. (II. 87)

To set such a doubt at rest, it is not mathematics that is required but therapy.

This, however, is not, today, the general attitude towards the significance of consistency results. Nobody seriously doubts the consistency of classical number-theory, for example; and everyone recognises that a proof of consistency is only as secure as the machinery in which it is embedded. Consistency proofs are interesting when this machinery is relatively weak—or when the consistency of a relatively powerful system of set-theory, for example, is shown to follow from that of a relatively weaker one. So we should be quite content to endorse the remarks at the end of II. 86. The interest of a consistency proof —its practical purpose—is indeed to give us reasons for the prediction that trouble of this sort is not going to originate in the object-system, but these reasons are no stronger than our confidence in the weaker system relative to which consistency is established and in the meta-mathematics in terms of which the result is achieved. Perhaps there were people who felt, or wrote as if they felt, about the matter in the way which Wittgenstein attacks; but in this case at least the attitude which he is attacking has already disappeared. The standpoint sketched in II. 81—

> Well, I shall go on. If I see a contradiction, that will be the time to do something about it.

—would not, I think, be vigorously opposed. Of course, it is *desirable* to have a consistency result; and the greater the disparity between the strength of the object-system and the weakness of the meta-theory—however these notions are to be explained—the better. But we should not in general regard it as irrational to proceed without such a result.

Even this low-key, pragmatic attitude, however, is under pressure from a Wittgensteinian point of view. For it still conceives of a (relative) consistency proof as *unearthing* something, as bringing to light something to do with the inherent potentialities of the object-system(s). We are thinking of a system as having certain determinate deductive capacities, settled by its axioms and rules of inference; the effect of a consistency result is conceived as showing us something about them relative to a certain hypothesis. We learn that it is not in the *nature* of the system to permit the generation of certain sorts of proof-structures unless the hypothesis is false. From this point of view, Wittgenstein's suspicion towards the desire to be able to avoid contradictions mechanically takes things the wrong way round. It is not just that we find the opposing picture, of continuing to work in an inconsistent system and attempting to avoid trouble by a kind of discretion, so difficult to swallow—because of the anxieties about truth and application already discussed. Rather, we conceive of a formal system as *already* a kind of mechanism, and of its capacity to avoid, or generate, contradiction as correspondingly an inherent mechanical capacity, comparable, say, to the incapacity of the first and third gears in a linear train of three to move in different directions, or of three pair-wise meshing gears to move at all. What we want to know is, as it were, what capacities of *movement* have we designed into the system?

In terms of this kind of figure—a favourite of Wittgenstein's[1]—an interesting consistency proof is like a demonstration which shows us that the terminal cog in one, relatively complex, gear-train will rotate as intended if that of another, relatively simple, gear-train will. And, of course, there seems no room for anything corresponding to *discretion*; if some of the gears are put together in such a way as to jam, we cannot, as it were, step in and force the others to move in the way we wanted without unmeshing them. But then we no longer have a *machine* with inherent capacities of movement determined by its design; whereas we wanted a structure that would be adequate to a certain task—transmitting the motion of the engine to the wheels, or generating valid principles of inference—without the need for continual interference. Now it is our use of this sort of picture which Wittgenstein is criticising when he talks (e.g. II. 83) of the misuse of the idea of a 'mechanical insurance' against contradiction. But it is not the *analogy* itself which is under attack. In Wittgenstein's view, error pervades our interpretation of both its sides. We mistake also the sense in which the capacity of the gear-train for a certain sort of movement is 'implicit' in the design. For, tritely, the components of the design do not themselves actually move; all we have is a diagram. The effect of the design, or of our interpretation of it, is that we form a certain conception of what can be done by a physical realisation of it. This act of 'recognition' is *very* like recognising that particular results cannot be achieved by correct application of particular rules to particular postulates. The effect of both is that we adopt certain rules of description: if the gearbox is actually built, and does not function as it 'ought', the rules enjoin us to look for a failure to implement the specification as an account of the matter, or for some instability in the materials, etc. The bare design gives us no guarantee that things will go well when the gear-train is constructed. All that is guaranteed is that if we are satisfied with the design, we shall describe failure in certain sorts of ways.

The relation between the basis of a formal system—the design-specification consisting in its axioms and rules of derivation—and its *practice* is the same. The system does not have an inherent nature which a proof of (relative) consistency brings to light. Its nature resides in our understanding of its rules and axioms, and that is conceived by a Wittgensteinian, motivated by the ideas about rule-following, as objectively indeterminate. The same shadow falls between the specification of a formal system and its practice as between the design of a machine and the performance of its physical realisation. We have the picture of an *ideal* machine which unquestionably would behave in the manner which we seem to be able to read off from the design; and a correspond-

[1] See, for example, *RFM* I. 119–22, 125; II. 20–1, 49; v. 51. Also *PI* I. 193–4; *PG* I. 141; *PB* 231; and *LFM*, lectures XVIII, pp. 178–9, XX, pp. 194–9.

ing notion of ideal practice of a formalised branch of mathematics, properties of whose range of accessible results we may, by a suitable pice of metamathematics, 'read off' from the formal specification. But there is vagueness in our concept of the *variety* of ways in which the physical realisation might fail to be ideal; so there is something peculiar about the claim to have *recognised* that this is how things must go provided we in no way default from the ideal.[1]

The specification of a formal system is, as it were, a highly schematised picture of its practice, and a proof of consistency part of our interpretation of the picture. On both sides of the analogy we have a representation—an engineer's diagram, a specification of axioms and rules —the effect of the relevant pieces of reasoning about which is not to be seen as our recognition of how things can, or cannot, go if the representation is properly implemented, but as persuading us to adopt a certain conception of when it is not. The reasoning gives us no *guarantees* that the implementation will go well—

> . . . a good angel will always be necessary. (v. 13)

All that is guaranteed is that if they do not, we shall not blame the design.

6.   It is, finally, Wittgenstein's scepticism about the objectivity of sameness of use which most fundamentally underlies his attitude towards 'hidden' contradictions.

> We went sleep-walking along the road between abysses. But even if we now say, 'Now we are awake', can we be certain that we shall not wake up one day and say, 'So we were asleep again'. Can we be certain that there are not abysses now which we do not see? But suppose I were to say: the abysses in a calculus are not there if I do not see them. (II. 78)

And previously:

> Let us suppose, however, that the game is such that whoever begins it can always win by a particularly simple trick; but this has not been realised so it is a game. Now someone draws our attention to it and it stops being a game. What turn can I give this to make it clear to myself? For I want to say: and it *stops* being a game, not 'now we see that it wasn't a game'. . . the other man did not *draw our attention* to anything; he taught us a different game in place of our own. (II. 77)

The point is not simply that suggested earlier: that an inconsistent system, like a defective game, may work as long as our innocence is preserved—as long as we do not know how to introduce disorder.

---

[1] The thought here is suggested by *RFM* I. 119–20, read in conjunction with: 'The proof is our new model of what it is like if nothing gets added and nothing taken away when we count correctly, etc. But these words show that I do not quite know what the proof is a model of' (II. 39).

Rather, the status of a calculus as inconsistent is for a Wittgensteinian, like any necessary statement, in need of explicit ratification. It is in this sense that the 'abysses' do not exist if we do not see them. The capacity, or incapacity, to generate contradiction is not something inherent in the system of which we may be in ignorance; nothing 'inheres' in the system in that sense. Before a putative derivation of a contradiction is presented to us, there is an indeterminacy about its status; not that it is undetermined (causally) whether or not we shall accept it—rather it is objectively open what will be the *right* thing to do; there is no verdict in investigation-independent conformity with correct understanding of the concepts involved.

XVII

Proof and Experiment

1.   No question receives more attention in *RFM* than that of the nature of the distinction between calculation, or proof in general, and experiment; to no question does Wittgenstein's thought revert more often. No special explanation is needed for this. The distinction embodies the most obvious, and intuitively most fundamental, contrast between pure mathematics and empirical science: the method of mathematics is the construction of proofs, that of empirical science the construction of theories and their experimental testing. So to account for the character of the distinction between proof and experiment is to be equipped to understand the distinction in character between mathematical and empirical statements—the cardinal problem of the philosophy of mathematics.

Dummett suggests, nevertheless, a more specific diagnosis of Wittgenstein's preoccupation with the question: a strong attraction towards, but a reluctance to accept, the traditional empiricist conception of the meaning of mathematical statements.[1] According to this conception, proposed by Mill but not without more recent adherents,[2] there *is* no essential distinction between mathematical statements and those of empirical science. In particular, mathematical statements do not enjoy a different kind of certainty to that possessed by scientific laws; they are certain in the same way but to the highest *degree*. The attractions of such a proposal are those of a general empiricist epistemology. The view promises both to liberate us from the mysterious aura which seems to envelop the traditional notion of necessity, and to

[1] 23 (2, p. 503).
[2] See, for example, Mackie 60; and of course, famously, Quine—*Locus Classicus: Two Dogmas of Empiricism* in 66.

facilitate an account of the empirical applicability of mathematics; if, for example, '27 + 14 = 31' actually states an empirical law about what happens when counting is done correctly, small wonder that when results seem not to conform to the law, we are almost always able to find a specific mistake.

Someone who was attracted to such an account would be obliged to feel a special concern with the apparent contrast between proof and experiment. If mathematical truths are to be assimilated to scientific laws, how is it that we assert them as the conclusions of *proofs*? What, now, *is* a proof? It looks as if the methods of mathematics might be maladapted to the actual character of mathematical statements. But that would be an intolerable conclusion, outweighing any advantages of the empiricist account. So the empiricist has to explain away the apparent anomaly. One obvious strategy would be to attempt not to account for the distinction between proof and experiment but to conflate it—to attempt to see a proof, or calculation, precisely as a kind of experiment, affording the same type of corroboration to its conclusion as any experiment affords to the general hypotheses for which it constitutes a test case. In the next two sections of this chapter I want briefly to consider the feasibility of an empiricist account. For, whether or not it is right to think that Wittgenstein was attracted to such a view, we shall achieve by exploring its limitations a better understanding of the nature of the position which he actually adopted. Wittgenstein, it seems to me, is to be seen as at one remove from empiricism; though the divide is still considerable.

2.   It was suggested that an empiricist might be tempted to collapse the distinction between proof and experiment. Whether this is so, however, depends both upon the scope of his empiricism and on the specific account suggested of the sense of those mathematical statements for which an empirical reading is proposed. Consider, first, an empiricism about elementary arithmetic of the sort glimpsed above, an empiricism which holds that arithmetical equalities are to be regarded as contingent generalisations about the results of correct counting: about the results which we obtain when groups of objects are amalgamated or partitioned, and the initial and resulting groups counted correctly by operational, that is, result-independent, criteria; and when, again by result-independent criteria, we are satisfied that nothing has been lost or added in the course of the partition or amalgamation. (There is some unclarity whether an explanation of the full range of application of arithmetical equalities is going to issue smoothly from such an account; the nature of the 'result-independent' criteria has to be explained for cases where we are concerned not with observationally discrete objects but with, for example, pooling distinct volumes of liquid. This sort of problem does not concern us now.)

Generalisations of this kind are obviously capable of corroboration by experiment. But the relevant experiments will be *counting-experiments*. It will be a matter of actually uniting or dividing groups of objects, satisfying ourselves that nothing has been lost or added, and then seeing what results we get when the resulting group(s) are, by operational criteria, correctly counted. The generalisation predicts the results of processes of counting. So the question is: what relation to such a generalisation could a *calculation*—an addition sum—have? We think of ourselves as working out the truth of arithmetical equalities by calculation; alternatively, we could supply proofs of them by appealing to the definitions of the integers involved in terms of zero and successor and the recursive definitions of arithmetical operations. But if arithmetical equalities are interpreted as described, what is the *relevance* of such procedures? The fact is that we do not have any direct experimental corroboration, of the sort which the present empiricist account suggests we ought primarily to rely on, for the majority of arithmetical equalities which we are actually prepared to accept. We accept them on the basis of calculations which have no obvious connections with experimentation of the appropriate kind.

Can an empiricist meet this difficulty? Seemingly the most promising strategy for doing so would involve precisely that he should deny that calculation is any sort of experiment at all. The sort of corroboration afforded to the conclusion of a calculation by the calculation is not that of an experiment bearing out a generalisation for which it is a test case. Calculation should rather be assimilated to making an *application* of a set of scientific hypotheses with a view to arriving at a prediction which is then capable of experimental corroboration. It is to be compared, for example, to the application of a hypothesis which postulates a functional interrelation between temperature, pressure and volume in a particular gas. Applied to data concerning, say, the pressure and volume of a sample of the gas, the hypothesis yields a prediction about the temperature of the sample. Similarly, on the suggested view, a calculation enables us to pass from a statement of initial counting data to a prediction about the results of a further counting experiment. If the prediction is successful, then what are corroborated are the general principles of calculation used in passing from the initial data to the prediction. It is these which correspond to the general hypothesis concerning the pressure, temperature and volume in the gas; and it is for these general principles that we have overwhelmingly powerful inductive support, so that we have confidence in the results to which they lead even where we lack direct experimental corroboration for those results. Hence our readiness to accept arithmetical equalities on the basis of calculation alone.

Suppose, for example, that we have calculated that $21 \times 37 = 777$. The general principles used have probably been certain appropriate equalities from the multiplication tables, and certain rules about the columns into which the resulting figures are to be placed and about carrying digits over. These rules, plus the equalities from the multiplication tables for 1–9, suffice for any multiplication. But while the equalities are capable of direct confirmation by counting experiments, the column-rules are not susceptible to that kind of confirmation at all. Asked to justify the latter, we should think it appropriate rather to supply a proof that the technique in which they are embedded yields the right results—that is, results coincident with those of determining the product of *m* and *n* by appealing to their definition in terms of zero and successor and the recursive definition of multiplication. For an arithmetical empiricist, however, the recursive definition would not be correctly regarded as definitional in character; it would rather be a general hypothesis, systematically associating the results of counting experiments as the gas laws systematically associate experimental determinations of pressure, temperature and volume.

As far as it goes, then, arithmetical empiricism of this kind provides no motivation for thinking of calculations as a kind of experiment; quite the contrary. The corroboration afforded to the conclusion of a calculation by its being so is that afforded to any lower level consequence of hypotheses which enjoy a high degree of inductive acceptability. A calculation, on this view, is not an experiment; it is a movement *within* the theory. It is to be compared to the derivation of any lower-level predictive hypothesis from theoretical laws.

Such an approach can obviously be extended to any single branch of pure mathematics. An empiricist account can be attempted of its axioms and theorems without there being any need to assimilate its proofs to any kind of experiment. The strategy, every time, will be to attempt to regard those of its theorems which have a direct physical application as hypotheses associating the empirically determinable parameters involved in their application, and the axioms as a theoretical codification of such hypotheses. A 'proof' of a particular hypothesis is then an application of the theory to envisaged experimental data.

The question arises, however, whether the motivation for seeking accounts along these lines is to be as originally suggested. It *could* be that someone held such a view for reasons which he thought were peculiar to arithmetic, say, so that it did not form part of a more general empiricism. On such a view, the situation would be merely that the boundary between a priori and empirical knowledge has been misdrawn; that what had been viewed as lying on one side was properly conceived as lying on the other. But the empiricism in which we are interested is implicitly *global*. There is to be no such boundary—everything comes on the empirical side. The central motif combines a general scepticism concerning a priori knowledge and necessity with a desire to allow that logic and mathematics deal in truths of some sort which we are capable of knowing in an ordinary sense. So the question concerns not the feasibility of an empiricist account of any single mathematical theory, but that of a simultaneous such account of the entirety of logic and mathematics—of everything, indeed, that is usually considered as known via proof or rational intuition.

The difficulties with the global view are immediate. We were to regard an arithmetical calculation not as an experiment but as the application of certain well-established laws to actual, or hypothetical counting data with a view to predicting the result of another counting process. An arithmetical equality is to be seen as a biconditional linking the results of distinct counting experiments. What account, then, is to be given of the statement that from the laws in question the biconditional follows? This is to be compared to the statement that the gas law yields such-and-such a value for pressure, given such-and-such values for temperature and volume. True statements of this sort would ordinarily be regarded as necessarily so. A global empiricism has thus to provide an account of their content which makes possible their assimilation to contingent statements of some sort while preserving their typical applications.

Generalised, the problem is that of giving an acceptable empiricist account of statements of logical consequence—or the associated conditionals: if Γ, then B. The empiricist wishes to allow that such statements can be known, or at least reasonably believed; what he wants to deny is that they are known a priori. The obvious suggestion would be that our knowledge of such a conditional is based on the empirical finding that whenever Γ is true, so is B. This account, however, would rule out the reasonable believability of the conditional where we *have* no experience of the associated truth of Γ and B, where the inference from Γ to B marks a new transition which we have never made before. So it seems that in such a case the situation must be assimilated to what was said above about the calculation of a new arithmetical equality: our confidence in the truth of 'if Γ, then B' is justified by its being a consequence of certain ulterior logical principles, Δ, for which there *is* powerful inductive support.

This move, however, is clearly no longer of any avail. For a justified confidence in Δ is not enough; we require, in addition, justified confidence in the further conditional, 'if Δ, then if Γ, then B'. Evidently our confidence in this cannot be justified in terms of the original suggestion, that our experience has been of an association in truth between Δ and 'if Γ, then B'—for, by hypothesis, we have no experience of the truth of 'if Γ, then B'. But, equally clearly, to attempt to justify our confidence in the new conditional by seeing it as a consequence of certain ulterior logical laws is simply to reintroduce the problem one stage further on. How, then, can our confidence in acceptable statements of logical consequence be regarded as empirically based at all? It would appear that in attempting to construe as empirical the knowledge which we have of logical and mathematical inferences, the empiricist has made it altogether impossible for himself to regard our confidence in them as justified. I think this signals an essential incoherence in the attempt to construe the validity of an inference as any kind of hypothesis. I shall come back to the matter.

3.   Wittgenstein spends time[1] considering a second empiricist account of the content of statements of logic and mathematics quite different to the sort of thing so far entertained. The suggestion would be that we regard arithmetical statements, for example, as *descriptive* of what tends to result when we are satisfied that calculations have been carried through properly; in general, that logic and pure mathematics are descriptive of our logical and pure mathematical practices. An obvious objection to such an interpretation, in contrast to the kind of account that we have been considering, is exactly that it obscures the *application* of logical and mathematical statements in ordinary practical contexts. The premisses, that there are exactly 7×7 marbles on the table and that we in general compute the product 7×7 to be 49—or find fault with the computation—do not appear to entail that there are 49 marbles on the table. The objections which Wittgenstein himself brings to bear on this account illuminate the respects of his departure from any empiricist view.

Wittgenstein repeatedly reminds himself that the existence of a con-census in calculation and inference, and so the possibility of general descriptive truths of the kind central to this second sort of empiricist interpretation, is of enormous importance.

> This consensus belongs to the essence of calculation, so much is certain . . . in a technique of *calculating,* prophecies must be possible. (II. 67)

> What would happen if we rather often had this: we do a calculation and find it correct; then we do it again and find it isn't right; we believe we overlooked something before—then, when we go over it again, our second calculation doesn't seem right, and so on. . . . I might say: where *this* uncertainty existed there would be no calculating. (II. 73)

And:

> Calculating would lose its point if *confusion* supervened. Just as the use of the words 'green' and 'blue' would lose its point. And yet it seems to be nonsense to say that a proposition of arithmetic *asserts* that there will not be confusion. (II. 75)

The consensus is a pre-condition of the possibility of calculation; unless there can be such a thing as 'a science of conditioned calculating reflexes' (II. 72)—a science successfully describing the way in which we calculate and so successfully predicting the results which we shall get

[1] *RFM* II. 65–6, 72–3; v. 36.

when we are content that we have calculated properly—there can be no such thing as calculation at all.

The experiments of such a science would actually be calculations. Its predictions would be borne out or refuted by what happened on particular occasions when we calculated and were satisfied that we had done so correctly. But would such a science be mathematics? Well, why would it not?

> Are the propositions of mathematics anthropological propositions saying how we men infer and calculate?—Is a statute book a work of anthropology telling how the people of this nation deal with a thief, etc.?—Could it be said: 'The judge looks up a book about anthropology and thereupon sentences the thief to a term of imprisonment'? Well, the judge does not USE the statute book as a manual of anthropology. (II. 65)

How, then, does he use it? Obviously, to *regulate* his performance of his office; the statutes are understood as directives.

This, prima facie, is the really crucial division between Wittgenstein's conception of the meaning of mathematical statements and any empiricist account. The empiricist, Wittgenstein implies, can give no satisfactory explanation of the *normative* character of mathematics (v. 40). If the statute book is regarded as a 'manual of anthropology' and then a judge proceeds to pass sentence in a manner inconsistent with its contents, it is the manual and not the judge who is wrong. There is no *contravening* an empirical generalisation; it is merely shown to be false. But this is not how either statutes of law or the statements of logic and mathematics are used; the eccentric judge is not an exception—he is guilty of malpractice. If mathematical statements are regarded as any sort of empirical generalisation, this normative character is lost. To think, for example, of arithmetical equalities, as earlier, as contingent biconditionals linking the results of distinct counting experiments is to sign away their capacity to sustain the inferences which, where results are not as they require, we always make: that either a miscount has taken place, or that the number of the group has somehow altered between the counting experiments (I. 156).

How can such inferences be justified if the content of the arithmetical equality is simply to describe what happens when such experiments are properly conducted? How can it be rational always to argue that the experiment was *not* properly conducted, that its initial conditions were not fulfilled, whenever it does not turn out as predicted? The same objection arises for the second sort of empiricist account. If '21 × 37 = 777' is a generalisation about the result which we get whenever we are finally satisfied about the correctness of the multiplication, how can we explain the attitude which we adopt to someone who does not get this result? For we do not allow that his calculation may be a rare exception—we say that he *must* have miscalculated.

This normativeness seems to come to this: where logical or

mathematical techniques are used to mediate an inference in a contingent context, we never allow as legitimate any other explanation of the falsity of the conclusion, should it prove to be false, then the falsity of the premises or a misapplication of the techniques. This point, for Wittgenstein, is at the heart of the nature of mathematics. Mathematics tells us what it is sense to say, what conclusions it is allowable to draw.

> I have not yet made the rôle of miscalculation clear. The rôle of the proposition: 'I must have miscalculated.'—

or must have misapplied some germane rule of inference

> —it really is the key to an understanding of the 'foundations' of mathematics. (II. 90).

We do not allow that correct application of logic and mathematics can steer us false—exceptions are not permitted. For Wittgenstein this normative character of the statements of logic and mathematics is something for which any adequate philosophy of mathematics must account; and it is his giving priority to the need for such account that motivates his departure from any form of empiricism. As with the judge, the only admissible explanation of a piece of mathematics providing an apparent exception to what is laid down in the mathematical manual is *malpractice,* of some sort or other.

But is it certain that an empiricist could in no way account for this putative feature of our use of mathematical statements? After all, the 'hypotheses' of mathematics are supposed to enjoy the highest degree of inductive certainty; and if an experiment is carried out whose results prima facie conflict with a theory which we regard as well established, we *are* inclined to say that the experiment must somehow have been misconducted. Getting results which conform with a well-established theory is accepted as a criterion for the correct conduct of schoolroom experiments; well-established scientific theories are, up to a point, used normatively—as criteria for non-observance of relevant conditions in a prima facie recalcitrant experiment.

But, it is tempting to reply, *only* up to a point. If the 'wrong' result is repeatable, and ordinary criteria suggest that the conditions of the experiment have been properly observed, we concede that the theory is under pressure in a way we never should with a mathematical statement. If matters turn out repeatedly other than as predicted by inferences mediated by accepted arithmetical statements, for example, we never allow that arithmetic is comparably under pressure. On the contrary, appealing precisely to arithmetic, we conclude that things *must* be other than as they seem—that there must have been a miscount, or some other peculiarity, even if non-mathematical criteria most strongly suggest that nothing of the sort has happened. We go on insisting that *something* of the sort must have happened, even if it means doubting the veracity of observations. If the phenomenon is repeatable, we shall look for a physical theory to explain why in these circumstances we are unable to locate the error, or other peculiarity, which must have taken place. And this is the crucial distinction. For we regard physical theory as the slave of observation; just the point at which we suppose it ceases to be rational to accept a physical theory is where its capacity to accommodate repeatable data requires the construction of a further theory to explain them away. With mathematics it is the other way about: the veracity of observations is subservient to their conformity with accepted mathematics.

An empiricist can apparently give no explanation of this treatment. From his point of view, it would amount to holding to mathematical theories as non-empirical *dogmas*. Of course, we are hardly ever called upon to display this attitude; but it is characteristic of our acceptance of something as a logical or mathematical truth that we *should* so hold to it—from an empiricist point of view, irrationally and dogmatically—if appropriate circumstances arose. A statement is used normatively if it is used unfalsifiably to prescribe certain forms of conduct—in the cases which interest us, to prescribe certain forms of description, revisions in our estimate of a situation, or modification to accepted beliefs. Physical theories are used normatively in proportion to their degree of inductive support; but mathematical statements are used *absolutely* normatively—and the explanation of that cannot have anything to do with inductive corroboration.

Appealing as this view is, it is open to doubt whether it is true. The precedent of Euclidean geometry suggests doubt. For we did *not* go on saying that astrophysicists must have taken faulty readings or miscalculated, etc. We modified our geometry. And why should the same thing not happen even to arithmetic? Would we really go on postulating that some sort of error, or other peculiarity, must be involved if our arithmetic repeatedly landed us in trouble, and the theories required to explain away the rogue data proved to be enormously complex? It is natural to protest that, if we did modify arithmetic, the inescapable price of doing so would be to change the meaning of our arithmetical language. But if that is meant to suggest that a modified arithmetic could somehow not *cohere* with the present interpretation, it is not a protest which the Wittgensteinian would want to make. In so far as it is acceptable, it is trivial; we should have changed the criteria which we used for the veracity of certain observations—certain antecedently normative statements would no longer have that status. It is unclear whether any statements are used absolutely normatively in the manner described above, if that is taken to require that their status if not revisable. Hence it is not a clear failing of empiricism that it cannot account for our assignment to logic and mathematics of an absolutely normative rôle, cannot account for this putative aspect of the way in which the statements of logic and mathematics are actually used. The justice in Wittgenstein's suggestion that there is in this area a point which the empiricist cannot accommodate has still to be brought out.

What, then, is the justice in this suggestion? It has to do, I believe, with the misgivings, voiced in the preceding section, about the *coherence* of the global empiricist's idea that the entirety of logic and mathematics is merely a network of well-established hypotheses. The problem for the empiricist is not to explain why logic and mathematics are used absolutely normatively; it is to explain how the practice of scientific theories is possible at all.

A natural objection to global empiricism, more immediate than that presented earlier, is that hypotheses become well-established precisely by yielding consequences corroborated by experience; so it seems that the status of something *as* a consequence cannot itself be hypothetical, or we could not verify that a particular hypothesis was well corroborated. But this implication, of course, is something which empiricists have been known to embrace. The hypotheses of a theory do indeed not receive confirmation individually; rather it is the whole network which is subject to conformation—the notion of confirmation is itself *holistic*. Thus it is right not to regard ourselves as capable of certainty that a particular hypothesis is well corroborated; confirmation essentially suffuses over the entire theoretic framework into which it is integrated.

The holistic conception, however, seems merely to postpone the difficulty: how can we know even of the entire theory that it is corroborated by, or inconsistent with, particular findings unless it is no mere hypothesis that it has the consequences which these findings verify or contradict? Even in terms of a holistic standpoint, it appears, it has to be possible to do better than hypothesise about what results conform with the theory as a whole—if it is to engage with reality in an ascertainable way. The price of the opposing conception, that the consistency, or inconsistency, of a theory with particular findings is always itself hypothetical, is that it is unclear what operational account can be given of when it is right to consider the theory to have proved inadequate. If it *was* itself a hypothesis that the theory is consistent, or inconsistent, with certain particular findings, an explanation would be mandatory of the conditions under which *this* hypothesis could rightly be considered as acceptable. Until we had such an explanation, we should not know what move to make. But nothing would constitute a satisfactory operational explanation of this which did not assign to this hypothesis in turn determinate conditions of acceptability and rejection. And after such an assignment is effected, the statement that, for example, the hypothesis is inconsistent with such-and-such states of affairs, whatever it is, is clearly no *hypothesis*.

The question is: how are we to get any purchase on how to apply and revise a theory if it is conceived in the global empiricist manner—if everything, including the underlying logic, is conceived as hypothetical in status? How, if the idea of what is compatible with the theory is itself conceived as a hypothesis integrated within the theory, can we get the concept of confirmation started at all? It may possibly be true that our concept of the conditions of justified acceptance or rejection of any particular hypothesis within any theory is irreducibly holistic; that hypotheses do not confront the world singly, as it were, but only by reference to other theoretical assumptions can the compatibility or incompatibility of a certain situation with a hypothesis be explained. Certainly, the view is plausible in at least two initial respects. First, even in what we think of as the most direct kind of falsification of a hypothesis, the *veracity* of the data is a theoretical notion—would require appeal to theory if questioned; and, secondly, the *bearing* of the data upon the hypothesis requires that no unmonitored variation should have taken place in any parameters whose relevance the hypothesis, or its background theory, postulates—which, again, is a question we have to approach armed with theory. Nevertheless, if a given theory is to be a structure which we understand how to modify in the light of experience, we shall require, sooner or later, not hypotheses concerning the findings which a theory does or does not tolerate, but *rules*. Let us grant, for argument's sake, that it is only relative to a fixed anchor of background theory that a particular situation may be conceived as conflicting with, or conforming to, a particular hypothesis; then the present suggestion is that the 'fixed anchor' must contain not only theory but directives. In particular, the underlying logic of a theory cannot be supposed to be itself under assessment when a key experiment is constructed; for it is only within the network of inferential and descriptive rules which the logic supplies that we can give sense to the idea of an *apparent* conflict of the results of the experiment with the theory.

This is not to hold that the logic must be altogether immune to revision; still less is it to allow that it has a unique epistemological status—that the fixed point supplied by logic is somehow apprehensibly absolute and imposed. It is simply that in any situation in which the confirmation or disconfirmation of the theory is in prospect, we will be in a position to arrive at an assessment of how the theory fares only if we can agree in assigning a normative status to certain statements articulating its consequences for that situation. Merely to argue for the correctness of a particular assessment is already to bring the underlying logic of the theory to bear, to accept certain of its theorems as not themselves here open to question but as regulative of what courses of action are open to us. Unless we are in a position to proffer argument of that kind, moreover, we do not know how to practise the theory; and not to know that is, in effect, not to have a theory at all.

The charge against the global empiricist, then, is not that he cannot explain why the 'hypotheses' of logic and mathematics have an absolutely normative rôle—why they are beyond revision in any conceivable circumstances; it has yet to be made out that they are. The charge is rather that, in his attempt to conceive of everything simultaneously as hypothetical in standing, the empiricist makes it impossible to explain how logic and mathematics, or any hypotheses, can *secure* the high degree of inductive support which would explain why they come so close to absolute normativeness. Normativeness cannot in general be explained by depth of inductive corroboration; rather, some statements have to be assigned a normative rôle as a precondition for the receipt by any hypotheses of inductive support. In Wittgenstein's view, we have assigned this rôle to the statements of logic and mathematics. Our acceptance of the soundness of inferences is not in general inductively supported—we have seen that it could not typically be so; rather, it is licensed by the application of norms of a kind whose currency is presupposed if scientific knowledge is to be possible at all.

The proposal is thus that an attempted global empiricism about necessity dialectically inflates, as it were, into something akin to Wittgenstein's conception. If we are to be equipped with the means to agree when a particular scientific theory licenses a certain expectation, or when modification of the theory is called for, then directives have to be supplied concerning such judgments, whose rôle will be normative of our assessment of a situation as of these or other relevant kinds. Wittgenstein's contention is that it is the apparent 'statements' of the logic and mathematics in which the theory is embedded which have this essential function; it is this which gives them their 'peculiar unassailability' which we are inclined mistakenly to interpret as necessary *truth*. But, to repeat, this picture of the matter does not preclude that it might prove expedient to revise the directives. What is precluded is that it could prove expedient simultaneously to revise them *all;* something has to be held constant in terms of which the inexpediency of the practice of inferring in accordance with the other directives can be appreciated. Suppose we have a theory, embedded in a certain logic and mathematics, which repeatedly yields predictions confounded by observation. It is at least a possibility that the simplest way to square the theory with those data, while preserving its domain of success, is to tamper with certain elements in the logic and mathematics rather than the theoretic basis. But the intelligibility of the supposition requires that it remains our ambition to avoid conflicts with the facts—that we are not to tolerate a situation in which the theory, fed with certain initial data, entitles us to assert certain statements which by other (observational) criteria we are entitled to deny. And in contemplating such a revision, evidently, we continue to hold it unquestioned that from the theoretic basis and certain initial data, and by the underlying logic and mathematics of the theory, the particular results follow which we cannot square with observation. That is, there is no means of understanding what it is for data to be *recalcitrant*—for them to put pressure on a theory plus its underlying logic—save by reference to judgments of the proof-theoretic capacities of that logic, and the principle that there is to be no simultaneous entitlement to assert inconsistent statements. A verdict of recalcitrance is intelligible only against a background of judgments and principles which are not called into question.

I suggested earlier that Wittgenstein could be seen as standing at one remove from the empiricist account. For an empiricist all knowledge and all rational relief is, ultimately, derived through the senses. Vague as this conception is, necessity immediately presents a problem to it. Anyone but a sceptic will allow that on the basis of sense experience we can know, or reasonably believe, how things are; it is, however, quite unclear how we could recognise by such means how things *must* be. The major point of similarity between empiricism and Wittgenstein's attitude is a rejection of the claim that we possess such knowledge. But whereas the empiricist succumbs to the temptation to try to interpret this apparent knowledge as empirical, Wittgenstein may be seen as embracing from within the framework of a general empiricist epistemology the other horn of the dilemma: this seeming-knowledge is not factual knowledge at all. The 'statements' of logic and mathematics are not fact-stating; their rôle is regulative—it is to supply the procedural guides which, in the sense we have tried to bring out, make factual knowledge possible.[1]

If these ideas are sound, a double insight is promised. First, we should not look to understand the distinctive unintelligibility, which we seem to sense in the suggestion that things might be other than as logical and mathematical principles seem to state, by attributing it to a rational intuition that they transcribe essential features of the world; it is rather, at least in part, occasioned by a loss of guidance as to the use of the statements which they enable us to link inferentially. Secondly, a suggestion begins to emerge about how a Wittgensteinian might meet the difficulty with which Chapter VI concluded: the difficulty of explaining the 'deep need for the convention', of explaining why there are any necessary statements at all. We shall return to this suggestion in Chapter XXI.

4.   If we adopt a Wittgensteinian conception of the nature of necessary 'statements', it is clear that we have to forego one traditional respect in which a proof, or calculation, *is* assimilated to an experiment: we can no longer see a proof as, like an experiment, a procedure whereby the truth of something—its conclusion or 'result' —is *found out*. Rather we are bound to attempt, as Wittgenstein commends, to see the effect of a proof as persuading us to concede to its conclusion the status of a rule. And we can no longer interpret our concession of this status to the conclusion as a product of our apprehension of its essential correctness.

[1] Cf. *RFM* v. 15; and *OC* 105, 210–12, 340–4, 509.

There has to be a disanalogy between proof and experiment just in the respect in which they are traditionally assimilated.

What Wittgenstein is trying to do in the repeated discussions of proof, or calculation, and experiment is to bring to light certain features of proof, of the mechanisms of proof and what happens when proofs are worked over and accepted, which count against the traditional account and suggest instead his own conception. How successful is he in this regard?

Wittgenstein grants that a proof always *can* be used to bring to light something in the same way as an experiment. Consider, for example, the drill that is done with the marbles (1. 36) as a proof that $10 \times 10 = 100$. The process could be regarded as experimentally demonstrating that the marbles can actually be moved around in certain ways, that they were not, say, glued to the table; or the experiment could be concerned with finding out which patterns we find memorable, or as investigating whether or not we are easily decieved if an extra marble is smuggled in (1. 75). The manoeuvres with the chain (1. 79) could be regarded as a geometric proof (of a highly complex proposition), like solving a metal-puzzle; but equally they could be part of an intelligence test, or an investigation into the mobility of the links or the tensile strength of the metal, etc. In the same way, a calculation can always be a psychological experiment into, for example, what strikes the subject as agreeing with paradigms (1. 156), or simply as determining whether he can calculate properly (II. 67). But the conclusions of such experiments are not the mathematical propositions which we regard these processes as demonstrating qua proofs or calculations; indeed, they are not *mathematical* propositions at all.

This seems unexceptionable. But now, what is the difference between these processes, taken in these various ways as experiments, and their being treated as proofs? Wittgenstein seems confident about what sort of distinction this is. 1. 160:

> Is a calculation an experiment? —is it an experiment for me to get out of bed in the morning? But might it not be an experiment, to show that I have the strength to raise myself up after so-and-so many hours sleep? And how does the action fall short of being this experiment? —merely by not being carried out with this purpose, i.e. in connection with an investigation of this kind. It is the use which is made of something that turns it into an experiment. . . . (161) If a proof is conceived as an experiment, at any rate the result of the experiment is not what is called the result of the proof. The result of the calculation is the proposition with which it concludes; the result of the experiment is that from these propositions, by means of these rules, I was led to this proposition.

But now, we want to ask, why is it not similarly the context of a particular *sort* of investigation which determines that the procedure is to be assessed as a proof? —an investigation, namely, into the mathematical facts. And why should we not, therefore, regard such procedures when viewed as proofs as bringing to light the truth of what is stated by their conclusions?

Wittgenstein's counter-arguments are various. Consider two of them. First (I. 55–7): often we draw the conclusion after only going through the proof *once*. Suppose, for example, this construction is given as a proof of the proposition that a rectangle can be made up of two like parallelograms and two triangles half their area:

(I. 50)

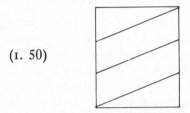

Now, how can carrying this construction out just once establish the general conclusion? Should we not perhaps just concede the more modest proposition '*This* rectangle consists of two like parellelograms and two triangles half their area'? The same question arises for calculations. How is it that calculating the product of 17 and 19 just once can establish that $17 \times 19 = 323$? If the calculation is an attempt to establish something general, ought we perhaps to insist that carrying it out just once is not enough? If the calculation is performed only once, should we not just concede the modest proposition that *this time* 17×19 proved to be 323? Strikingly, however, if we understand the 'more modest' propositions in such a way that they at least can have been demonstrated, even if the generalisations have not, then that is to understand them as something other than instantiations of those generalisations which we should ordinarily take the construction, or the calculation, as establishing. In the geometrical case, indeed, we may not have established Wittgenstein's 'more modest' proposition at all; we may very well not have verified that the figure really is a rectangle or that two of its constituents really are parallelograms. They may very likely not actually be so; the figure may be badly drawn without jeopardising its status as proof. So it cannot be claimed that we have at least established the particular case.

The last point seems not to apply to the calculation. There is no practical doubt about whether 323 resulted, comparable to a doubt about whether the resultant figure really is a rectangle. But generalisation of what is certain, viz. that on this occasion a seemingly correct multiplication of 17×19 produced the result 323, yields merely the proposition that, whenever 17 and 19 are apparently correctly multiplied, 323 will result —and *that* is not the mathematical proposition that $17 \times 19 = 323$; nor is the geometrical proposition which the above diagram, somewhat laconically, establishes, equivalent to the generalisation that wherever what appear to be two like parallelograms and two appropriate triangles half their area are arranged in this manner, the resultant figure will impress as a rectangle. When the diagram, and the calculation, are viewed as proofs they do not build on any apprehension of what has happened in the particular case more secure than the generalisations which they demonstrate. The propositions which they prove are not generalisations of any proposition which is in this way relatively secure—which could reasonably be accepted while they were held in doubt. So the essential rôle of the diagram, or the written calculation, is other than to be recognised to *exemplify* the proved generalisation.

The second argument is this. We talk of 'unfolding' the properties of a mathematical structure, of 'drawing someone's attention' to what can be done with the hundred marbles, for example, or with the chain. But we should not speak of 'unfolding' the melting point of a bar of iron, or 'drawing someone's attention' to something which it takes an experiment to bring to light; we can only draw his attention to what he is already in a position to notice for himself. These nuances signpost something important: proof and calculation can take place in the medium of *imagination* (1. 80, 98). This is not to say that proof is fundamentally a private mental activity, as the intuitionists historically thought. It could be insisted that the imagery be capable of some concrete realisation —a diagram or a film, say (1. 36). The point is that it does not matter whether we carry out the actual drill with the hundred marbles, or whether we draw a cartoon film of the process, or represent it in some other way. The 'mathematically essential' thing is apparent in any case. What comes to the same thing: there can be no 'rigging' of the film in the way in which we could rig a film so as to suggest, for example, that iron melted in boiling water. Not that the film could not be doctored so as to distort what actually happened with the hundred marbles; but if it shows 100 spots going without loss or remainder into 10 groups of 10, then, whatever actually happened with the hundred marbles, that is good enough. Now, how can we get to know any property of the *real* world just by imagining, or drawing cartoons?

The natural response to these considerations is as follows. They may have some power as directed against an empiricist conception of mathematical fact and the notion of proof as experimental corroboration, but they altogether pass by the idea of a mathematical fact which we should actually want to endorse and the conception of a proof which corresponds to it. If 'a rectangle can be constructed out of two like parallelograms . . . etc.' is thought of as a general hypothesis, and the construction carried out in the proof as an attempt at its experimental confirmation, then it is indeed obscure why we accept a single, badly drawn construction as entitling us to assert it. And it is indeed obscure how an act of imagination can establish an empirical generalisation, how drawing a cartoon can teach us something about the real marbles. But in terms of the ordinary, non-empiricist conception of proof and mathematical fact, are these matters so difficult to understand? According to the ordinary conception, pure mathematics deals in a special category of purely conceptual truth—truth dictated by the rational understanding alone. And a proof of a mathematical statement is a procedure which calls our attention to aspects and implications of our understanding of it, which brings the character of the involved concepts into view in such a way that we are enabled to discern that what is stated must be true. How this is done varies; the statement may be exhibited as the terminus of truth-preserving inferences, driving us to admit that it is so understood as to have its truth a consequence of that of previously accepted necessary truths. But this, obviously, cannot be the *general* form of recognition of necessity. In other cases we feel that we need no proof; or better, the proof is just to think about the sentence—we seem to have an immediate apprehension of the implications of the way in which the sentence is understood. In others, again, like the example above, intuition is jogged by the use of a diagram. Here we tend to think of the figure merely as an heuristic tool; it is used to focus someone's attention on features of his understanding of certain concepts, though the same end could likely be achieved in all sorts of ways.

Thinking of mathematical truth, and necessity in general, in this sort of way encourages us to see no difficulty in the idea that the *same* necessary statement can have various proofs; there seems to be no reason why we should not be lead to the same feat of conceptual cognition in different ways. It is suggested, too, that our need for proofs is a function of our intellectual limitations; that a being is conceivable to whom the apprehension of necessity was *always* immediate. But this is only suggested, not required. We could repudiate the idea of a *direct* apprehension of the truth of, say, Fermat's 'theorem', while still allowing that the effect of a proof of that statement would be to articulate and display the character of our understanding of the concepts involved, and thereby to reveal a crucial implication of that understanding—the statement's necessary truth.

In terms of this view, we are inclined to think that the facts to which Wittgenstein called attention, while anomalous for an empiricist, are not so for us. There is no longer any obvious puzzle why it can suffice to carry through a construction just once in order to establish a mathematical generalisation, or why what is empirically certain about the particular construction is not an instance of the proved generalisation. For we want to suggest that the way a proof leads us to an apprehension of conceptual generality does not require empirical knowledge of the actual character of the constructed particular case. Equally, there is no obvious puzzle why a proof can take place in the medium of imagination; the rôle of the proof is to display to us the character of our understanding of certain concepts, and this can be done as well by working with an imagined realisation of them as with an actual one. What is important is that we should see the particular construction *as a* realisation of the relevant concepts —which, in geometrical cases, we are sometimes able to do even if the figure is overtly badly drawn. The proof works by calling our attention in terms of this vision to the essential relations of these concepts. No separable, non-general conclusion has to be demonstrated about the particular figure or calculation; they serve only to mediate our apprehension of the essential character of the concepts, 'rectangle', 'parallelogram', 'correct multiplication of 19 by 17', etc. It seems, then, that nothing has yet been done to disrupt the conception of proof in mathematics as a means of discovering facts. Rather the two points adduced by Wittgenstein seem to be reflections of the special character of the facts which mathematical proofs lead us to recognise.

We should want to make a similar point about another disanalogy which Wittgenstein stresses: the difference between the concepts of repeating a proof and repeating an experiment (II. 55; v. 6). Nothing counts as repetition of a proof which does not get the same result. From an empiricist point of view, this would look like rank prejudice—as if we were only prepared to countenance data which are convenient. But we should wish to say that this disanalogy is just a reflection of the fact that it is the business of the proof to demonstrate an *essential* conceptual connection, and such connections preclude variation in the outcome. Only one outcome is *possible* when 17 is correctly multiplied by 19, for the truth or falsity of any statement of the form '$17 \times 19 = n$' is settled just by the character of the concept of correct multiplication and the identity of the integers involved. The rôle of the calculation is to display an essential connection between these concepts, and thus to tell us what the outcome has to be.

5.   So easily does this way of looking at things come to us that we are apt to mistrust any opposition to it. But the truth is that it makes a mystery out of the proving-power of a wide class of proofs. How exactly, in particular, is a calculation supposed to *display* an essential connection? We noticed that the mathematical proposition which it proves is not to be seen as a generalisation of anything about the particular construction of which we are empirically certain. What is empirically certain is, for example, that no error is apparent in this calculation and that its result is 323; but the mathematical proposition is not that whenever we compute the product of 17 and 19 and can find no error, we shall have got the result 323—it is that whenever the calculation is done *right*, we shall get that result. Yet how do we see any more in the calculation than the aspects of which we can be empirically certain?

The question applies to any proof involving a process of construction; the proof succeeds only if it is demonstrated that the process essentially culminates in a certain outcome—though this is not necessarily the proposition which we are trying to prove. *How* is this recognised? And what is the rôle in its recognition of our empirical awareness of the properties of the particular construction? The ordinary conception comes close to suggesting that this empirical awareness has no *essential* rôle; the particular construction is viewed just as an aid to apprehension of a connection of concepts. But whatever plausibility this has for a proof consisting of a single, possibly badly drawn figure, surely it cannot be right when the proof consists of a *process*. For then we are not satisfied with the proof until we have checked it over; and if this is not an empirical investigation of the proof, what is it?

Let us try to track down this suspicion by reverting to an issue on which we touched towards the end of Chapter VI. Wittgenstein talks of the 'equivalence of process and result' in mathematics, of the method 'presupposing' the outcome (I. 82–7; III. 40, 50–1).[1] We want to say *both* that we learn what the product of 19 and 17 is by doing the multiplication, that the calculation teaches us the mathematical fact, *and* that the correct multiplication can only yield one result—to get anything else would be a criterion for having miscalculated. Wittgenstein seems to think that these claims are in tension. Recall the previous analogy: 'Jones has it in him, if suitably trained, to become a great artist.' It was pointed out that there can only be evidence for this statement if an account can be given of what a 'suitable training' is in this context which does not presuppose the outcome, the emergence of a great artist. But what is meant by 'presuppose' here? If the example is to be a significant empirical conjecture, it has to be possible to set up the conditions of the experiment, viz. Jones' receiving the suitable training, without settling in any but a causal sense what the outcome will be; so it must not be a conceptually necessary condition for Jones' receipt of a suitable training that he should emerge as a great artist. If it is, the result has been incorporated into the concept of the correct application of the method; and that means that the conditional can no longer be confirmed by experience—there is no finding out that it is true, or false, of Jones by implementing the antecedent and seeing what the outcome is.

Now consider the conditional 'If the product of 17 and 19 is correctly computed, the result will be 323'. We are inclined to think that it is possible to be in ignorance of the truth of this, and to find out that it is true by doing the calculation and seeing. But can this be coherently maintained if getting 323 is a conceptually necessary condition of correct execution of the multiplication—if nothing counts as doing the multiplication correctly unless the result is 323? There would be no difficulty if, with the mathematical empiricist, we regarded the conditional as an empirical conjecture—for then there would be no question of its being a conceptually necessary condition of the correctness of the multiplication that 323 should result. The problem is that we do *not*

[1] See also *RFM* I. 112; *PG* II. 39 (p. 457); *LFM*, lectures v, p. 73, VIII, p. 79. And *T* 6.1261.

regard the conditional as empirical, but we do regard it as capable of something in some respects *akin* to experimental corroboration: doing the calculation correctly is like setting up the conditions of an experiment, and then we see what the outcome is. Is there any room for this attitude? By regarding the outcome as implicit in the process, have we perhaps left ourselves no scope for even this partial analogy with experimentation, and so for the idea that the calculation informs us of something?

We are confronting the following dilema. If we allow that it is *possible* to do the calculation correctly and not get 323 —that is, if it is not a conceptually necessary condition for the multiplication's being correct that any particular outcome is obtained—then while it now makes sense to think in terms of experimental corroboration of the conditional, we can no longer suppose that the calculation teaches us something of the *essential* character of the concepts involved. But if we do not allow that it is possible to do the multiplication correctly without getting 323 as the result, then in order to assure ourselves of the correctness of the multiplication, we have to verify that we have got this outcome. And now, it seems, as with Jones' potential artistic greatness, we cannot *learn* anything about what happens when the multiplication is done correctly just by doing it, for its status as correct—cf. the status of the training as 'suitable' —cannot be allowed unless we get the presupposed result.

The ordinary view that the calculation informs us of the necessary truth of the conditional is purportedly squeezed by this dilemma into a vanishing space between Wittgenstein's suggestion, that its effect is rather to persuade us to adopt a new rule, and mathematical empiricism. And the latter is no option in general if the thought of the earlier sections of this chapter is sound. There is, however, no doubt how a defender of the ordinary view would want to respond to the dilemma. He will claim that its second part contains a non sequitur. It does not follow from its being a conceptually necessary condition of the correctness of the calculation that 323 should result that in order to assure ourselves of the correctness of the calculation, we have to verify, in the sense presupposed in the dilemma, that the result is the 'right' one. The reason why it is possible to *discover* the product of correctly multiplying 17 by 19 is precisely that the notion of correct multiplication is not *explained* by reference to any particular outcome in this case; so though it is indeed a conceptually necessary condition of the correctness of the multiplication that 323 should result, it is, antecedently, no *criterion*. There is an epistemic independence between the concept 'correctly multiplying 19 by 17' and the concept 'getting the result, 323'. The idea of a correct multiplication of 19 by 17 is understood in terms of a quite general account of how to multiply correctly of which we come to recognise it as an essential implication that 323 should result if 19 and 17 are so multiplied. It is this epistemic independence which saves the possibility of discovery here. In contrast with the situation of the conditional about Jones, we can give an account of how to implement the antecedent of the conditional about the multiplication without any reference to the outcome at all. Discovery is always possible that when such-and-such is done, so-and-so is the outcome, provided that the criteria for having done such-and-such do not *explicitly* involve that so-and-so should result. What is idiosyncratic in the example about Jones is that this is not so—if it were so, it would be possible to have evidence for the truth of that conditional. In short, by the use to which he is trying to put the notion of 'presupposition' in this context, the inventor of the concept of a criterion as a guideline in judgment confuses it with the concept of a logically necessary condition.

Wittgenstein, it seems to me, succeeds in drawing attention to two grounds for dissatisfaction with this reply. The first is suggested in this passage:

> . . . I want to say that we have no right to say: though we may be in doubt about the correct reverse of e.g. a long word, still we *know* that the word has only *one* reverse. 'Yes, but if it is supposed to be a reverse in this sense, there can be only one'. Does 'in this sense' here mean 'by these rules' or 'with this physiognomy'? In the first case the proposition would be tautological; in the second, it need not be true. (III. 50)

The point, if I interpret it correctly, is this. We obviously have to concede that once getting 323 is inbuilt epistemically, in the above sense, into our concept of a correct multiplication of 19 by 17, there can be no quasi-experimental learning by calculation what is the product of 19 and 17. But so long as the concepts remain epistemically independent, that is, so long as it is not yet a criterion for the correctness of the multiplication that we should get that particular result, it will be conceivable that a multiplication of 19 by 17 should *seem* correct—that it should have the 'physiognomy' of correctness—while giving some other result. When we do a calculation which we have never done before, we are ready for any result falling within a certain vague range. It is antecedently intelligible to us that any of these integers should be the result; that is, so long as we get one of them, we will not find fault with the calculation on sole grounds of the result; and it may, of course, be that the calculation seems in other respects correct. The question is, how are we to reconcile these obvious facts with the view that the calculation enables us to discern something of the *essential* character of correct multiplication? We are supposed to learn that a correct multiplication of 19 by 17 essentially produces 323; but we have to allow that before we know what we are *supposed* to get, a computation of 17×19 may seem correct to us while giving a different result. How, at that stage, do we make the transition from knowing the result which we have obtained by a seemingly-correct multiplication to knowing that it is the result which a genuinely correct multiplication of these integers *has* to give?

This brings us back to our earlier suspicion: it really contains no explanation at all of the mechanism of our recognition of a proof to think of the particular construction as a mere aid to our apprehension of an essential conceptual connection. With the kind of proof of which the calculation is illustrative—an ordered array of formal transformations—it seems that an empirical awareness of the character of the particular construction must have an essential part to play. In contrast with the badly drawn geometric figure which somehow succeeds as a proof, we *do* investigate such constructions and their status as proofs turns on the result of the investigation. But it seems that the most such an investigation can tell us is that starting from such-and-such a basis, so-and-so resulted and no error is apparent in the intermediary steps. How then do we make the transition to recognition of necessity?

The point here is not the sceptical one that an unnoticed error may always remain, that a seemingly correct calculation may never be known really to be so. It is that while in any ordinary sense we can certainly claim to know that the multiplication is correct, the criteria for this knowledge seem to have nothing to do with an intellectual perception of the essential nature of the concept of 'correctly multiplying 17 by 19'. What actually happens, if the transition from seeming correct to being correct is called into question, is that we go over the calculation several times. If no error comes to light and we go on getting the same result, then we are satisfied—and, if somebody is not satisfied, we do not know how to satisfy him. But if the distinction between a seemingly correct and a genuinely correct multiplication is drawn in this ordinary way, then—the point on which a sceptical doubt would fasten—it is consistent with our being rightly satisfied by ordinary criteria that the calculation is correct that it should actually not be so. How, then, by doing the calculation and checking it over can we learn something about what the result *has* to be if the calculation is *really* correct? We can learn that repeated, apparently correct calculations all give the same result; but how can we learn that this is the result which we *must* get if the calculation really is correct? The ordinary view is that our knowledge of this conditional is based on an apprehension of the 'essential nature' of a correct multiplication of 17 by 19. This view sits very oddly alongside the plain fact that we discover this 'essential nature' quasi-experimentally by constructing an instance of that concept for whose status as such we have no absolutely non-defeasible means of recognition. Indeed, just because a sceptical doubt is possible, it begins to look as though our treatment of arithmetical conditionals of this degree of simplicity as absolutely certain and guaranteed is unwarranted. Wittgenstein's suggestion is rather that the warrant is not wholly cognitive in source.

The point applies to any quasi-experimental 'discovery' of the necessity of any proposition of the form 'x results from y by such-and-such a transformation'. We regard ourselves as having recognised the necessity of such a proposition whenever we accept a proof consisting of an ordered array of sentences. But so long as getting y is not incorporated epistemically into our concept of such-and-such a transformation on x, we shall find it conceivable that a seeming-instance of the transformation should produce any among a variety of results; so how does carrying the transformation out, and repeatedly getting y when it seems correct, teach us that this is what must result if it really has been carried out correctly? Is it not rather that we have, so to speak, prepared ourselves for the obtaining of an internal relation here, and then pick on what proves to be experimentally the most convenient candidate (cf. 1 inch = 2.54 cm)? On the other hand, once process and result are epistemically interwoven, there is no confirming the connection by carrying out the process.

The second ground for dissatisfaction, at least for a Wittgensteinian, in the intuitive rejoinder to the original dilemma issues from the rule-following considerations. It has to do with the supposition that the character of someone's understanding of certain concepts can be opaque to him in the manner which the rejoinder presupposed. The Wittgensteinian will not accept that the invoked idea of epistemic dependence may legitimately be regarded as a narrower notion than the conceptual dependence with which it was contrasted. This distinction will not be accepted because of its implicit play with the possibility of unacknowledged internal relations. Unless we accept that such relations can obtain unacknowledged, we have no access to the ordinary notion that any so-far untried calculation has a unique proper outcome —something which we do not yet know of, but which is the only possible result if the calculation is carried out correctly. Not that, if we accept a proof, there is any reason why we should not say that that shows that we were understanding the concepts involved in such a way as to admit the established connections; mutatis mutandis, if we reject it. But the substance to this familiar way of talking is that our agreement about the proof is a product of a similar training, that further explanations are not in general required to secure agreement. What has to be dismissed is the picture which we tend to associate with such talk: that of an objective constraint imposed by the character of our understanding. There are no investigation-independent facts about the way in which we ought to apply certain concepts if we are to remain faithful to our understanding of them. As a result, there is no sense in the idea that rules of language or, more specifically, of inference already determine certain connections which proof and calculation can draw to our attention, but whose status does not depend upon that event.

The first argument underlies, and the second is explicit in this passage:

Might I not say: if you do a multiplication, in no case do you find the mathematical fact, but you do find the mathematical proposition? For what you *find* is the non-mathematical fact, and in this way the mathematical proposition. For a mathematical proposition is the determination of a concept, following upon a discovery. . . . The concept is altered so that this *had* to be the result. I find, not the result, but that I reach it. And it is not this route's beginning here and ending here that is an empirical fact, but my having gone this road, or some road to this end. (48) But might it not be said that the *rules* lead this way, even if no-one went it? For that is what one would like to say—and here we see the mathematical machine, which, driven by the rules themselves, obeys only mathematical laws and not physical ones. I want to say: the working of the mathematical machine is only the *picture* of the working of a machine. The rule does not do work, for whatever happens according to the rule is an interpretation of the rule. (II. 47–8)

XVIII

'True Purely in Virtue of Meaning'

1.   Dummett[1] formulates the philosophical problem of necessity as twofold: what is the source of necessity, and how do we recognise it? The formulation is compelling, and its being so casts light on some of our assumptions about the issue: we think of necessity as something of which we have a cognitive apprehension, and an apprehension, very often, of a strikingly certain sort —at any rate, when contrasted with most empirical knowledge. We are not inclined to ask: *Do* we recognise necessity? Or: Is recognition of such a thing even possible? If an answer to Dummett's first question were given which called the possibility of our apprehension of necessity into question—or even merely made it difficult to explain how we recognise it—our inclination would be to regard that fact as an objection.

We do not want to accept any account of necessity which does not put it in a place where we can find it and be legitimately certain that we have done so. Indeed, in sufficiently simple cases at least, we regard our knowledge of necessary truths as beyond question. So when we ask for an explanation of the source of necessity, while what is primarily wanted is an account which makes apparent how this knowledge is possible, this is not sought in a spirit of justification. That such knowledge is possible we do not doubt. If a philosopher arrives, as Quine did, at a position from which it is difficult to see that there can be a genuine distinction between necessary and contingent statements, our initial reaction is to suspect some inadequacy in his premises.

In this chapter I want first to elaborate some of the difficulties in the conception of necessity which, in the middle part of this century, has more or less prevailed among those philosophers who have not rejected the notion altogether; and then briefly to compare the standpoint of Wittgenstein with that of Quine.

2.   The prevailing conception of necessity, as Dummett remarks, has been, broadly speaking, linguistic. According to it, necessity is always a matter of the analyticity of a sentence; where 'analyticity' is understood not in any of the variant, narrow senses given to it by Kant[2] —nor even, as in Frege's writings,[3] as marking the status of

---

[1] 23 (2, p. 494).   [2] *Locus classicus:* 49, *Introduction,* section IV.
[3] 35, §3.

sentences which can be shown to express truths by means solely of logic and definitional transformations —but, more generally, as characteris- ing the circumstance that the sentence's expressing a truth is settled by its sense alone. It is a measure of the appeal of this conception that we find it difficult to conjure any other account which seems remotely plausible. It did not seem preposterous, for example, that Quine, in calling into question the acceptability of the idea of such a species of truth, was held to be calling into question—and intended to be so held—the whole distinction between necessary and contingent state- ments.

The attraction of the conception is that it seems neither to belittle the status of necessity nor impose on us an over-exotic picture of our capacity to recognise it. The conception presents no obvious obstacle to the possibility that necessary truths may be subtle, or elaborate, and their discovery an exercise in intelligence and ingenuity. Equally, there is no obvious mystery how necessary statements, so conceived, can describe the non-linguistic world: if we have so fixed the senses of 'red' and 'green' that correctly to describe a surface as 'red all over' precludes its simultaneous correct description as 'green all over', then there is no way a surface can actually *be* simultaneously red and green all over; nothing counts as its being so. Our recognition of necessity, in these terms, is made possible by a general capacity of reflective language- mastery; it is furnished by our ability to know what we mean by the expressions which we use, and to discern the implications of meaning them in such-and-such ways. We are already inclined to credit ourselves with the ability not merely to use a particular expression correctly but to know in a general, self-conscious way in what its correct use consists. If we have such knowledge, it therefore seems readily intelligible that we should be able to recognise in a particular case that, for example, the correct use of one predicate requires, or excludes, that another may be correctly applied to the same object.

If we take a sufficiently superficial view, it is almost possible to comprehend why such a picture of the matter might have led some philosophers to think of necessity as conventional. Necessity is seen as originating in meanings, and meanings are conventional things. The picture leads naturally to the 'modified conventionalism' discussed by Dummett.[1] Meanings are established in a variety of ways: ostensively, by explicit definition, by explanation of criteria, etc. But another thing which we seem sometimes to do is simply to postulate as analytic certain sentences involving the expression whose sense is being determined, which nevertheless do not amount to straight explicit definitions of its sense. Such is a natural way of looking at, for example, the mathemati- cal axioms for the notion of a *group*. If this is the only kind of explanation which the expression receives, it is usually said to be *implicitly* defined by such postulates. More usually, however, postulates of this sort

[1] 23 (2, p. 494).

supplement explanations of other kinds. An example of something which could feature both as an explicit definition and in this latter supplementary way would be the stipulation, '3 feet = 1 yard'. This could be used to explain the concept of a yard to someone familiar with the practice of measuring with rulers calibrated in feet; it might be all the explanation of the concept of a yard which he received, and it would be a *complete* explanation—there is nothing left to know about the concept of the yard ignorance of which would not also be ignorance of aspects of the concept of a foot. But if both 'yard' and 'foot' had already been explained to him in operational terms, the stipulation would still be supplementary to what he already knew; for all he could verify in terms of his original training would be that 3 feet approximate *very closely* to 1 yard. Either way, then, '3 feet = 1 yard' is an explicit convention, holding necessarily because its rôle is precisely mutually to determine the concepts of a foot and a yard.

It seems different with '27 feet = 9 yards' or '9 sq. feet = 1 sq. yard'. The character of the necessity of these statements is exactly the same, but it does not depend upon their explicit postulation. Rather, it is natural to suppose that in stipulating that 3 feet = 1 yard we have implicitly adopted all its consequences.

Now, according to modified conventionalism, all necessary statements are divisible into two such classes: direct registers of conventions, which may take the form of explicit definitions, implicitly definitional postulates, or supplementary postulates of the sort just illustrated; and registers of their more or less distant *implications*. The view thus effects a straightforward extension of Frege's concept of analyticity; we are to include within the materials from which an exhibition of analyticity by logical derivation may proceed not merely explicit definitions but stipulations in the other two categories as well. And to recognise necessity, according to modified conventionalism, is to recognise either that the rôle of the statement is that of direct convention is one of the three stated ways, or that it is a consequence of such.

A paradigm of such recognition would be the construction of a truth-table exhibition of a tautology of classical propositional logic. The status of the tautology is shown to follow from conventions concerning the senses of some of its contained expressions—the matrices for the propositional connectives. It is usual to present these in tabular form; but of course their content may just as well be expressed propositionally:

'$\sim P$' is true if 'P' is false; otherwise false.

'$P \lor Q$' is true if at least one of 'P', 'Q' is true; otherwise false.

'$P \supset Q$' is true if it is false that 'P' is true and 'Q' is false; otherwise, false.

and so on.

Recognition that a statement is a tautology may be conceived as recognition that its truth follows from such conventions and the further convention that for each statement truth and falsity are the only possibilities.

3.   The most immediate objection to this modified conventionalism is that it is obviously incapable of explaining every kind of example to which the general notion of analyticity would naturally be applied —every kind of example of which we should intuitively want to say that truth, or falsity, was consequent on meaning alone. The above paradigm is striking for its contrast with the situation in other cases. One such is our old faithful: 'nothing can simultaneously be red and green all over.' We want to say that, in contrast with the situation of '3 feet = 1 yard' even after 'foot' and 'yard' have been explained operationally—or, Dummett suggests, with that of 'nothing can simultaneously be blue and green all over'[1]—the analyticity of this sentence is an immediate consequence of the nature of the meanings of 'green' and 'red' as determined by appropriate ostensive training. It has no supplementary rôle—there is no indeterminacy left by the ostensive training which it serves to resolve. If a trainee who understands the kind of thing which the sentence is meant to exclude does not grasp it as analytic, that shows he must somehow or other have missed the point of the ostensive samples.

How, in terms of the paradigm, are we to understand the 'flow' of truth from meaning in such a case? In order to assimilate the example to the paradigmatic type, we have to articulate certain explicit conventions concerning the senses of 'red' and 'green' of which 'nothing can simultaneously be red and green all over' can be apprehended as a consequence. These ulterior conventions must somehow be recognised to codify information imparted by the ostensive training. But the suggestion that we might *search* for such conventions would obviously be a muddle in this context. Of course, metalinguistically, the information given by the ostensive training is perfectly expressed as that 'green' may truly be applied to all and only green things, and 'red' to red ones; but no means is apparent for *articulating* how the truth of ' "simultaneously red and green all over" is not truly applicable to anything' is a logical consequence of this information. And the whole point of the example is that no conventions need have been *pronounced* in the course of the training from which that sentence follows by accepted logical techniques. The applicability of the paradigm cannot be allowed to become

[1] 23 (2, p. 494): '. . . is a direct register of a convention, since there is nothing in the ostensive training we give in the use of colour words which shows that we are not to call something on the borderline between green and blue "both green and blue".'

a matter of our somehow recognising that certain sentences, from which the example so follows, encapsulate information given by (some aspect of) typical ostensive training, for it is of this very species of recognition that the paradigm is supposed to supply an explanatory model.

The difficulty, in summary, is that the paradigm is appropriate only in cases where semantic information has been imparted by means of explicit conventions presented in sentential form. In other cases, even if we find it easy to see how to draw up what we take to be the essential import of the training in the form of such conventions, the model of modified conventionalism fails. It fails because as soon as it becomes necessary to attempt to *impose* the paradigmatic form on our apprehension of the analyticity of a sentence, we face the problem of explaining our recognition that the conventions which we devise for the task codify information given by the explanations which we actually—historically—received. For this recognition just *is* recognition of truth flowing from meanings unregistered in the form of explicit declarative conventions. We have, admittedly, been thinking of the modified conventionalist paradigm as working with metalinguistic premises; but our recognition that such a premise holds good in virtue of the meaning which we attach to the expressions of which it makes mention is the same feat of recognition—if, *pace* Wittgenstein, we may be allowed to talk in such terms—as recognition of the analyticity of the appropriate object-language analogue.

Obviously, it would be futile for modified conventionalism to respond to such problem cases by attempting to extend the explanatory model backwards; and not merely because it is certain that this cannot in general be done, that we cannot go on indefinitely devising suitable ulterior conventions. In addition, because we are looking for a model of what we take to be a feat of recognition, it is required that any viable candidate supply us with a determinate starting point—yield a finite structure. This is exactly what the play with direct, explicit conventions is supposed to achieve; and it cannot be achieved if, in the attempt to assimilate a particular example to the paradigm, we are forced to project the model backwards indefinitely.

4.   Dummett brings a different complaint against modified conventionalism. His point concerns what it is for something to be a *consequence* of certain direct conventions.[1] Is it, or is it not, a matter of convention whether one convention is a consequence of another which we have explicitly adopted? If it is supposed that it is, it becomes unclear what has been achieved by the 'modification'; unclear, that is, how the resulting position differs essentially from the *radical* conventionalism, attributed by Dummett to Wittgenstein, according to which every necessary statement achieves that status by being primitively an

[1] 23 (2, p. 494–5).

explicit register of a linguistic convention. There is no longer any important difference between the status of initial conventions and that of what we are pleased to regard as their 'consequences'; it is merely that there are two forms of election to that status—for that we do *count* some conventions as consequences of others the radical conventionalist has no reason to deny. So it seems that the modified conventionalist must deny that when one convention follows from another, it is a mere convention that it does so. But then, what account is he to give of the necessity of the statement that it does? For correct statements of logical consequence are par excellence necessary.

The argument captures one respect in which the superficial idea, that to think of necessity as originating in meaning somehow enjoins that it is conventional, has always tended to receive criticism. The idea might pass examination when we are concerned with the initial conventions that form the foundation of the modified conventionalist view, the conventions whose rôle is precisely to establish or refine meanings. But unless he is prepared to take the further step into radical conventionalism, someone attracted to such a view would seem to have to concede that when we are concerned with the consequences of conventions, and the flow is in the opposite direction—from pre-established meanings to necessity—the fact of such a flow is nothing conventional; it is something objective and imposed.

Nevertheless, this conclusion was contested by Bennett,[1] whose view I want now to consider. Bennett thought that the truth about necessity lies with a form of modified conventionalism; he contended that a genuinely conventionalist view is still available which avoids the 'radical' extreme. Dummett's question was: what is the nature of the necessity of 'C implies Q', where C is a set of initial conventions and Q a consequence of them? If we say that such a statement is itself merely a convention, the resulting position is tantamount to a version of radical conventionalism; it is just that our adoption of the convention Q takes the form 'Let Q be a consequence of C'. But if we say that the statement is not a convention, how can we avoid the conclusion that its necessity consists in its description of a non-conventional conceptual connection? Bennett's answer was that the necessity of the statement of consequence is to be regarded in neither of these ways; rather it is *itself* a consequence of certain ulterior conventions whose rôle is to fix the notion of logical consequence. We immediately want to protest, 'But then, what is the status of *this* statement of consequence?' But now the same answer is forthcoming again: it is a consequence of conventions for the notion of logical consequence. And so on.

It is not straightforward to see what this move amounts to. Consider this picture. A committee is getting down to the business of establishing a network of necessary statements for an emergent society of which we are members. Its task is to compile two volumes of initial explicit

[1] See 3.

conventions: Vol. II will cover the notion of logical consequence, and Vol. I everything else. Then our problem is whether, when the committee's work is done, it has *thereby* been settled what are the consequences of everything in the two volumes by the conventions in Vol. II as rules. If we say 'No', we are adopting a radical conventionalism; we are saying that the committee only completes the first stage of an ongoing task which is then taken up by the society as a whole, that further conventions will always be needed—there is no finally settling the extent of what is conventionally true. If we say 'Yes', we are granting that it does not need further decisions by society at large to determine that something follows from items in the two volumes by items in Vol. II; but then it is no convention that it follows, and it seems that what can be derived from and by any particular conventions is something objective, antedating our choice of any particular set. So which answer is Bennett giving?

Clearly, he has to be interpreted as refusing to answer, as rejecting the viewpoint from which the dilemma seems to arise. The dilemma tries, as it were, to stand outside the conventionally chosen system of necessary statements and ask something about its character. The point of the picture of a committee, settling on such a system while the rest of society goes about its business, is precisely to suggest the intelligibility of such an exterior standpoint. It suggests that there is already a conceptual framework in terms of which such questions can be raised. Bennett's answer is then, presumably, that there simply is no such intelligible exterior position; we can operate only within the framework which the committee lays down. There is thus no intelligible exterior question whether the necessity of '*C* implies *Q*', *Q* a convention on which they did not explicitly decide, depends upon a further explicit decision by society or whether it is imposed upon us and needs no explicit ratification. The truth, and so the necessity, of '*C* implies *Q*' is settled by showing that it can be derived from items in Vol. II by items in Vol. II. But that is not to concede that the possibility of such a correct derivation was, in the sense essayed by the dilemma, something independent of our conventions. The only understanding which we can have of the necessity of '*C* implies *Q*' is that it is possible to derive it from conventions which we use by conventions which we use; if it is then asked, do we not therefore have to grant that it is a non-conventional, objective fact that *this* derivation is possible, the answer is again that the only understanding which we can have of this fact is that that Vol. II implies that *C* implies *Q* is itself derivable from items in Vol. II, using items in Vol. II as principles of derivation. So all necessity is conventional; every necessary statement is either an explicit convention or a derivable consequence of such. And statements of logical consequence in particular are neither in general explicit conventions nor expressions of non-conventional facts—they are themselves consequences of conventions, and so conventional in character.

The natural reaction to the foregoing is that it is just an evasion. Bennett intends his conventionalism to provide answers to Dummett's twofold formulation of the problem: all necessary statements, including statements of logical consequence, are either directly adopted conventions or consequences of such; and our recognition of the necessity of a statement is just recognition that it comes into one of these two categories. But it is an essential part of the account that there can be no *explanation* of the character of the constraint upon us to accept the consequences of our adopted conventions. We want to know, in what sense *must* we accept the consequences—or is it perhaps that we have a freedom which we did not suspect? But no answer can issue from Bennett's account. If someone asks, 'Why must I accept Q if I accept C?', the answer is that Q is a consequence of C. But if he then asks, 'What is the nature of this "consequence"—what sort of statement is it that Q is a consequence of C, what do I ignore if I repudiate it?', no explanation is available which does not invoke directly the concept of consequence itself. All there is to say about the character of the truth of 'C implies Q' is that it is itself a consequence of certain ulterior conventions. The account, indeed, is *designed* so that that is all there is to say. If Bennett were to admit that a further, deeper question arose, he would have to confront the original dilemma. Not surprisingly, then, he seems to accept the implication that the idea that there might be some deeper-reaching explanation of necessity, going beyond what can be accounted for by modified conventionalism, is a mistake:

> I therefore conclude that the aim of the —[Bennett's]—paper is to answer Dummett's questions; or if it is not, then the questions admit of no answer.[1]

The problem was, if necessity is conceived as flowing from meaning, what is the nature of the fact that it does so in any particular case—or, equivalently, the character of the necessity of a *statement* of the fact? Bennett's answer was: 'The same.' The statement of the fact of the flow is itself true in virtue of meaning. And, surely, there *can be* no other answer if it is in terms of truth in virtue of meaning that all necessity is to be explained. So no one who is inclined to that conception of necessity ought to object to Bennett's answer. Bennett may be right, moreover, to contest the suggestion of the original dilemma that there are no alternatives to interpreting statements of logical consequence either as freely adopted explicit conventions or as reflections of a non-conventional conceptual reality—a question with which we shall be exercised in subsequent chapters. He may even be right to suppose that the desire for a deeper explanation of necessity than his account is going to allow, is folly. What *is* questionable, however, is whether the particular play which Bennett makes with the notions of explicit convention and logical

[1] 3, p. 20.

consequence can provide a satisfactory explanation of the manner in which, in other then explicitly conventional cases, the necessity of statements of logical consequence can be apprehended.

Doubt arises as soon as we consider more carefully the model which Bennett seems to be proposing of how we apprehend the necessity of any particular such statement Q. Prima facie the model is straightforward: apprehending the necessity of Q is deriving it from certain initial explicit conventions C for the notion of logical consequence. But, of course, it is not enough to construct any old derivation; it has to be recognised that Q is derivable from C by *correct* application of the conventions for logical consequence—something which, if so, is necessarily so. Accordingly, we have either to find an explicit convention to the effect that Q is so derivable, or to recognise its being so as a more remote consequence of such conventions. But we can discount the first possibility. The conventions governing the notion of logical consequence will be couched in purely general terms, and we are now concerned with an application of them to a particular example. Our recognition that Q is correctly derivable from C has thus itself to be the conclusion of a derivation; we have to recognise that that Q is correctly derivable from C is itself correctly derivable from the initial conventions for logical consequence. And this, if true, is a *further* necessary statement which, because it deals with a specific example, has to be apprehended by constructing a derivation. So we have a regress.

Why should a regress of this sort matter? Because it appears to bring out that the model imposes an infinitistic structure upon our recognition of the status of any non-initial convention Q. In order to recognise the necessity of Q, we have to recognise (i): that Q can be correctly derived from certain initial conventions. But (i), if true, is itself a necessary statement and one which will not feature in a set of purely general logical consequence conventions. In order to recognise the necessity of (i), then, we have to recognise that of (ii): that (i) can be correctly derived from certain initial conventions. And so on indefinitely.

A natural immediate complaint is that this reasoning conflates recognition that such-and-such is a consequence of certain conventions with recognition that it is necessary that it is; we do not need to recognise the necessity of (i) and (ii), etc., but only their truth. Surely, however, there is, in this context, no material distinction to conflate; what would it be to recognise the truth of the statement 'Q is correctly derivable from C', where such recognition was not eo ipso recognition of necessity? And, in any case, truth is enough. Since the statements (i), (ii), etc., are, if true, recognisably so only and purely in virtue of the character of the concept of correct inference, recognition of their truth must proceed in terms of a derivation of them from conventions codifying that character. For that is the essence of the view which we are considering: *all* necessity is truth in virtue of meaning; and for a statement to be true in virtue of meaning is for it to be a—perhaps totally trivially—derivable consequence of explicit meaning-determining conventions.

The model thus appears to require that in order to recognise the status of any consequence of initial logical consequence conventions, we have to recognise the same of infinitely many statements. This is obviously an impossible requirement, unless it can be supposed that infinitely many such feats of recognition take place simultaneously. But is that so extravagant a supposition? Suppose I have derived Q from the consequence conventions C, using the latter as rules; then it ought to be straightforward to transform the derivation into one of the conditional 'if C then Q', from C using C as rules; and it ought similarly to be straightforward to transform this derivation into one of the conditional 'if C then if C then Q' from C using C as rules; etc. All this need assume is that the consequence conventions are sufficiently rich to yield the principles:

$$\frac{\Gamma \vdash B}{\Gamma, \Delta \vdash B} \quad \text{and} \quad \frac{\Gamma, A \vdash B}{\Gamma \vdash A \rightarrow B}$$

If we have these principles, then the failure of any of the derivations requires the failure of its predecessor. So it is not implausible to claim that the proof of Q from C using C as rules is simultaneously a proof of all the statements of consequence, taking these as the appropriate conditionals, which we appear constrained to regard as involved in our recognition of the necessity of Q. It is not implausible because we know that if the set C is sufficiently rich, then once we have a derivation of Q from C we can construct the derivation of any of the conditionals in a uniform manner. Is the regress, then, harmless?

I think not. The foregoing reasoning invites us to recognise that a correct derivation of each of the conditionals is possible. But the statement, R, that such a derivation is possible for each of the conditionals, while itself necessary, will be—for the same reason as before—no explicit convention. Accordingly we have, in terms of the model, to recognise that R is itself correctly derivable from C. It is, admittedly, unnatural to try to view in that light the reasoning sketched on behalf of R. But the important point is that to attempt to apply the model here will be to generate a further regress. And a similar regress will occur at any higher level at which the model attempts to cope with a similar infinitary recognitional feat. Not that any question is raised by these considerations about our ability to perform such a feat; the question in effect concerns the capacity of modified conventionalism to accommodate that ability—indeed, to accommodate our ability to recognise logical consequence at all. For it appears to be able to supply no finite model of such recognition.

5.   The intuitive idea was that all necessity is truth in virtue of meaning. And modified conventionalism is essentially a gloss on this intuitive idea, an attempt to be concrete about 'in virtue of'. But what we have learned, assuming the correctness of the foregoing, is that if we are to have a finite model of our recognition of necessity, requisite if such recognition is to be completable, we cannot in general coherently conceive of the 'flow' of truth from meaning in terms of the modified conventionalist model of premises and logical consequences. We have learned two reasons for this. First, the model cannot apply if the premises have not been given in an explicit declarative form as part of the explanation of the character of the meanings involved, but have themselves to be gleaned from semantic information given another way. For our recognition of their correctness is then just a further instance of recognition of truth in virtue of meaning. If we attempt to explain *this* recognition in terms of the model, the model threatens to extend backwards indefinitely, and we never reach the direct, initial conventions supposedly at the source of necessity. And now, secondly, it appears that even where explicit initial conventions are available, modified conventionalism still cannot provide a finite model of our recognition of their consequences. For it conceives that, in order to recognise that something is a consequence of conventions, it is necessary to recognise that that it is so is a consequence of conventions—and now there is no end to what we have to recognise. Only if we can recognise simultaneously of infinitely many statements of consequence that they are each consequences of the conventions for consequence can the process be completed. So, indeed, intuitively we can; but there is no way of explaining *this* recognition as in general other than the product of an inference, if it is viewed in terms of the premises/consequences model of modified conventionalism; so the difficulty recurs. The route from the initial conventions to recognition of their consequences thus becomes a transfinitely extended path of infinite paths; so a route which we cannot follow.

Modified conventionalism thus seems a totally unsatisfactory exegesis of the original intuitive conception. The flow of necessity from meaning cannot in general be conceived in terms of the notion of logical consequence. Our recognition of the necessity of a statement, while perhaps informed and guided by a self-conscious awareness of the character of our understanding of it, cannot always be thought of as recognition that its truth follows from premises stipulative of the character of that understanding. Rather, if the intuitive conception that all necessity is truth in virtue of meaning is to be retained at all, recognition of the necessity of statements which are not explicit conventions has at least sometimes to be direct and unmediated by inference. There is no other way of granting us access to premises for the modified conventionalist model, where these are not available in the form of initial conventions, correctly regarded as indispensable in an adequate explanation of the concepts involved. And there is no other way of doing justice to our ability to recognise finitely that something follows from such premisses.

The interest of Bennett's proposal derives entirely from its bearing upon our intuitive conception of analyticity. For we do think of our admission of certain sentences as analytic as imposed on us by the nature of their syntax and the senses of the expressions which they contain. We think that no other course is open to us, and, if someone does not see as much, then he betrays either confusion or a divergent understanding of the sentence. No other course than admission of the analyticity of the sentence is *consistent*, we want to say, with the way in which we understand the sentence. But if modified conventionalism is inadequate, what account are we to give of this invocation of the notion of *consistency*? For inconsistency is exactly that relation which holds between statements *P* and *Q*, just in case the negation of *Q* is a logical consequence of *P*; and what we have just noted is the unsuitability of a premisses/consequence model for our recognition of analyticity in general. Recognition of analyticity cannot in general be conceived, therefore, as recognition either that the sentence is an initial convention or that its negation is inconsistent with sentences which follow from such conventions. The notions of logical consequence and inconsistency as propositional relations are not suitable for the task of explaining the intuitive idea that we are coerced by our understanding of certain sentences to regard them as analytic, that no other course is consistent with that meaning. So the effect of all this is that we are not in a position to resist the charge that the notions of consequence and consistency at work in the intuitive idea are as much metaphorical as that of 'flow'. And that, if admitted, seriously detracts from the standing of the intuitive idea. For it seems quite unsatisfactory that the source of necessity and the attendant epistemology can *only* be conveyed by metaphor; and Bennett's account is the only obvious strategy for cashing the metaphor out. Whatever the content of the idea that all necessity is 'truth purely in virtue of meaning', it cannot be held to involve that in order to recognise the necessity of a statement, it is necessary and sufficient validly to derive it from meaning-determining postulates.

6.   A supporter of the idea that all necessity is truth in virtue of meaning is unlikely to be too dismayed by the failure of the above sort of modified conventionalism. For he will have felt all along that a distortion was involved, in at least a wide class of cases, of the essential phenomenology of recognising necessity. Recognising the necessity of something which is not an explicit convention does not have to involve constructing a proof. The real motivation for thinking of necessity as truth in virtue of meaning derives from the fact that, in simple cases at least, we want to describe ourselves as reflecting on the content of a

sentence and *thereby* coming to see that it cannot but be true. In such
cases necessity is recognised by *the light of* understanding, even if the
way the understanding casts its light cannot be explained in terms of a
premisses/consequences model. We want to attribute to ourselves a
capacity reflectively to apprehend impositions and constraints which
the manner in which we understand particular expressions places upon
us. It may, or may not, be possible philosophically to explain how such
a genre of reflective cognition is possible; but possible it unquestionably is—what else could explain our agreement about the analyticity of
sentences which we have never thought about before and which are too
simple, or fundamental, to admit of coercive proof?

Wittgenstein's ideas have an enormously destructive effect on this
conception. But in order to see their bearing, it is necessary to notice
that the kind of reflective grasp of meaning appealed to is essentially
idiolectic—it is a matter of each of us discerning the character of his own
understanding of expressions. There is no temptation to claim a reflective knowledge of features of *others'* understanding of a particular
expression—except against the background of the hypothesis that it
coincides with one's own.[1] Indeed, the special certainty which 'the light
of understanding' is supposed to bestow on the truth of sufficiently
obviously analytic sentences would be in jeopardy if the claim to have
grasped that truth was defeasible by a lack of communal sympathy; an
inductive scepticism would be possible, whereas the certainty of sufficiently simple 'truths of reason'—if not absolutely sceptic-proof—is
supposed at least to be proof against scepticism of that sort. It is always
conceivable that we may have an intractable disagreement about a
putatively analytic sentence; that we may be unable to locate anything
involved in our respective judgments which we can agree was a mistake
or oversight. In that case, if I stand alone, I shall have made the
discovery that my understanding of the sentence is deviant. But that
still seems to leave available the idea that I can know that to accept it as
analytic conforms with *my* understanding of the sentence. Indeed, it
seems that it must be possible for me to know facts of this sort, if I am to
be in a position to recognise analyticity in the reflective manner which
the present conception believes in. For the response called for when a
putatively analytic sentence is put to me is a response from me. It is on
the character of my understanding of the sentence that my response will
be held to bear; if, therefore, that response is viewed as an expression of
knowledge which I possess, this knowledge must concern the character
of my understanding of the sentence. The capacity to discern analyticity reflectively is thus, for the present conception, essentially a capacity
to discern the character of one's own understanding; it is a supplementary presupposition that this understanding is shared.

According to the intuitive conception, recognising necessity is a
matter of recognising certain relations between the types of use of

---

[1] Or with one's understanding of another specified expression.

particular expressions which conform with the meanings which one attaches to them. To recognise the necessity of 'Nothing can simultaneously be red and green all over' is to recognise a relation between the kinds of use of 'red' and 'green' which conform to one's understanding of those predicates; and to recognise the necessity of a statement of logical consequence is to recognise a relation between the kinds of use of the premisses and conclusion which conform to one's understanding of those statements. It is thus essential that it be possible to defend the conception that each of us can know—barring error about the facts—whether a particular application of an expression exemplifies the kind of use constitutive of—or required by—the sense he attaches to it. If this idea proves indefensible, there will be nothing in which a reflective knowledge of the general character of idiolectically correct use of an expression can consist; for it makes no sense to attribute to someone knowledge of the *kind* of use of an expression which conforms to his understanding of it unless he is also supposed to know under what circumstances a particular use is of that kind. And without appeal to the possibility of such reflective knowledge, there will be no justification for the idea that we can discern reflectively those relations between such patterns of use whose recognition is supposed to constitute our recognition of necessity.

Wittgenstein repudiates the view that each of us may regard himself as knowing reflectively what kind of application of an expression conforms to the meaning he attaches to it; that we can have such egocentric, reflective knowledge of sense. His ground for doing so, elaborated in Chapter II, is simply that *whatever* sincere applications I make of a particular expression, when I have paid due heed to the situation, will seem to me to conform with my understanding of it. There is no scope for a distinction here between the fact of an application's seeming to me to conform with the way in which I understand it and the fact of its really doing so. If such a distinction could be made out, there would indeed, be an insuperable difficulty about explaining how I could reasonably be sure whether, idiolectically, I was using an expression correctly. But because the distinction cannot be drawn, there is no possibility of an account of what my seeming-knowledge here is knowledge *of*, no possibility of an account of what the fact of the idiolectic correctness of my use of the expression consists in—representing it as something which I am especially well-placed reflectively to discern. There is, that is to say, no *objectivity*[1] in the supposition that a particular sincere use of an expression conforms with the understanding which I have of it, if this conformity is to be something which I am to be in a uniquely favourable position to recognise. But facts, it is unnecessary to stress, have to be by their very nature objective. We

---

[1] The notion of objectivity relevant here is, of course, neither the transcendent objectivity of realism nor investigation-independence. It is the absolutely minimal requirement: the existence of operational standards of right and wrong judgment.

employ the notion precisely to mark a contrast between mere impression, opinion, and attitude and the obtaining of something exterior which vindicates them. So the concept of the *fact* of my idiolectically correct use of an expression, if this is to be something, barring misapprehension of other facts, of which I am to be capable of a special reflective certainty, encounters insurmountable obstacles. It follows that I have no right to claim a reflective awareness of the nature of my own understanding of an expression; there is nothing to be the object of such awareness. Neither, therefore, am I capable of anything properly described as the 'discovery', either by an immediate act of reflection on the character of my understanding of a sentence or by a series of operations sanctioned in the medium of such reflective understanding, that the sentence in question must express a truth if it is meant in the way in which I mean it. In particular, proofs—whatever their degree of formality or rigour—cannot be conceived as instruments of such discovery. Whatever goes on when I seem to myself to discern just by thinking, or reasoning, about meanings that a certain sentence cannot but express a truth, these matters are not illuminatingly—or even in the end intelligibly—to be viewed in these traditional ways.

This idea of Wittgenstein's is, of course, most familiar in the context of the celebrated 'private language' argument. And to some philosophers it has seemed less than cogent. For it is felt that the difficulty in drawing a distinction between its seeming to me that I am using an expression in an idiolectically correct way and my really doing so points to a quite different conclusion—can be explained equally well by the conception that I have a unique and certain cognitive access to such facts. Applied more generally in the context of the philosophy of mind, this idea issues in the ancient doctrine that we have an 'incorrigible' awareness of the nature of our mental 'contents'—in particular, of the character of our sensations. This alleged awareness has, indeed, repeatedly been conceived as the possible foundation of an entire anti-sceptical reconstruction of the architecture of our knowledge. But ask, what *reason* could be given in favour of slanting the situation this way? Why say that something is *known* here rather than, less tendentiously, that we find ourselves with the disposition to re-apply a certain concept? The only possible justification for preferring the former account would be if it could be explained in what the fact supposedly known—the fact of my use of an expression conforming with my idiolectic understanding of it—consisted; and this in such a way that both its objective standing and the possibility of certain knowledge of it were *consequences* of the account. Such an account is not ruled out immediately; it is not to be insisted that objectivity requires an ineliminable possibility of a misapprehension. What it requires is that the facts do not *consist* in one's having a certain impression of them—that there is something for my impression to be an impression *of*. For this an ineliminable possibility of a misapprehension is sufficient but not necessary. But in the present case an account of this kind is not going to be forthcoming. The order of things is the other way about. It is not that we have such a concept of these alleged facts that the possibility of a privileged, egocentric access to them is a consequence; it is that, faced with the impossibility of intelligibly explaining how things might be *other* than as they seem—how my sincere use of an expression might in fact be idiolectically *incorrect*—we dignify our inability to do so into the discovery that here things cannot be other than what they seem, that here at least sceptic-proof knowledge is available. We keep the facts and the impressions apart, and then pretend to prove that, though apart, they coincide. We engineer our conception of the facts to cohere with the impossibility of making intelligible a certain distinction; when the truth is that we have no business thinking in terms of 'facts' here at all unless we can explain their character in such a way as to give them objective stature.

In summary: we have been concerned with the intuitive view that the phenomenon of necessity reduces to that of the occurrence of analytic sentences, the analyticity of a sentence consisting in the circumstance that it expresses a truth just in virtue of the senses of its constituent expressions and its syntax. That a sentence has this status, it is conceived, may result either from its being explicitly stipulated to do so or as a result of explanations of meaning in which it played no explicit rôle. The central idea of modified conventionalism was to build an explanation of necessity upon the foundation of analytic sentences of the first kind, those of the second kind being conceived as derivable consequences of them. And the failure of modified conventionalism is a failure of the most natural attempt to substantiate the original, intuitive view. It leaves us with the bare notion of a postulated capacity to recognise directly aspects and consequences of the character of our understanding of particular expressions. If, now, we accept the foregoing Wittgensteinian considerations, we shall reject even this bare notion. The character of my understanding of an expression is nothing of whose aspects and consequences I have a capacity for a privileged, reflective awareness. Nor is it therefore in point to request a philosophical explanation, or model, of such a cognitive ability. The point may be expressed like this: if in the traditional notion of analyticity we had hold of the germ of an account of what necessity consists in, of its source, it would immediately have to be conceded that there could be no answer to the second problem of necssity—that of explaining its epistemology. If the analyticity of a sentence is thought of as something objective, originating in the way in which the expressions involved are understood, then analyticity is something which we are empowered to discern only if 'the way in which an expression is understood' is a possible object of our knowledge. But this conception slides irresistibly into the idea that we are each empowered reflectively to apprehend the character of our *own* understanding of particular expressions; and this idea proves impossible to make good. No objective notion of correctness of use can be made good of such a sort that each master of the language may be credited with a special, reflective capacity to know in what, for a particular expression, it consists. A solution to the twofold problem of necessity must take some quite other point of departure.

7.   It might be thought that the appropriate course for someone who accepts these reflections would be, rather than looking for some other solution to the problem, to abandon the notion of necessity altogether. We might, I suppose, abandon the use of the word 'analytic', if this is thought too closely wedded to a discredited idea of 'truth purely in virtue of meaning'. But this was not the most general sense of the word employed by Quine in *Two Dogmas of Empiricism*:

> . . . it becomes folly to seek a boundary between synthetic statements, which hold contingently on experience, and analytic statements, which *hold come what may*.[1] (my italics)

What is the relation between the considerations with which we have just been concerned and the central argument in *Two Dogmas*?

Famously, Quine does propose to reject the analytic/synthetic distinction altogether. There are no statements 'true in all possible worlds', no statements which 'hold come what may'. What there is is a distinction of degree; the degree, namely, of dispensability of various sorts of sentence, measured as the inverse of the degree of our reluctance to abandon them in the face of 'recalcitrant experience'. There are sentences which we may abandon within the context of a particular special theory; sentences abandoning which will constitute revision of the theory; sentences whose abandonment will involve abandoning a particular branch of theory; sentences the abandoning of which will amount to abandoning several branches of theory, etc. But since we have a vested interest in finding the world intelligible and predictable, we are reluctant to abandon our theories; so, confronted with a recalcitrant experience, we incline to abandon those sentences whose release has the smallest revisionary impact on the edifice of the theory while preserving its capacity to predict the actual course of phenomena. Analyticity/syntheticity is thus a measure of the depth of entrenchment of a sentence in the corpus of current theories, a measure of how much is at stake for us in the sentence—the extent of our commitment to it. But this, for Quine, is just to say that analyticity as an absolute is a fiction; there are no absolutely indispensable sentences contrasting with in principle dispensable ones:

> Any statement can be held true come what may if we make drastic enough adjustments elsewhere in the system . . . even a statement close to the

[1] 66, p. 43.

periphery can be held true in the face of recalcitrant experience by pleading hallucination, or by emending certain statements of the kind called logical laws. Conversely, by the same token, no statement is immune to revision.[1]

This is a generalised empiricism about necessity of the sort discussed in the previous chapter, provoking several immediate queries. How is the idea of 'recalcitrance' to be explained—in particular, can Quine's account coherently be applied to the true statement that such-and-such observations are recalcitrant for our total theory? What determines the *extent* of the revisionary impact of rejection of a particular statement? And why should it follow that there are no absolutely indispensable statements—statements which will feature in anything intelligible to us as a theory at all? In the previous chapter it was suggested that we could not make sense of the practice of a theory unless certain state- ments—precisely, those comprising its underlying logic—were assigned a normative, or directive rôle. We have to be instructed what moves to make, how to apply and revise a theory; and the statements codifying these instructions, while perhaps in principle revisable, can- not coherently be regarded as confronting experience in the rôle of *hypotheses,* even supremely deeply entrenched ones, without making it impossible to understand how it is reasonable to accept their applica- tions to particular inferential contexts. So, I argued, an empiricist has at least to accept the analytic/synthetic distinction in something akin to Wittgenstein's interpretation of it. I shall revert to this suggestion in Chapter XXI; our present interest is rather in any parallel between the considerations which lead Quine to scepticism about analyticity and points to be derived from Wittgenstein.

Quine's central argument is extremely simple: the concept of analy- ticity as traditionally conceived cannot be satisfactorily explained. Clearly such an argument, if upheld, would be decisive. If there is no satisfactory explanation of the notion, there is no way whereby we can have been introduced to it; so unless the concept of analyticity is to be something of which we have an a priori grasp, we cannot be supposed to understand what class of statements it allegedly marks off. But it would be natural, having been promised this dramatic conclusion, to be puzzled by the way in which the argument is actually developed. For Quine seems to concentrate on just one sort of analytic state- ment—those arrived at by substitution of synonyms for synonyms in logical laws—and then proceeds to develop an attack on the concept of synonymy. The analyticity of the statement in which the substitution is made, and that of certain statements which are neither logical laws nor derivable from them by substitution of synonyms, is thus, it appears, untouched by the argument. The most of Quine is going to be able to show is that the status of *one* kind of traditionally conceived analytic statement is unclear.

[1] 66, p. 43.

A supporter of Quine would be expected to disavow this limitation. The validity of a logical schema consists in that of its instances; and their validity will turn on continuity of interpretation both for any multiplicity of occurrences of a given logical constant within them and for the expressions with which a particular schematic letter has been replaced. So the idea of a valid logical schema presupposes the intelligibility of that of sameness of sense among token-expressions.[1] And a similar point applies to analytic statements, like 'Everything red is coloured', which are not derivable from logical laws by substitution of synonyms. The invitation to consider whether such a statement is analytic is only intelligible in terms of the presupposition that we are to assign to the expressions involved their standard senses, that we are to understand this occurrence of 'red' in the same way as previous ones. So a successful attack on the notion of synonymy really is going to call into question the status of all classes of so-styled analytic statement.

The puzzled reader of *Two Dogmas*, however, is liable to reply that such an attack would do much more than that; it would put in jeopardy the very idea of assessing the *truth-value* of any statement. If the intelligibility of judgments of synonymy among token-expressions is held to be presupposed in judgments of logical truth and non-substitutional analyticity, then it is presupposed universally. *Every* judgment of the truth-value of a sentence may be construed, comparably, as requiring that it be intelligible to assume that the contained tokens have the same (or in cases of intended ambiguity, different) senses to other tokens of the same type either within the same sentence or in previously uttered ones.

Quine's attack upon the notion of synonymy is actually upon the relation conceived as pre-established in the language between expressions of distinct types; whereas the proposed generalisation of his argument concerns the idea of synonymy among tokens of the same type. But this reflection merely puts the puzzled reader in position to set up the following dilemma for Quine's supporter: if the intelligibility of token-synonymy stands or falls with that of type-synonymy, then a successful attack by Quine on the latter undermines the distinction between truth and falsity; but if token-synonymy is intelligible in the absence of an account of type-synonymy, then no way is apparent to generalise Quine's argument to the other classes of analytic statement.

That Quine's argument will bear on all classes of analytic statement is however, on reflection, clear. It is true that recognition of type-synonymy seems to play an explicit rôle in our recognition of only one class of analytic statement. But the fact remains that it is a cardinal feature of the traditional concept of analyticity that the sense of an expression is a possible object of intellectual scrutiny; it is to be possible to carry out an investigation into the essential characteristics of senses with the upshot that, for example, a pair of predicates turn out to be

[1] The point is used *against* Quine by Strawson in 72.

essentially co-extensive. Such an investigation is what, as traditionally conceived, would constitute the *discovery* that '$(x) (Fx \leftrightarrow Gx)$' is analytic. It is precisely because such an investigation is conceived as possible that it seems we have to be able to make sense of the idea of an initial datum, or informational state, into which the investigation is made; and thus it seems there must be some intelligible notion of sameness and distinctness among such states—or we could not be held to know what we were investigating. If there is to be scope for the idea of an investigation, perhaps by inference or just by direct reflection, into the characteristics of senses, of which the outcome is recognition of analyticity, there has to be *a* notion of sameness of sense. If there is not, the whole conception of recognition of analyticity as the product of scrutiny of sense becomes mystifying. So it seems to me that Quine's strategy of argument is entirely appropriate to his purpose. To recognise 'truth purely in virtue of meaning' we have to know *what* we are scrutinising; and this means, have a concept of the identity of what we are scrutinising. If we have such a notion, it must be possible to give some sort of account of it—though whether this ought to take the form of a statement of necessary and sufficient conditions for sameness of sense is a different matter. So, if no satisfactory account is possible, we do *not* possess any such notion, nor, therefore, the concept of analyticity as traditionally conceived.

The arguments of Quine and Wittgenstein thus seem entirely complementary. Wittgenstein's argument—one of the cluster of rule-following considerations—is that the traditional notion of analyticity involves construal of the meanings, scrutiny of which is to enable us to recognise analytically, as fundamentally idiolectic; with the consequence that even the most minimal notion of objectivity cannot be introduced into the sort of 'investigation' which the traditional notion conceives us as capable of. And Quine's essential point is that a further presupposition of the possibility of such an investigation—our possession of a satisfactory conception of *what* we are investigating—cannot be made good. Quine's tendency to present his doubt as one concerning the legitimacy of 'reifying' meanings has probably obscured its true force—as though the issue were purely one of the ontological taxonomy of meaning. The real question concerns the *particularity* of the information an investigation into which may terminate, on the traditional picture, in recognition of 'truth purely in virtue of meaning'. Whether the particular critique of the notion of synonymy which Quine advances achieves this end is a another question.

What is questionable, however, is the right with which Quine passes between the idea of statements true purely in virtue of meaning—statements whose truth is settled just by the 'linguistic component' as he expresses the matter —and the idea of statements which 'hold true come what may' or however we want to paraphrase necessity, Why, if we reject one picture of what necessity is, ought we to reject the concept altogether? For the dilemma set by the puzzled reader above has a point: the same notion of particularity of content can be argued to be involved in our ability to make the distinction between true and false statements. It is correct, and harmless, to say that in order to assess the truth-value of any statement, we usually need to know what it means. And now it seems that unless a satisfactory account can be given of the nature of, and so of identity and distinctness among, such states of information, we cannot conceive of ourselves as in a position ever to make such an assessment. But this will not do. When someone does not know what a statement means, the information sought is exactly an explanation of how, ideally at least, to decide its truth-value; and we are able to give and to understand such explanations. So if Quine is right in thinking that the idea of sameness of meaning eludes satisfactory explanation, the proper conclusion is that neither its mastery nor its intelligibility is presupposed by our capacity to assess truth-value. But then we ought to be prepared to allow something similar in the case of necessity.

Thus, if Quine is right, we must reject the distinction between statements whose meanings settle their truth-values and those whose truth or falsity depends upon an additional worldly component; for the antecedent conception of particularity of meaning cannot be properly explained. But what is thereby rejected is just a certain picture of the source and epistemology of the analytic/synthetic distinction. For, as Grice and Strawson emphasised,[1] it is foolish to deny that we possess a concept when a capacity is manifest to agree about its application to new cases. The hard fact is that our receiving the linguistic training which we do receive just does secure our assent to certain previously unencountered statements without the need for anything that it is natural to describe as an 'empirical' investigation; and that, where such assent is ultimately not forthcoming from someone, we are able to make the situation intelligible to ourselves only via a reinterpretation on his behalf of certain of the expressions involved. It is in these facts that the substance of the image of 'truth purely in virtue of meaning' resides. What Quine—and Wittgenstein—are to be taken as criticising is the elevation of the image into something literal, a schematic representation of the source of necessity with attendant implications for what is involved in its recognition. But nothing here yet warrants the dismissal of the concept of necessity. And for Wittgenstein, unlike Quine, the distinction between necessity and contingency is as genuine as—and very much akin to—the ordinary contrast between the rules of a game and the moves made in the course of playing it.

The traditional picture is that of a pair-wise correlation of sentences and contents—*propositions* in the sense of Moore.[2] But the distinction then seems to be imposed on us between sentences whose repudiation is and is not respectively compatible with preservation of the pairing;

---

[1] 41.
[2] See, for example, 'Propositions and Truth', in 64, pp. 132–52.

imposed on us because we regard the rejection of certain sentences as a criterion for misunderstanding of their contained expressions or syntax, irrespective of the sensory environment in which the rejection is made. So now it seems that there must be some propositions rejection of whose sentential expressions is possible only via a break in the pairing. There is thus no way of coherently rejecting the proposition *as such*. Of course the status of any *sentence* may be revised; but with formerly analytic sentences that is just to assign to them a new partner. It is certain that much of the antipathy provoked by Quine's discussion of this topic derives from an attempt to interpret his suggestions within the framework of this picture. So interpreted, he appears simply to 'fudge' the question how we might reject a necessary *proposition*. But it is the picture—the idea of sense as something objective and apprehensible as such—on which his discussion bears. Quine himself, however, did not—in *Two Dogmas* at least—seem clearly to appreciate that necessity does not stand or fall with this background notion of sense; it is just that, if the background notion falls, we lose our preferred 'explanation' of what necessity is—the long-standing explanation, that is, which dignifies the forms of interpretation we place upon sincere rejection of an analytic sentence into a defective metaphysic of the source and epistemology of necessity.

XIX

Radical Conventionalism

Sources

'Decision':
RFM: II. 27; III. 23
BGM: VI. 7, 24
  PI: I. 186
  OC: 368
  BrB: 5 (p. 143)
LFM: lectures II, pp. 30–1; v, p. 56; xi, p. 109; xiii,
        pp. 124–5; xxiii, p. 226

1.   Someone who had thought, in the venerable and inchoate way, of all necessity as 'truth purely in virtue of meaning' need not have meant to endorse any kind of conventionalist account. It would be possible, and natural, for him to embrace the second 'prong' of Dummett's dilemma: to grant that a wide class of necessary statements, including correct ascriptions of logical consequence, record members of a special category of fact which, when pointed out to us, we have no rational option but to acknowledge. The novelty of Bennett's account was that it attempted to combine the attractions of the venerable and inchoate view with a global conventionalism: all necessary statements, including ascriptions of logical consequence, are both true purely in virtue of meaning and conventional; for they are all either explicit meaning-determining conventions or derivable consequences of such. But we saw that in order to explain how a finite recognition of logical consequence was ever to be possible, Bennett would be constrained to invoke the notion of a convention-informed recognition of necessity which was not achieved via the construction of a derivation from anterior conventions. So no real advance is achieved beyond the venerable and inchoate view: no clear sense emerges in which the necessity of logical consequence is conventional, no clear middle path between the prongs of Dummett's dilema is opened to view, and the account is no better placed to handle the objections raised against the venerable view by Wittgenstein and Quine.

    Dummett's dilemma, then, still stands. Can there be a global conventionalism which does not embrace the extreme of radical conventionalism, the conception that all necessity is freely adopted explicit convention? The principal task of this chapter is to consider whether a negative answer to this question would entail that no satisfactory conventionalist account of necessity can be given. But I want first to consider the problems which conventionalism looks likely to encounter with the construal of reiterated modalities.

2.   We have not, so far, enquired in any detail what positive account of necessity the conventionalist should offer. Of course, we know the kind of thing that Wittgenstein had in mind. Dominating his thought on this topic is precisely the figure of the *language-game*; and

the laws of logical inference are rules of the language-game. (v. 28)

The figure of a language-game naturally encourages the prominence in our thinking about language as a whole of the ordinary contrast between the rules of a game and the moves made in the course of playing it. In the language-game, however, a 'move' is itself a linguistic act. Necessity originates in the circumstance that a statement is upgraded from the status of a conceivable move to that of a rule, prescribing how the admissibility of certain moves is to be associated with that of others.

This idea gives an immediate point to withholding the concept of truth from necessary statements. It is the incongruity of thinking of any game-rule as literally true. The rule is, rather, merely accepted—we play by it. To drop it, or never to have used it, are options we could have exercised without ignoring any worldly fact. For Wittgenstein, the contrast between necessary statements and others is to be located in their respective *rôles* in the language-game, the kinds of use to which they are respectively put. The traditional conception of necessity inflates this sort of contrast in rôle into a spurious distinction between two kinds of truth. It is Wittgenstein's rejection of that traditional idea which signals him as a conventionalist archetype. For the essence of any conventionalism about necessity worth the title is just this negative point: the suggestion is to be repudiated that necessary statements state a priori *facts* whose acknowledgment constitutes our recognition of necessity, and failure to acknowledge which is a kind of worldly ignorance. The leading conventionalist thesis has therefore to be that there is, admittedly, a fundamental contrast between necessary and contingent statements, but one against which the ordinary notion of necessary truth is already prejudicial. Not that there is no sense at all in describing necessary statements as 'true'; for they are *quasi*-assertable, and so, like all quasi-assertions, yield a well-formed sentence when prefixed by 'it is true that'. But this analogy, for a conventionalist, must somehow be all there is to the notion of necessary truth—though not, of course, all there is to the notion of necessity.

What follows from this thesis is not that ignorance of the necessity of a particular statement can be ignorance of no fact, properly so called, but that it cannot be ignorance of a fact stated by any necessary statement. So the following question arises. Since some statements are necessary and some are not, not to know of a statement which is necessary that it is so must, it seems, be interpreted as *some sort* of worldly ignorance. There thus appears to be a danger that for a conventionalist 'it is true that' and 'it is necessary that' will no longer commute as prefixes. 'It is true that it is necessary that *P*' would appear to state something of which I may be in ignorance; while it is unobvious that 'it is necessary that it is true that *P*' can now be regarded as *stating* anything at all. For if we accept that where *P* is necessary, so is: it is necessary that *P*—the distinctive axiom of Lewis's S4[1]—the latter quoted statement is itself liable to be necessary and so, for the conventionalist, non- fact-stating.

It might be supposed that a conventionalist will have to reject the S4 principle anyway. For, on his account, the necessity of *P* will consist in its having been assigned a certain sort of rôle; and the whole point is that this assignment is not *imposed* on us—it is a matter of convention whether any particular statement receives this rôle. So while there is a disanalogy of function between inner and outer occurrences of 'it is true that' in the kinds of context illustrated—the inner occurrence is quasi-assertive license, the outer substantial—there is no reason why we should not regard both 'it is true that it is necessary that *P*' and 'it is necessary that it is true that *P*' as fact-stating, indeed as stating the same fact: that *P* has been assigned a normative rôle, or whatever rôle the conventionalist proposes to be distinctive of necessity.

Now, though, any good reason for cleaving to the S4 principle becomes an objection to conventionalism. One intuitively powerful such reason is as follows. Suppose it is merely contingent that it is necessary that, for example, there are no cubic spheres. Then there are logically possible circumstances under which it would not be necessary that there are no cubic spheres. Like any logically possible circumstances, these circumstances might have obtained—and then the existence of cubic spheres would have been a contingent matter. But it seems impossible to give any intelligible account of what such circumstances might be—circumstances in which a cubic sphere might actually be constructable. The kind of possibility to which the conventionalist account will point—that, for example, we might have used the statement that there are no cubic spheres in a non-normative way—seems to do nothing to help us understand the character of the kind of circumstances in question; circumstances in which the claim to have constructed a cubic sphere would not be an absurd one. It does nothing to help us understand this because we seem unable to interpret the suggestion as anything other than that the *words* 'cubic', 'sphere', etc., might have had a different meaning; and while the conditional,

[1] See 55, pp. 490–500.

If 'cubic' had meant what 'yellow' now means, there could have been objects correctly described as 'cubic spheres',

is presumably true, this version:

If 'cubic' had meant what 'yellow' now means, there could have been cubic spheres,

seems unacceptable. Or better: there is *a* reading under which the latter is unacceptable, namely when the hypothesis of difference in sense is taken not to affect the content of what is asserted now, *using* the expressions in question. And it is this which must be the reading relevant to assessment of the S4 principle; for the construction of reiterated modalities does not require us to *mention* any expressions.

The difficulty has arisen because the conventionalist has been interpreted as attempting an account of the truth-conditions of statements of the form: it is necessary that P. And surely, we are tempted to think, any adequate account of necessity must do at least that. If we are to have an account of what the necessity of a statement consists in, this must be an explanation of what kind of thing makes true the claim that a statement is necessary. How then are we to interpret the statement, that it is necessary that P, as anything other than the claim that this kind of thing obtains for the case of P? And how, on a conventionalist account, can the obtaining of this kind of thing be anything other than a contingency?—contrary to the argument.

It seems to me that a possible conventionalist reply to the difficulty proceeds in two stages. To begin with, the conventionalist should concede that there is a natural reading of the conditional, 'if "yellow" had meant what "cubic" now means, there could have been cubic spheres', in terms of which it is unacceptable. But this, he should claim, is because the convention, that nothing is to count as a cubic sphere, is being appealed to in its assessment. And it is on an appeal to the same convention that the plausibility depends of holding the proposition that it is necessary that there are no cubic spheres to be itself necessary. What the intuitive argument essentially did was to collapse the double negation in '$\Diamond--\Diamond-P$', hypothesise that the 'outer' possibility obtained, and then assess the envisaged circumstances in terms of the very convention that nothing is to count as a cubic sphere. To be sure, there is no relevant reason to quarrel with the double negation elimination or with the elision of the outer '$\Diamond$'; it must be coherent to hypothesise as actual any logically possible circumstance. Nor does the conventionalist have any well-founded complaint against the argument on any grounds of circularity; for, presumably, the assessment of any possible circumstances has to proceed in terms of the conventions of correct description which we actually use. What he ought to argue is that, since the proscription of 'cubic sphere' as a permissible characterisation of anything is actually implicitly appealed to in the argument, nothing is done to establish that the acceptance of that convention is imposed on us in any *deep* sense, that to admit the necessity of 'It is necessary that there are no cubic spheres' would preclude him from regarding the necessity of 'There are no cubic spheres' as conventional. On the contrary, if both are necessary, then the necessity of both is attributable to the same convention. We may still want to complain that the conventionalist misinterprets the total unintelligibility—to which the intuitive argument appeals—of the fiction of circumstances in which the claim to have constructed a cubic sphere would not be an absurd one, that it is not just a matter of our having proscribed the description 'cubic sphere'. But we are likely, in any case, to want to level complaint against the conventionalist's treatment of the necessity of 'there are no cubic spheres'—whatever treatment he offers of reiterated modalities, it will not give *extra* offence.

The second stage of the reply is this. To allow the S4 principle is to allow that reiterations of 'it is necessary that' always yield a necessary statement when applied to one. The proper conclusion of the intuitive argument, according to the central thesis of conventionalism, is therefore to deny that the results of such reiterations are properly regarded as statements at all. Rather, what that argument illustrates is how to construct a case for assigning a normative rôle—if that is what is distinctive of necessary statements—to 'it is necessary that P' whenever such a rôle has been granted to P. What kind of thing, then, do such higher order conventions prescribe? Exactly that nothing is to count as a coherent elucidation of circumstances under which the existence of counterinstances to P (cubic spheres) would be contingent. The conventionalist, that is to say, now rejects the presupposition which led to the difficulty. It is not part of the brief of an adequate account of necessity to explain the *truth-conditions* of 'it is necessary that P'. If the validity of the S4 principle is allowed, such 'statements' actually have no truth-conditions; 'it is necessary that P' can no longer be regarded as stating that facts exist wherein the necessity of P consists. The initial problem of commutativity is thereby resolved. Both inner and outer occurrences of 'it is true that' are to be read as grammatically permissible tags, enclosing quasi-assertions.

What, in that case, is to *be* the canonical form of a statement of the facts in which necessity consists? Presumably it is to be in terms of whatever account the conventionalist gives us of the truth-conditions of: 'P' is used in the manner distinctive of a necessary statement. It is of the obtaining of such conditions that 'worldly ignorance' is to be possible. We misapprehend the problem if we think of it as the task of explaining the conditions of correct application of a predicate to whatever we take to be the referent of a that-clause. The problem is rather to characterise the use of *sentences* which we mistakenly conceive as effecting statements of necessary truth. The proper form for an expression of the facts in which necessity consists is a statement which mentions *P* and attributes to it a certain sort of use; so such a fact cannot be stated by a statement which, mentioning no sentence, makes use of *P*.

The impression, then, that a conventionalist must either reject the validity of the S4 principle or wind up imposing some pretty strange behaviour on the operations 'it is true that' and 'it is necessary that' may be mistaken. 'It is necessary that' will uniformly be conceived as a function from conventions to conventions; and 'it is true that' uniformly as a mere grammatical filler when applied to such expressions. But other difficulties certainly remain. For one, is it not just *false* that it is necessary that there are no yellow spheres? More generally, if 'it is necessary that' expresses a function from conventions to conventions, what account is to be given of its value when, as just illustrated, its argument is not a convention at all? A conventionalist could be expected to reply that actually the 'statement' in question is not merely false. Rather, we have a *permissive* convention that it is possible that there are yellow spheres, that it is sense to claim, for example, to have drawn one. 'It is false that it is necessary that there are no yellow spheres' is itself a quasi-assertion, giving expression to this permissive convention.

There seems, however, to be no prospect of restricting the admissible arguments of 'it is possible that' to conventions alone: it has to be applicable to what may be genuine statements, for what the permissive convention, 'it is possible that *P*', permits is the *truth* of *P*. This suggests that in order to achieve a general account of the rôle of modal operators, the conventionalist should actually conceive of that to which they are applied as some sort of statement-radical and the results as determined according as a radical has the rôle of genuine statement, convention, negation of convention, or whatever categories he might find it meet to introduce. But I shall not pursue this issue here. It cannot be ruled out that the conventionalist's leading thesis may well make it very difficult to devise a satisfactory general account of contexts of this sort. I have wanted to suggest only that he is not bereft of all approach. In particular, the kind of debate we might consider it appropriate to have about reiterated modalities is not obviously pre-empted by a conventionalist outlook.

3.   According to radical conventionalism:

> That a given statement is necessary always consists in our having expressly decided to treat that very statement as unassailable; it cannot rest on our having adopted certain other conventions which are found to involve our treating it so.[1]

The position is certainly unattractive; it plays havoc with our

[1] 23 (2, p. 495).

impression of what happens to us when we work over proofs, or contemplate statements which we regard as obviously necessary. It pays no heed to the sense of constraint recognised by our description of sound argumentation as 'cogent'. Stroud[1] attributes to Dummett a successful attack on this implausible view. The attack, as I read it, is twofold. The first part is the argument discussed above in Chapter V. It cannot be correct to regard us as free to lay down as necessary any statement whatever—at least, if it involves only concepts already in currency—since to do so is to grant it the status of an admissible rule of inference; and to give it this status is to allow that if correct application of it seems to take us from true premises to a false conclusion, some error must be involved either in the assessment of the premises as true or of the conclusion as false. But the rule will not tell us *what* error; and if the claim that some error has occurred is to be correct, then some more specific claim about the *nature* of the particular error must also be correct. So much is absolutely constitutive of the meaning of 'some'. But if such a more specific claim is to be correct, it must be possible—or at least have been possible—to disclose the occurrence of a particular error in the independent assessments of the truth-values of the premises and conclusion. How can this possibility be guaranteed if the status of the rule is simply that of an arbitrary convention?

Let me recapitulate why, as it was suggested, this argument does not succeed against a Wittgensteinian. The suggestion is that a radical conventionalist account of necessity cannot meet certain responsibilities to the senses of contingent statements. But, for Wittgenstein, statements of the type, 'some specific error has occurred', are not, in the manner needed to make good the objection, predeterminate in truth-value. That is, we ought not to think of the result-independent criteria for the occurrence of specific mistakes as having a life of their own, as determining as errors things which we have yet to recognise as such, so that the claim that error has occurred is answerable to the range of verdicts which they autonomously yield. The status of any error as such by these criteria requires a judgment by us. There is thus no way in which we can *fail* in our alleged responsibility to the sense of 'some' (as Dummett actually presents the argument, to the sense of 'or'). If, appealing to the validity of an arbitrarily picked rule which appears to have led us to a false conclusion from true premises, we insisted on saying that some error must have occurred in our assessment of these truth-values, we should run no risk of conflict with the facts. Of course, we cannot guarantee that we shall *actually* be able to find anything which we are content to regard as an error; but this is something which we cannot guarantee anyway, even if we conceive of ourselves as having codified in our principles of inference a system whose correctness goes beyond anything which a conventionalist is in a position to allow. The guarantee is rather that a specific error is *there* to be found. And this is

[1] 74, p. 510.

scarcely a precarious guarantee—it has the indefeasibility characteristic of open existentials. No doubt we think we know the guarantee to be sound. But ask: how *could* we know such a thing? And do not answer, 'By recognising the validity of our principles of inference'. For it is in an apprehension of their capacity to issue such guarantees, sound in some objective way, that, for a non-conventionalist, our apprehension of the validity of principles of inference will presumably consist.

For a Wittgensteinian, on the other hand, there is no such recognisably sound guarantee. The distinction in point of superiority between contrasting systems of inference should be sought, rather, in the respective frequencies with which they bid us postulate such errors and, among such cases, the frequency with which such postulation proves in practice to be capable of substantiation.

Dummett seems to anticipate something like the reply just sketched;[1] at least, he realises that a Wittgensteinian will not allow him the kind of idea of truth which the objection requires for the allegation that, for example, a miscount has taken place. But his second, and main complaint is rather that

> if Wittgenstein

—i.e. the radical conventionalist—

> were right, . . . communication would be in constant danger of simply breaking down.

The objection has to do with the absolute freedom which we are supposed to enjoy in approaching the question of the necessity of any given statement—

> It is all very well to say, 'Say what you like once you know what the facts are': but how are we to be sure that we can tell anyone what the facts are if it may be that the form of words we use to tell him the facts has for him a different sense as a result of his adopting some logical law which we do not accept?[2]

I have often heard in discussion the complaint that Dummett here simply misunderstands Wittgenstein's insistence that there is in the end no foundation for human agreement in linguistic practice; that there is nothing more primitive than our disposition to agree, and that it is improper to look for a philosophical account of how agreement is possible or to fault just on that ground a philosophical position which provides no possibility of such an account. That Wittgenstein held such a view is uncontentious; it is one of the strands in the discussion of following a rule. Sense is given to the idea that a particular description of particular circumstances is correct only, ultimately, in terms of our

[1] 23 (2, p. 503).
[2] 23 (2, pp. 501–2).

capacity to agree on the matter. In particular, it is futile to conceive of
agreement as always informed, or made possible, by shared apprehension of universals, or whatever. To think of agreement about the
application of an expression as possible because we are informed by a
shared conception of its correct application fails of explanatory force
since it is a necessary condition for possession of the relevant equipment
that we do in fact agree—if it were not, the picture would not explain the
agreement—and we have no *other* criteria for affirming that the equipment is shared. All that is achieved is to embroider the phenomenon of
agreement with a misleading terminology. This is not, of course, to say
that there can be in no sense an explanation of our ability to agree in
classifications; in terms, for example, of shared neurophysiology. But
that is not the deep explanation sought for. No doubt there are causal
pre-conditions of the possibility of communication. The goal, however,
was rather the elucidation, for each particular concept, of a state of
information—'grasp' of a universal—which successful training would
bestow and which, once shared, would guarantee, *modulo* sensory
divergencies and human prejudice, the possibility of ultimate agreement about the application of a concept. Wittgenstein perceived that
this project is chimerical. There *is* no deep explanation of why communication does not break down. Causal types of theory provide no
guarantees, certainly; a theory cannot guarantee its own adequacy.

It is not clear, however, that this meets Dummett's point. Dummett's objection concerns the rôle of *decision-making* in our acceptance
of new statements as necessary. He is arguing that matters cannot be as
conceived by radical conventionalism. The importance of the point is
that, if a global conventionalism is to be possible, decision must always
have *some* part to play; necessity must always be conferred—it cannot be
recognised, or imposed on us. The protest just considered does nothing
to illustrate the part played by deciding—its intention is merely to stress
that communal agreement founds judgments of necessity like all others.
But the essence of conventionalism is precisely that these 'judgments'
cannot be assimilated to all others; that here decision has a special rôle.

Why can this rôle not be interpreted as by radical conventionalism?
Because if that view is correct, there is not merely no *foundation* for our
agreement in linguistic practice but no practical reason for expecting
agreement at all. The distinction relevant here is that illustrated by the
ordinary expectation, to which we are entitled on inductive grounds,
that we shall be able to agree, say, about the expansion of the series of
natural numbers at any stage, and the expectation, to which we are *not*
entitled, that if invited severally to construct a series of arbitrary
choices, we should agree in the choices which we actually make. What
entitles us to the first expectation is just that similar training tends to
produce harmony in applications; we are entitled to suppose that each
of us knows how to count. To be sure, there are misconceptions
concerning the character of this knowledge which Wittgenstein is keen

to explode; *we* have still to construct the path which the integers follow—it is not settled for us by the criteria of correct continuation. But the fact is that we *need* the training to get agreement here, and that we shall appeal to the criteria if divergences arise. With a series of arbitrary choices, on the other hand, there is, in the relevant sense, no such thing as knowing how to continue, no sense in the idea of disagreement about how the series *ought* to be continued. How to continue is literally a matter for individual decision. And this is the conception of our admission of each new necessary statement which Dummett is associating with Wittgenstein's supposed radical conventionalism. There is no sense in which our acceptance of a new statement as necessary can be conceived as criterially *governed*; there is nothing—not nothing 'ultimately', but nothing at all—in terms of which a dispute about the matter could be resolved. The very idea of a dispute, indeed, is misappropriated; for there is not anything to dispute, any more than we could significantly dispute the right continuation of a series of arbitrary choices.

Surely, the 'risk' to the possibility of communication involved in such a picture is obvious enough. Consider an analogy with a board-game again. Suppose we have, by ordinary standards, a clear and complete explanation of the rules. These rules correspond to initial conventions; they are adopted by fiat, we all agree to play by them. What, however, if the radical conventionalist is right, is the *substance* to this agreement? For in order to play by the rules, we shall have to make inferences from them. The trouble is not just that the game may be sufficiently elaborate to require that, in order to adjudicate a particular position, we have to appeal to a rule which was not explicitly stated as such, which is rather a derived rule. If this is so, and the acceptability of such a rule is always a matter for a further explicit decision, we are bound to say that the initial list of rules was not after all a complete explanation of the practice of the game, that there is perhaps no *complete* explanation. That is certainly counterintuitive in the face of the fact that we probably will not need more than an 'incomplete' explanation to secure agreement about admissible play. But, more important, it is now obscure how a presentation of the rules could be taken as a genuine *explanation* of the practice of the game at all. For even in situations which we should ordinarily regard as falling directly within the province of an expressly stated rule, its application will still be a matter of inference. The rule will be a general statement, designed to adjudicate all situations of a certain general sort. It will need application to the particular cases, and agreement about the character of this application needs more than agreement about the nature of a particular case; it needs agreement about the directive resulting from instantiation of the rule to that particular case. However, since this directive was not explicitly issued, its status on the present picture—or rather, *what* directive is actually involved—is a matter for explicit free decision. So it becomes impossible to think of the original general rule as part of a genuine elucidation of how to play. Our agreement about the original rules, our willingness to be party to them, comes to seem totally insubstantial. It was not agreement about how to *do* anything; when it comes to action, further arbitrary decisions are needed at every turn. But if conventions are to guide a certain practice, there must be implicit agreement about what their consequences are; so explicit judgments about those consequences cannot require further, totally free decisions.

Matters are actually worse with the 'language-game', since it is not a co-operative activity in the same way as, say, Chess. If language is to be used to inform, we cannot be required to scrutinise the origins of every assertoric 'move'—the responsibility for the 'decisions' involved has to be delegated to individuals. But the transmission of information can be secured only if there is implicit agreement among speakers about the conditions under which particular assertions are warranted. If the acceptance of new statements as necessary were a matter of the free decisions of individuals, it is impossible to see how there could be this implicit agreement; for an uncontentious rôle of necessary statements is precisely to contribute towards the determination of the assertibility-conditions of contingent ones. How am I to *receive* the assertion that P if there is every possibility that the assertor's willingness to make it is the product of an arbitrarily made inferential decision? Even if there is verbal agreement about the principles of inference which are relevant, how is this agreement any more substantial than that involved in accepting the initial rules for the board game? The assertor may have decided to accept different instantiations of the principles to the particular assertoric situation than those which I should decide on. The effect of such a conception of the 'rules of the language-game' is that they surrender the function of rules, that they are deprived of exactly the normative rôle which is supposedly essential. Wherever the warranted assertability of a contingent statement is the product of an inference, the idea of 'warrant' drops out altogether; and then there is no way of intelligently receiving the assertion.

Radical conventionalism is thus an incoherent position. It cannot be supposed that the status of consequences of conventions is a matter of arbitrary decision, or we can give no account of the capacity of the conventions to inform linguistic practice. The whole point of having conventions is to secure conformity of conduct in certain areas. To be sure, we have no guarantees that this objective will be secured. But it *cannot* be secured if the manner of application of the convention to a particular situation is a matter of arbitrary choice. For if anything is a 'very general fact of daily experience' (1. 118), it is that when human beings are given free rein to make decisions arbitrarily, they do not in general choose the same. The application of general conventions is irreducibly a matter of implicit inference. The question, what are the consequences of a convention, comes up every time the convention is put to any work. And the status of something as a consequence cannot, therefore, be viewed as by radical conventionalism. Once more: it is idle to attempt to remedy its shortcomings by appeal to the basic phenomenon of human agreement in judgment. There *is* no such phenomenon when we are told to do as we like.

No doubt Wittgenstein is misinterpreted if taken to be advancing this view. But it was politically justified for Dummett to interpret him in this way. For one thing Wittgenstein *does* introduce into the description of our admission of proofs and ratification of necessity—in general, indeed, into the description of our following of rules[1]—the idea of 'decision'; if it was not meant as in radical conventionalism, none the less it is there. But even if he had not done so, the fact remains that if a generalised conventionalism about necessity is to be essayed, the notion of decision must have *some* part to play every time we accept a statement as necessary. Conventions are things which we decide to adopt; not that decisions have always to be taken explicitly—a conventional practice may simply develop. Still, the point of describing the character of the practice as 'conventional' is just to suggest that it is not imposed upon us, that is something on which we could have decided explicitly and is at most answerable to considerations of practical convenience. Only two kinds of consideration enter into the justification of conventions—those of manageability and of utility. It is unmistakable that Wittgenstein wants to say something similar about our principles of inference, in particular that he wants to deny that their validity is ever simply *recognised*.[2] So it seems that he is committed to granting to the concept of decision some part to play in our acceptance of every new case. The question is not whether Wittgenstein intended radical conventionalism; it is rather, if the rôle of decision-making in the admission of proofs and the application of conventions is not to be the absurd one implicit in radical conventionalism, what is it to be? Is a conventionalism which avoids this untenable extreme possible, if it is not to be beset by the difficulties of the 'modified' brand? Has Wittgenstein supplied the materials for a more satisfactory account?

4.   Wittgenstein's position, as so far interpreted in this book, has five facets. First, he wants to deny that the concept of truth is properly applicable to necessary statements. Their function is normative; it is to direct and regulate the use of statements of which truth and falsity are genuinely predicable.

Secondly, there is no ultimate justification for the principles of inference which we use. There are only the considerations of utility and manageability appropriate to any conventional practice. In particular,

---

[1] See, however, *PI* i, 219; *LFM*, lecture xxiv, p. 226.

[2] That is, discovered. 'Recognise' sometimes carries the sense of voluntary acknowledgment; but never in this book.

the apparent constraint that they be truth-preserving in application is not a real constraint. For our rules of inference are antecedent to truth; that is, they are among the criteria of truth for the statements to which they are applied. There is no possibility of their clashing with other observational, or operational, criteria for the acceptability of such statements. Criteria do not issue verdicts autonomously; their results require our explicit ratification, and this ratification will be dominated by our principles of inference. To accept a statement as necessary cannot be out of accord with the use of the expressions which it contains, or to which it may be applied. For to accept it as necessary is further to determine the correct use of those expressions.

Thirdly, perhaps because the standing of these principles is answerable to no 'ultraphysical' considerations, because there are no objective truth-liaisons among contingent statements which it is their responsibility to codify, Wittgenstein claims that the concepts of *correct* inference, calculation, etc., have no more in them than is defined by the rules which we actually employ—just as the concept of playing Chess correctly goes no further than playing in accordance with the rules which we accept as the rules of Chess.

Fourthly, as these considerations seem to suggest, we can conceive of the possibility of alternative systems of calculation and inference. If our principles of inference are antecedent to truth, then there seems no reason why we should not have used different principles.

Finally, we ought not to think of our subscription to particular principles of inference as a rigid contract in understanding, predetermining what in any particular case is to count as their correct application if we are to remain faithful to their content. It is our ever-expanding network of applications of them which determines their content, rather than the other way about—though this way of talking, too, is misleading. Wittgenstein commends as an analogy the relation between behaviour and character:[1] it is the man's ever-expanding tapestry of actions and responses which reveals his character—and *all* that is revealed are further actions and responses. The ground for this conception of meaning is the central conclusion to emerge from the discussion of following a rule: to conceive of the content of a rule as something finally established and stable, which each successive application objectively either implements or violates, it to conceive of grasp of such a content as something which we can recognise neither in others nor ourselves. We cannot significantly give our sincere inclination to make *this* application of *this* rule such a dignity. There is, in the end, only the inclination. (In a comparable sense, a man cannot act out of character.) Here is where it is important to emphasise the phenomenon of a consensus in calculation, of agreement in judgments in general. Agreement cannot be explained by appeal to a grasp of the putative unity of those applications of an expression which correct understand-

[1] *RFM* I. 13.

ing of it enjoins—for such grasp can manifest itself only in the disposition to agree.

In Chapter IV we rejected the third and fourth elements in this résumé. Wittgenstein seemed to be paying insufficient heed to the way in which the concept of a certain sort of inference, say, calculation, connects with the kind of application made of its results. This connection is such that 'correct inference' cannot be regarded as defined purely by the principles of inference which we happen to allow. Indeed:

> It is the use outside mathematics and so the *meaning* of the signs that makes the sign-game into mathematics. (IV. 2)

The criteria for whether or not a tribe has a system of calculation do not coincide with those for whether or not it has *our* system of calculation.

The reasons for saying so need no further elaboration at this stage. We had occasion to sketch them again in Chapter XVI. But perhaps it is worth while making a little more explicit what is acceptable and what is unacceptable in the hypothesis that alternative systems of inference and calculation are possible. What criterion ought a field anthropologist to use for saying in any particular case that inferential procedures are being employed other than his own? The claim makes sense only if there is a plausible interpretation for at least a large proportion of statements made by the studied people in terms of which he feels he understands why they make them when they do and what their interest in making them is. But his confidence in the interpretation will vanish if they proceed to make, as it seems to him, invalid inferences among these statements and are not disconcerted by the disharmony of their conclusions with what he had interpreted as their objectives in employing them. He will lose confidence that he does after all know what the conclusions mean, and look for another interpretation.

Just because the essential rôle of inferential principles, on Wittgenstein's view, is to further determine the assertability-conditions of contingent statements, even where these have received an antecedent explanation, there is no possibility of identifying as such a community who infer statements of which we have a correct interpretation by means of alternative, and doubtless—as it seems to us—invalid rules; *unless* they are suitably often disconcerted by the results and feel compelled to look for errors in their assessments, direct or theoretically inspired, of the truth-values of premises and conclusions. If they do, we can retain some confidence that we understand the statements which their principles of inference take them from and to; but the effect of this confidence is that we shall describe them as inferring invalidly. The price of the confidence is *commensurability*. We shall feel that we ought to be able to step in and explain to the community the source of the disharmony of their objectives with their inferential techniques, and persuade them to change them. Of course, it is not guaranteed that we shall be able to do so. But the community's behaviour is still liable to reasonable criticism; for there are alternative inferential methods which there is every inductive reason to think will serve their purposes better. If, on the other hand, they are not suitably often disconcerted, and all appears to conform with their expectations, then there *is* no basis for attributing alternative inferential practices to them—we are still awaiting an interpretation of their language in this area.

All the examples by which Wittgenstein attempts to give the fourth strand concretion—the soft-ruler men, the wood-sellers, the dividers by zero, the men who 'count' taking objects twice over, those who 'fancy' that $(a + b)^2 = a^2 + b^2$, those who infer via contradictions—turn out, if we press for detail, to de-stabilise in one of these two directions depending on how the fuller story goes. It will not do blandly to suggest, like Stroud, that we can understand the possibility of alternative methods of inference non-constructively; that is, without it being in principle possible to construct intelligible examples.[1] If Wittgenstein thought we could understand the suggestion in that way, it is hard to see what his objection is to the claim, for example, that it is a possibility both that '550' does not occur in π and that no demonstration can be given; or that our perceptions of colour could differ in radical but behaviourally inconsequential ways.[2]

Now, the two important points here are these. First, the principal thesis of Wittgenstein's conventionalism, that necessary statements function regulatively and are in no sense imposed, does not *commit* him to the idea that we could use totally different, rationally incommensurable techniques of inference. Secondly, it is nevertheless open to a Wittgensteinian to take a sympathetic view of the suggestion that not only the hypotheses of a theory, but its underlying logic may be emended as practical convenience may dictate. After all, that is precisely what a tribe of zero-dividers can be expected to do—or at least to be capable of being persuaded to do—if our confidence in interpreting certain of their statements as contingent arithmetical judgments is warranted. The point here is not the trivial one that people may always agree to change the interpretation, and so the use, of any particular sentence or sentential schema. Rather, in order for something to be intelligible as the employment of inferential techniques other than those which we should now endorse—whether it is a matter of identifying another community as doing so or of interpreting our own former practice—there has to be overlap (or continuity) of interpretation for a large cluster of statements transitions among which are mediated by both sets of putatively contrasting inferential rules. Naturally, a change in inferential procedures will involve a change in the use—the circumstances under which we are prepared to assert—the statements in question. What it must not involve, if it is to be intelligible as such, is a

---

[1] 74, p. 513.
[2] See *PB*, 41; *BlB*, pp. 60–1; *PI* i. 272–9

change in our conception of the conditions of *ultimate* justification for their assertion—in the case of contingent, numerical statements, for example, the idea of a correct count as decisive. The change (or contrast) in inferential patterns has to be interpretable as a change (or contrast) in our (respective) conception(s) of the conditions under which other, *basic* criteria—counting, measurement, etc.—for the use of the statements inferred from and to have been correctly applied. The root reason why a Wittgensteinian can sympathise with Quine's ideas is just this: correct application of such basic criteria is always in principle a defeasible claim; nothing counts as a decisive verification of it—there is always a distinction between seeming to have made no slip or error and actually not having done so. Rules of inference applied in practical contexts play on this hiatus, determine our conception of when the transition has been successfully negotiated. And this conception is, in principle, indefinitely revisable and re-revisable as practical considerations may dictate.

Our present problem, however, is to determine whether Wittgenstein has supplied the materials for a genuinely global conventionalism, beset by the difficulties of neither the radical nor 'modified' versions. And in the context of this problem, the loss of the third and fourth strands is inconsequential. It might have been thought otherwise; if we had been able to make ready sense of the availability of alternative, rationally incommensurable techniques of inference, would not that have supplied a sense in which our acceptance of such-and-such a consequence of certain conventions involved a decision?—for we might have had quite different canons of valid inference. But this mistakes the character of the problem; the conventionalist has to locate a decision in our every acceptance of a statement as necessary *within* the framework of the rules which we actually use. The first strand, moreover, is really just an expression of the faith that a global conventionalism is possible. So the question is: are we in a position to indicate the requisite element of decision in our acceptance of any new statement as necessary by deployment of the remaining two—the second and fifth—strands?

5.   Obviously, what is amiss with radical conventionalism is the suggestion that the decisions involved in our ratification of new necessary statements are wholly arbitrary. If decisions are involved at all, they must be construed as guided in some way. The conventionalist must hold that we are not *required* —by the constraint of conformity with the facts, or whatever—to accept the new necessary statements which we allow; but neither do we have a freedom intelligible in terms of the idea of arbitrary choice. This is actually the picture of the matter which Stroud commends as exegesis of Wittgenstein's intent:

Logical necessity is not like rails that stretch to infinity which we must follow in only one way, but neither is it the case that we are not compelled at all. There are the rails we have already travelled, and we can extend them beyond the present point only by depending on those that already exist; for the sake of navigability they must be extended in smooth natural ways—how they are to be continued is to that extent determined by the route of the rails already there. [1]

A critic will ask: what is this idea of guidance being contrasted with on the other side—do we ever *not* have the freedom which is supposed to originate in the gap between guide and verdict here? We could give an affirmative answer to this question, indeed the situation would fit Stroud's picture quite well, if the concept of valid inference had a measure of *flexibility*; the kind of situation the picture brings to mind is that of a judge aiming at a fair sentence. Our judgment of the validity of an inference would be a judgment in the same sense—a guided choice with the force of further precedent for future occasions. We should thus have a responsibility to ensure that it cohered with previous judgments, that it combined with them to form a manageable set of precedents. Such a picture, however, founders on the simple fact, acknowledgment of which is a precondition of talking sense about necessity, that when we make such a 'judgment', we cannot in general point to respects in which our verdict seems *discretionary*. If we could, the inference would seem to us to fall short of full cogency at just the places where discretion operated. There seems to be no vagueness in the idea of valid inference—at least for inferences mediated by specified rules. Nor, therefore, is the freedom, which the conventionalist needs to indicate, to be understood in terms of the model of guidance by inexact criteria. It cannot be plausibly claimed that the concept of a valid inference has the same sort of flexibility as that of a fair verdict.

A natural proposal at this point is that what Wittgenstein really wants to call into question is not the very idea of logical cogency but just a standard picture of the nature of the constraint —a misinterpretation of the character of the 'must' in conditionals like, 'if you want to infer in accordance with the rules, you must accept this step', or any conditional articulating what is required of us if we are to conform with particular rules. Nowhere in *RFM* is there a denial that there are such requirements; nowhere does Wittgenstein say anything to suggest that all such conditionals are actually *false*—as, presumably, they would strictly be if the judicial analogy were acceptable. It is just that accord with the rules does not have a settled, investigation-independent character—the conditionals do not state investigation-independent truths. Wittgenstein need not be rejecting our warrant to assert such conditionals, but only the idea that it derives from recognition that to do so is to pay due heed to the objective character of the practices which they concern.

This is undoubtedly part of Wittgenstein's view, but it cannot be all

[1] 74, p. 518.

of it. Since the sort of conditional in question will contain a demonstrative element —'this step', etc. —it will not, strictly speaking, be necessary. But its method of assessment will be that of a necessary statement; we have only to identify the demonstrated element and apply canons of inference to it. It is no contingent property *of* the step, or move, that it conforms to or breaches the rules. Any positive account of necessity therefore owes us an explanation of the character of our warrant to assert such conditionals; it cannot rest content with a mere denial that the warrant is as we standardly conceive it. And it is unmistakable that Wittgenstein was trying to work towards a positive account.

In fact, the would-be conventionalist's predicament crystallises around conditional statements of this sort. For these are the statements about which our implicit agreement has to be secured; it is conditionals of this kind which serve to articulate the character of conformity to a general practice as codified and explained by more general conventions. Even where it is right to regard the latter as issuing from arbitrary choice, we cannot extend this conception to these conditionals—or our agreement about the general conventions will become dislocated from agreement in their implementation. If general conventions are to have a normative rôle at all, our acceptance of these conditionals must be warranted in some other way than that offered by radical conventionalism; and if any global version of conventionalism is correct, it must be an error to suppose that we ever simply recognise the correctness of such a conditional. So the would-be conventionalist needs an account which steers between these rocks. There must be every practical reason to expect agreement about whether a particular such conditional is acceptable; implicit agreement at the very least, and explicit agreement if the people whose practice is informed by the general conventions are sufficiently articulate. But the account must avoid the conclusion that we *recognise* such statements' truth. There has to be some notion of warranted assertibility for them falling short of the warrant bestowed by recognition of the facts. The final step, as it were, must still be a decision, but a decision concurrence in which is no less secure than our general disposition to agree in linguistic practice. Wittgenstein, I believe, wanted to offer such an account; but so far no route to such an account is in view.

6.   There is a way of misunderstanding the discussion of private languages which might lead one to suppose that Wittgenstein cannot countenance any notion of *warranted* assertibility for these statements. For is there not in the end no interesting difference between our ratification of such a conditional and someone's application of a 'private' sensation word? There is, according to Wittgenstein, no substantial sense in which, when I sincerely apply S to *this* sensation *now*, I may be held to be using S rightly or wrongly. Whatever seems to me right is

going to be right and apparently 'that only means that here we can't talk about "right" '. But what important distinction is there between my situation here and *our* situation when we ratify, 'if Black is to conform to the rules here, he must move the king to QB1 or resign'? How is there an any more substantial sense in which we have here applied the notion of conformity with the rules of Chess rightly than the 'rightness' of my sincere use of *S*? What does publicity matter? For, in the same way, we cannot in the end draw a distinction between the conditional's seeming to us to be correct and its really being so. Is there not finally just the disposition to assent both in the public and private cases? How does it effect a difference in point of *correctness* of assent if we are concerned with a case which *we*, rather than I alone, can judge?

The tempting reply is that it is only in terms of a community of assent that we can make sense of the idea of correct use of an expression at all, that there is not such a thing as a correct description unless agreement can be elicited about its correctness. But this, as it stands, just begs the point at issue. For I agree with myself, as it were, about the application of *S*. We have to ask, rather, what differences are *consequent* upon the fact that we are operating in the public sector, so to speak? When a judgment is capable of communal ratification, what are we able to do, other than agree with each other, which we cannot do in the private case?

The obvious answer is: to correct our own individual judgments. If I find my assessment of a situation out of line with others', and am unable to change matters by calling attention to what I consider to be the relevant aspects, there are two possibilities: either they can draw my attention to features which I failed to take account of, as a result of which I concede my former assessment to be wrong; or they cannot— that either we mutually fail to see the relevance of the considerations which have respectively decided us, or if the judgment is sufficiently simple—of primary colour, say—then I simply cannot understand why others make the assessment which they do. In that case, supposing no perceptual divergence to be the root of the trouble, I shall have to face the fact that I have misunderstood the meaning of the sentence by which I gave my assessment. Either way, I have to conclude that I did not assess the situation *correctly*—if it is allowed that there is no residual sense in which I may have assessed it correctly relative to my own understanding. And we can take it that that much is allowed. For the doubt was whether, if there is no substantial sense in which I can recognise the dictates of my own understanding, there can be a more substantial sense of correctness for *our* judgment that Black must move the king to QB1 or resign.

Because there is in general a process of assessment for such judgments, we can make sense of the idea of an oversight. Black replies, for example, 'Not at all; by R to Q4 I can block the check, and simultaneously discover a check on White by the KB.' Publicity makes it possible to negotiate and revise our individual judgments. A judgment which seemed correct may come to be regarded as otherwise. This leads into a more subtle point. Ask now, why should the same not happen with a sincere use of S? Is not a judgment involving S revisable also? At first the sensation seems to me to be S again, but after a short while it strikes me as different and I want to revise my assessment. Clearly, however, other than in cases where I regard myself simply as having failed to say what I meant—a slip of the diarist's pen—I can have no possible reason for saying that the original description was wrong, rather than that the sensation itself has changed, that a sensation that was S has merged into one which is not. This distinction makes no *phenomenological* sense at all; and no one else can draw it for me. It is just here that the publicity of a calculation or inference (or chess position)—its capacity to be exhibited orally or on paper as a public object —is of decisive importance. For it is this which gives sense to the idea of a temporal examination of it and so to that of the discovery of features which an original inspection did not detect. Its publicity makes it a structure which we can return our attention to with a significant presumption that it has not changed in the meantime. Of course, there is still room for a *sceptical* doubt whether, say, a calculation might not originally have been correct and that in which we now disclose an error is not an unnoticed mutation of it. But this doubt will likely be sceptical because it will very probably be right to say, by ordinary criteria, that no change has occurred—none of the members of the class noticed any mysterious movement or transformation among the formulae on the board. It is because we have such ordinary criteria that there is sense in the idea of an investigation of the calculation. In contrast, a doubt about the enduring sameness of the sensation is not sceptical—which is far from saying that it might in a practical sense be real. Rather, the judgment that the sensation has not changed, if it is to be compatible with an admission that one originally mistook its character, is something for which there are no criteria, ordinary or otherwise. It is not intelligible that things should even *seem* to be so.

There is thus room for the idea of an investigation into the rights and wrongs of our assent to a conditional of the germane type. This idea has no counterpart in the situation of a self-ascription of a sensation, thought of as vindicated by the character of a private object and the meaning of a private word. The point applies even where we are concerned with the most simple, direct applications of a simple rule—the simplest kind of ingredient judgments in our ratification of a proof. A written representation of such a step remains as a public object which we can re-attend to and re-appraise, satisfied by ordinary criteria that its physiognomy has not changed. We can revise our assessment of such a step without having to acknowledge as a possibility which we have no way of defeating by ordinary (non-sceptic-proof) standards the supposition that, should we find ourselves wanting to make a revision, perhaps it is rather the step itself which has changed. It is a natural conjecture that this contrast characterises the distinction between those mental objects —sensations and images —whose exact nature we are inclined to think of as essentially incommunicable and those—especially thoughts —which we should ordinarily suppose capable of full public characterisation. (Traditionally the intuitionists seem to have conceived of proofs and mathematical objects generally as mental entities coming within the first category.[1] This mistake would provide ample motivation for rejecting a realist conception of truth for statements about them; but it would also totally obscure the possibility of a significant investigation of their constructive properties.)

There is thus more to the 'private-language argument', applied to sensations, than an application of the rule-following considerations to an egocentric situation. Our ratification of the conditional about Black, if it does not have the objectivity of accord with antecedent facts, has at least the quasi-objectivity associated with the possibility of a significant investigation, and, in consequence, of reassessment.

7.   So much for a possible misunderstanding. The situation, then, at which we have arrived is this. Suppose we have freely chosen certain general rules for a game; and let us grant, in accordance with Wittgenstein's second conventionalist strand, that thinking of the matter in terms of a game is not prejudicial—that there is not with rules of inference a problem of fidelity to the truth-values of contingent statements with no analogue in the situation of game construction. The would-be conventionalist's problem is that if the rules are to be capable of normative function, training in them must secure agreement in practice; more especially, if we have a language in which we can adequately describe the situations that may develop in the course of the game, agreement must be secured in articulate judgments about practice, about the rights and wrongs of particular moves in particular situations. So what account is to be given of the correct assertability of statements —the sort of conditional we have been considering —to such effect? For a conventionalist it can be a matter neither of arbitrary decision not recognition of the dictates of our understanding; for the essence of conventionalism is to deny that the rôle of necessary statements is actually to state anything. And the same ought to go for the statements in question, even if we quibble over their description as 'necessary', since they wholly concern 'internal' characteristics.

Earlier it was proposed that we need look no further than the suggestion that Wittgenstein's position here is just an application of the fifth strand. Of course we have criteria for the correct assertion of these statements; but we are not to conceive of these criteria as running the road ahead of us, and ourselves as following in their wake and finding

[1] For example, Heyting 46, pp. 8–9.

out what they have determined for us. To revert to the image of the committee: it is the case *neither* that by choosing the content of the two volumes they *thereby* settle implicitly all their correct applications and consequences, *nor* that the ongoing task of the community is the same in character as that of the committee—a matter for free decisions. We *construct* the road whose foundations the committee has laid, but we do so not by decision but by judgment.

Hacker and Baker[1] talk of Wittgenstein's conception of the 'auto-nomy of grammar'. The doctrine is twofold: grammatical propositions do not have to answer to any facts, and are mutually independent. There is, perhaps, no absolutely explicit statement to this effect in either the *Grammar* or *RFM*;[2] but it seems that a conventionalist must hold some such view. There can be no genuine truth-conditions for conventions or their acceptability will not be conventional; and if a global conventionalism is the object, there can be no ratification-independent liaisons between conventions either. So what we actually have as exegesis of this idea so far are simply the second and fifth strands respectively; grammatical propositions are not answerable to the truth-values of the statements inferences among which they mediate, and are mutually independent in the sense that our judgments concern-ing their consequences and applications are likewise mutually independent—we construct the connections between them and, in doing so, we add to grammar. Our understanding of the correct applica-tions of conventions, and their logical consequences in general, determines the route of no untravelled roads.

Now, why precisely will this account of the inter-independence of necessary statements not serve conventionalist purposes? Well, what the conventionalist needed to make out was a notion of guided decision. The whole point of invoking the concept of decision is lost if there is nothing to contrast it with. It has to be the case that acceptance of the conditional articulating Black's options is something which we *choose*, although the choice is guided and informed. But this conception is not made good just by pointing out that our 'choice' is answerable to no ratification-independent facts concerning correct application of the rules of Chess. For the same point, if we accept the rule-following considerations in general, holds for absolutely every judgment which we make. Every contingent judgment is in this way a *new* judgment: there are no objective truths as yet unrecognised by us dealing in the judgments which we ought ideally to make if we correctly apprehend the world and apply our concepts properly. The mutual independence of grammatical propositions, so interpreted, is just a special case of the mutual independence of *all* judgment. But now we have no significant contrast between our acceptance of the conditional about Black and our

---

[1] 44, pp. 278–9. See also Hacker 45, pp. 150–66.
[2] Various passages in the later texts come very close. See, for example, *PB*, 7; *PG* I. 133; II. 1, 15 (esp. p. 313); *Z*, 320, 331.

agreement about any new contingent judgment. If either is a guided decision, it seems, then both are. The notion of decision is idling.

We thus have as yet no purchase on the fundamental thesis of conventionalism that no necessary statement describes anything; that they are mere quasi-assertions whose prescriptive rôle is not to be explained in terms of the general prescription that our judgments and practices should conform to the facts. The rule-following considerations certainly bear down hard on the platonist 'alchemy' of objective conceptual structures; but, unsupplemented, they seem to leave intact for necessary statements whatever notion of recognition is now appropriate for contingent ones. So no genuine conventionalism issues from the rule-following considerations alone; and a careful reader of *RFM* can come to no other conclusion than that Wittgenstein was strongly attracted to a genuinely conventionalist account of necessity.

XX

Necessity

Sources

PI: 1. 371–3
LFM: lectures xv; xviii, pp. 172–4; xxxi
WWK: part i, p. 63. (*Geometrie als Syntax* ii, concluding remarks)

1.   The question for this and the remaining chapters is whether Wittgenstein's ideas (can be developed so as to) substantiate a globally conventionalist theory of necessity. The first section of this chapter will be given over to a rehearsal of our understanding of the situation up to this point. Accordingly, those who need no rehearsal should pass directly to §2.

During the debate provoked by the Logical Positivists' conventionalist views, A. C. Ewing,[1] identifying as one element in his opponents' motivation the idea that the truth of necessary statements 'depends wholly on the meaning of the terms used', wrote:

> This is true in a sense, for an *a priori* proposition does follow necessarily if the meaning of the terms used is given, and therefore its truth, unlike that of an empirical proposition, does depend wholly upon the meaning of the terms used. But this admission is compatible with my view as well as with the conventionalist view. If the terms used stand for characteristics, one of which entails or logically excludes the other, the truth of the proposition must follow from their meaning; and if they do not stand for such characteristics, the proposition is not *a priori*. It does not in the least follow that an *a priori* proposition is made true simply by the speaker, or most educated people, using words in a certain way. That the truth of an *a priori* proposition 'depends wholly on the meanings of the terms used' is indeed not inconsistent with any view about *a priori* propositions—except the view of Mill that they are based on empirical generalisations concerning the objects to which they refer. The question is whether they depend on the meaning of the terms in the sense of being a necessary consequence of the nature of the characteristics for which the terms stand, or whether they depend not on the objective nature of what they mean but on arbitrary rules of language which forbid us to combine certain words.

[1] 33, pp. 231–2, 239. See also Ayer 1, Broad 8, Kneale 52, Malcolm 61, and countless other papers of the period.

And later:

> . . . when I recognise the truth of an *a priori* proposition, I am seeing that
> a certain characteristic by its inherent nature necessarily involves or
> excludes another, and not merely learning that certain rules of language
> hold or ought to hold. In order to decide whether a proposition is
> necessarily true, I have not to consider the usage of words but the
> characteristics for which the words stand.

Ewing went further than he needed, in order to resist conventionalism, by his suggestion that our knowledge of these 'characteristics' is, in
some sense, direct, or unmediated by linguistic competence. All he
needed to maintain—perhaps, indeed, all he meant to maintain—is that
we possess a capacity so to reflect on our use of language, so to
extrapolate just what justifies the use of a particular expression from the
training in its use which we are given or which we consider paradigmatically adequate, as to be enabled to see that its correct use—keeping to
the kind of example he seems to have had in mind—requires, or
prohibits, justified use of some other particular expression. However, it
is clear that Ewing is right in the contention that to think of necessity as
depending 'wholly upon the meanings of the terms used' ought to
provide no impulsion whatever in the direction of conventionalism.

In the preface to the 1946 edition of *Language, Truth and Logic*, Ayer
in effect conceded Ewing's point, shorn of its platonist ornamentation,
and moved towards 'modified' conventionalism:[1]

> Just as it is a mistake to identify *a priori* propositions with empirical
> propositions about language, so I now think it is a mistake to say that they
> are themselves linguistic rules. For apart from the fact that they can
> properly be said to be true, which linguistic rules cannot, they are
> distinguished also by being necessary, whereas linguistic rules are arbi-
> trary. At the same time, if they are necessary, it is only because the
> relevant linguistic rules are presupposed. Thus it is a contingent, empiri-
> cal fact that the word, 'earlier', is used in English to mean earlier; and it is
> an arbitrary though convenient [*sic*] rule of language that words which
> stand for temporal relations are to be used transitively. But *given* this rule,
> the proposition that if *a* is earlier than *b* and *b* is earlier than *c*, *a* is earlier
> than *c* becomes a necessary truth. Similarly, in Russell and Whitehead's
> system of logic, it is a contingent, empirical fact that the sign '$\supset$', should
> have been given the meaning that it has; and the rules which govern the
> use of this sign are conventions which themselves are neither true nor
> false. But *given* these rules, the *a priori* proposition, $q \supset (p \supset q)$, is
> necessarily true. Being *a priori*, this proposition gives no information in
> the ordinary sense in which an empirical proposition may be said to give
> information; nor does it itself prescribe how the logical constant '$\supset$', is to
> be used. What it does is to elucidate the proper use of this logical constant,
> and it is in this way that it is informative.

Ayer's 'it is and it isn't' attitude to the question whether or not such

[1] 1, p. 17.

propositions are *informative* presumably harks back to the extremely restricted notion of verification with which the positivists originally tended to work. The fact is, nevertheless, that he now allows that necessary statements can be acceptable purely on the basis of, while not themselves amounting to expressions of, certain ulterior linguistic conventions. All that is preventing him from granting outright that necessary statements record objectively essential characteristics of universals, that is of the patterns of usage which the underlying conventions establish, is the core dogma of ten years previously: that the only, properly speaking, *factual* statements are those capable of verification purely by sense-experiential means. Ewing's point was in essence that it is a confusion to suppose that, just because it is conventional what meaning an expression has, it is also conventional what properties meanings have—or that their having particular properties merely depends upon aspects of *actual* linguistic usage. This is what Ayer is now conceding. By doing so, he in effect abandons altogether his former conventionalism. The original idea was that necessary statements were not *fact-stating* at all; that was the whole point of the play with the analogy of linguistic rules. But now it is allowed that their acceptability, at least in some cases, is an objective matter, settled by our 'conventions' for the use of the expressions which they contain; exactly the modified conventionalist standpoint.

We have found that we cannot rest content with anything like the conception of necessity illustrated in these passages. For the idea that meanings, or 'characteristics', can be determinate in such a way as to settle that some statements are incapable of conflict with contingent fact, irrespective of whether we ever appreciate as much, is simply unavailable to anyone who grasps the central point of Wittgenstein's discussion of rule-following. The idea is unavailable because we simply cannot give content to the sort of objective conceptual stability needed to sustain it.

To summarise the reasons why not. Objective conceptual stability means: objectivity in the idea of when an expression is applied in the same way, or to the same kind of circumstances, as previously. However:

(i) There is no such thing as a *unilateral* recognition that, by allowing such-and-such a use of it in such-and-such circumstances, the rest of the community has broken faith with its pattern of use for a particular expression—save in the irrelevant sense that, like the twelfth jury man, it may be possible to bring the others round; that is, contrary to the hypothesis of unilateralness. The proper conclusion for someone in such a position would rather be that he had just found out he did not understand the expression in question.

(ii) No residual sense remains in which someone in such a situation may be supposed to recognise that he has applied the expression perfectly properly vis-à-vis his *own* understanding of it; the understanding which, as it has just turned out, is not shared by the rest of the community. For he cannot make operational a distinction between its seeming to him that he is so applying the expression and his really doing so; between the judgment and the fact which, if correct, the judgment records. In order to operate such a distinction he needs a *community* of practitioners of his idiolect.

In isolation, then, we severally possess nothing properly described as a capacity of *recognition* of when we, or others, keep faith with or breach prior patterns of linguistic usage; we cannot prise our inclinations, on the one hand, and the supposed facts of the matter, on the other, far enough apart to have it make sense to speak of a correspondence between them.

(iii) However, neither can we, as part of a community, ultimately make sense of the idea of a correspondence between our consensus verdict about the correct use of a particular expression in a situation and the 'true facts' of the matter. What goes for the ideolect user goes, writ larger, for the community as a whole; when we are agreed about the description of some new situation, we have no ulterior purchase on the idea of its being *right* for us to agree in that particular verdict. We cannot give sense to the idea that our communal speech habits pursue objective tracks which we laid it down as our intention to follow. Certainly, as members of a linguistic community, we can distinguish between correct and incorrect usages of language, or between continuing to use an expression in the same way as previously and inventing a new use; but the intelligibility of these distinctions cannot presuppose that of a comparison between what we judge to be a correct description of a new situation and what *really* is, independently, its correct description in our language. The substance to these distinctions must rather be sought in the practical criteria on the basis of which we make them.

The main moral of the three points is this. We would readily grant that a Martian, radically translating our language, would be bound, by and large, to accept our use of it as correct *whatever* we do; there are, in terms of Wittgenstein's analogy, the same limitations on our ability to misuse our language as on a man's ability to act out of character. But we have a tendency to think that *we* have access to knowledge of the nature of our own, or of the communal, understanding of our language of a superior sort, inaccessible to the radical translator who can only observe our practice; so that we conclusively know, or are at any rate capable of reflectively discerning, what we mean by any particular expression—and are thereby capable of seeing to it that, by and large, we really do use our language correctly. To develop the analogy, it is as if, knowing myself as I do, I could ensure that most of the time I really do act in character. Such a conception of self-knowledge is immediately open to the challenge to explain in what consists this supposed privileged purchase on the ins and outs of my own character which enables me to make a *comparison*; and the effect of the rule-following considerations is that the corresponding challenge in the case of linguistic understanding cannot satisfactorily be met.

Ewing said that when he recognised the truth of an a priori proposition, he did so just by 'seeing that a certain characteristic by its inherent nature necessarily involves or excludes another'. But it is now evident that there is nothing to which he could have appealed to vindicate that flamboyant description of the matter, rather than the more austere account that he found himself disposed to assent to the sentence and could not envisage how it could be false. All being well, such will be the general response to any particular ('obviously analytic') sentence. But that there is, in any particular case, such a communal response falls far short of giving sense to the idea that to treat the sentence as necessary is no more than we *owe* to it in view of its syntax and the semantic properties of its constituent expression—or the nature of the 'characteristics for which the terms stand'.

The most striking corollary of these ideas is that we are obliged to reject the ordinary notion that internal relations can obtain *unacknowledged*; in general, that statements exist which are predeterminately necessary, but whose necessity we have yet to recognise. We must reject this notion precisely because no sense sufficient to sustain it can now be given to the belief that when we admit a new statement as necessary, there is a further objective issue whether our judgment is truly at one with the character of the concepts in play in it; that is, whether to admit the statement as necessary conforms to patterns of use for the expressions involved which it was the community's antecedent intention to preserve. An unacknowledged necessary statement is a statement which, while not yet used as though it was necessary—whatever that may mean—is such that to admit that it was defeasible, that there might be counterinstances to it, would be in conflict, albeit a conflict so far unrecognised by us, with the character of our understanding of the statement. However if we cannot give content to the idea of a sameness of use which overrides our successive consensus judgments about what constitutes it in successive cases, neither can we give content to a corresponding, objective non-sameness; nor therefore to the idea that to deny the necessity of the statement would conflict with, although we do not yet know as much, the meanings of the expressions which it contains.

These reflections are naturally taken to suggest a general constructivism about necessity. It appears, if we accept them, that we ought to think of ourselves as bringing necessity *into being* when we judge a statement to have that status; it seems no longer allowable to suppose that it was necessary all along, that our judgment is recognition of a prior fact. All the necessity of a statement can now consist in is its being *used* as such. But when a new statement is admitted as necessary, it will not have been used in the appropriate way before its admission; and we appear to be at a loss to make good any objective sense in which nevertheless it *ought* to have been so used before, or: admitting it as necessary was the *right* thing to do. So invention, it appears, is really the mother of necessity.

Thus at this point it appears that Wittgenstein's ideas, beyond calling attention to the naivety of the position advocated by Ewing and Ayer, are actually a powerful argument on the side of the opposite camp in the original dispute during the thirties and forties.[1] What any conventional-ist requires is some sort of prima facie viable interpretation of the Autonomy of Grammar: grammatical propositions—what we ordinarily take for necessary truths—must all be in some sense both independent of each other and incapable of conflict with fact. We have seen that a full-blown conventionalism, a conventionalism purporting to explain *all* necessity, must endorse some such conception. Not that the conven-tionalist has to deny that logical relations obtain among conventions: what has to be maintained is that such relations are themselves consti-tuted by convention. For if they are not, at least some necessary statements will be of a non-conventional character, contrary to the conventionalist's intention to give an absolutely general account of what necessity is. The second constraint, that statements dignified as neces-sary be somehow immune to conflict with empirical fact, arises similarly: if such conflicts were possible, that would be a non-conventional constraint on which statements we could soundly admit as necessary.

On the face of it, the rule-following considerations supply an interpretation of both elements. As an interpretation of the mutual independence of all grammatical propositions, we have: the non-objectivity of logical consequence. The, objectively speaking, open character of our understanding of any particular grammatical propos-ition means that it has no predeterminate consequence connections; there is no good sense in which by adopting one convention we *commit* ourselves to others. Rather, the existence of any such relation awaits our explicit ratification in order to bring it into being. And for an interpreta-

---

[1] Note one aspect of that debate which these ideas certainly do not touch on: the question whether there are any *synthetic a priori* statements. One thing which the 'linguistic' camp felt obliged to affirm, and their opponents to deny, was that all necessary statements were analytic. Part of the motivation for the denial was certainly the desire that necessary statements be capable of *depth*; for example, that the sorts of things which a philosopher might discover about Mind, Language, Reality, etc., could have this depth. Ewing complained (op. cit.) that the view of his opponents would completely trivialise philosophical enquiry. However it is evident that what is essential to the position which Ewing wanted to hold is that we be capable of discovering predeterminate internal relations between concepts. If this idea is allowed at all, it will surely be extended to propositions traditionally regarded as analytic; that is, propositions whose demonstra-tion needs recourse only to logic, definitional transformations, and implicitly definitional postulates. Once it is allowed that our concepts can have features which it may take subtle and ingenious argumentation to bring to light, no further obstacle is posed to the depth of mathema-tics, and indeed philosophy, by the conception—however implausible on other counts—that all their true statements are analytic. Thus, to be sure, Wittgenstein's ideas issue a challenge to something which the Ewing camp wanted to maintain: the deeply informative character of at least some necessary statements. But in order to maintain that thesis, it was quite unnecessary to defend the synthetic a priori; nor would it have been sufficient to provide an account of this supposed category of statement in order to meet the thrust of the challenge posed by Wittgenstein's ideas to the whole notion of discovery in mathematics—and in philosophy, if we conceive it as dealing comparably in essential truths.

tion of the second element, the impossibility of a conflict between grammatical propositions and empirical fact, we have Wittgenstein's conception of necessity as 'antecedent to truth'. This idea, too, ultimately rested on the rule-following considerations; in particular, on a rejection of the investigation-independence of statements alleging the existence of errors or misapplications in procedures like counting and measurements. So need we look any further? Do we not have here both the heart of a generalised conventionalism about necessity, as compelling as the anti-realism which underlies it, and an exegesis of Wittgenstein's own conventionalist sympathies?

Reasons for dissatisfaction with this suggestion began to crystallise towards the end of the last chapter. However natural a general constructivism is as a pictorial expression of the import of the rule-following considerations, the fact is that the manner in which unacknowledged internal relations are ruled out of bounds has yet to be made out to be anything other than a special case of a general rejection of investigation-independence argued to be implicit in anti-realism. If constructivism about necessity is suggested by the rule-following considerations, then it is suggested about everything: facts in general are not antecedent to our ratification of them. For, in order for them to be so, there will have to be ratification-independent truths about how we ought to assess them *if* we do so in accordance with correct understanding of any of the judgments which we might consider making; and since there is no objectivity in the idea of 'accord' there, neither can the truth of such a conditional statement have the requisite objectivity. In short, within the framework of the rule-following considerations nothing has so far been done to distinguish the situation of new *contingent* statements from that of new necessary ones. If we want to say, on the basis of these ideas, that statements only become necessary when ratified as such—that we thereby create the conceptual connections which on the ordinary view we regard ourselves as bringing to light—then we ought to be prepared to say something comparable about new contingent truths: when we accept a new contingent statement as true, we ought to regard ourselves as, in a sense, *creating* the truth of that statement; for there is, for just the same reasons, no investigation-independent fact about how the circumstances which we judge to verify the statement in question *ought* to be judged to bear on that statement if we judge in accordance with correct understanding of it.

I suggested previously that it would be best if we could resist the temptation to lard the rule-following ideas with constructivist imagery. The imagery does nothing to make the underlying arguments any clearer; and it risks coming into conflict with deeply entrenched linguistic habits, for example the ordinary tense-links, in ways of which it is difficult to command a clear view. Rather, the real force of the rule-following ideas is that the ordinary notion of objectivity which sustains the idea of investigation-independence is to be rejected for *all* statements, both necessary and contingent. It therefore so far remains an open possibility that, after we have adjusted our thinking to that rejection, whatever notion of recognition *remains* appropriate for new contingent truths will, as far as the rule-following considerations are concerned, remain appropriate for new necessary statements also. Expressions in the family of 'recognise', 'verify', 'find out', are not going to be banished from the language; it is just that we must refashion our thinking about contexts in which we typically employ them.

A rough analogy would be this. If, following Wittgenstein's suggestions in the *Investigations*, we abandon the conception of sensation as an *accompaniment* to behaviour which expresses it, we do not thereby negate the difference between, for example, pain behaviour which is expressive of pain and pain behaviour which is simulated. What we have to do is rather to find some other account of the substance of that distinction than that which the 'accompaniment' picture suggests. Likewise, in the present situation, if we abandon the idea of the status of a fact recognised, or verified, or unearthed, as something independent of our investigation, presentation with which *constrains* us to describe it in just the way in which we do if we are to keep faith with the way in which we understand the expressions involved in proposed descriptions, we do not thereby undermine all distinction between finding things out on the one hand and, on the other, stipulating, or deciding, that a proposition is to be counted as true. If the only interpretation which we have of the Autonomy of Grammar is as sketched above, then the independence of grammatical propositions from each other, and their non-answerability to empirical fact, become no more than implications of the independence of *each* of our judgments of every other. Nothing special will so far have been said about necessity. No theory of necessity will have yet been proposed; all that will have been achieved will be the exposure of error in one such theory.

This fundamental aspect of Wittgenstein's later thought, then, does indeed war with the conception of necessity of which I quoted Ewing and the 1946 Ayer as representatives. But, unsupplemented, it is insufficient to prompt a ruling in favour of the conventionalism which they opposed. The essence of a conventionalist view is that we are mistakenly inclined to *over-assimilate* necessary and contingent statements; that we seduce ourselves into thinking of necessary statements as a special category of descriptive truth, when it is actually almost a pun to describe them as 'statements' at all. The rule-following considerations suggest no such disanalogy. For that reason, they cannot supply a complete interpretation of Wittgenstein's own conventionalist streak. Wittgenstein wanted to suggest not merely that there is error in the way in which we ordinarily interpret our recognition of necessity, the cogency of proof and the absolute 'hardness' of the logical 'must'—error involved in the presumption of an illicit notion of objectivity—but more: the whole notion of recognition, of discovery, is *misappropriated* in the case of necessary statements. Beyond anything which digestion of the rule-following ideas requires, there remains the fact that Wittgenstein wanted to say that there is a disanalogy between necessary and contingent statements of so fundamental a sort that necessary statements are not, properly viewed, to be regarded as descriptive of anything; and that there is therefore no question of anything properly described as recognition that, as descriptions, they are correct. Rather, their rôle is to *pre*scribe; and their manner of acceptance involves not recognition but adoption. The rule-following considerations continue to supply the means for beating off certain natural objections to any conventionalist account; but for an adequate motivation for such an account we must look elsewhere.

2.   In order to clarify the view of logic and mathematics which Wittgenstein (on the whole) favoured, we have to go deeper into the roots of his hostility to the whole idea of necessity as a property of genuine *statements*, of necessity as necessary *truth*. The task is twofold: first, we must look more directly for justification of the view that *recognition* is categorically inappropriate to necessary statements. We have to see whether, beyond our inability to make out, in their case or any other, the presumption of objectivity with which the idea of recognition is ordinarily associated, some further reason remains for describing mathematicians, in *contrast* with other scientists, as inventors rather than discoverers. More sharply, the question is whether there really is always an element of *decision* in our acceptance of a new statement as necessary with no counterpart in contingent cases.

Secondly, we have to determine whether there are aspects of the *use* of necessary statements which tell against our ordinary conception of them as a kind of truth. Wittgenstein's idea was that necessary statements essentially function *normatively*; what is distinctive of a statement's having the status of necessity is wholly its possession of a certain sort of prescriptive rôle. In particular, this rôle is not a *consequence* of necessity. This is the idea, of course, which motivates the repeated analogy with imperatives. So if, by calling attention to facts about the use of necessary statements, we can substantiate that analogy, then that will constitute an argument for supposing that there is the same incongruity in applying the notion of truth to necessary statements as there is in applying it to any imperative, or other sort of directive. If that were to prove to be the situation, then we should immediately have an answer to the first line of enquiry also; we should immediately have a sound motive for discarding the notions of recognition, verification, etc., as descriptive of what really happens when it is our inclination to say that the necessity of a statement has been discovered. Accordingly, it is to this second line of enquiry that I shall devote the remainder of this chapter and the next one.

3.   Wittgenstein's suggestion that necessary statements possess a distinctive normative rôle actually has a double function. Suppose investigation into the first question just sketched turned out satisfactorily from a conventionalist point of view; that is, we proved able to vindicate the claim that some element of decision always has a part to play in our admission of a new statement as necessary. The conventionalist would them immediately have a responsibility to explain what, in that case, our decision in such situations is a decision to *do*. Since the correct answer is: to treat the statement in question as necessary, what is being called for is in effect an explanation of what necessity now consists in. If we are to abandon the picture of the necessity of a statement as consisting in its recording a special category of truth, constituted as a by-product of its syntax and the meanings of its constituent expressions, we are owed a new account of what necessity is. What is needed is exactly an explanation of what is distinctive of necessary statements in use. Wittgenstein's suggestion, that they are essentially normative of correct practice of the 'language-game', is, in addition to being intended to undermine our inclination to describe such statements as a kind of truth, meant as the germ of such an explanation.

We have thus to consider two sub-questions. First, can we so explicate the idea of normativeness that the possession of such a rôle by necessary statements conflicts with their capacity for truth, their status as genuine assertions? And secondly, can we so explicate the idea of normativeness as to characterise what, in use, is the hallmark of *necessity?*

That statements accepted as necessary are (at least potentially) in some way regulative of correct linguistic practice seems clear. But when the suggestion is that possession of such a rôle is what is distinctive of necessary statements, this clarity comes to seem superficial. What exactly is it that admitting a statement as necessary is supposed to require of us; what exactly does the statement then prescribe? The answer suggested by the 'rules of the language-game' idea, viz. a certain co-ordination in forms of description hitherto not so related, looks to be much too simple. The Chess theorem which states that Black's only conceivably sound defensive strategy in such-and-such a position is so-and-so, surely prescribes *action*, if it prescribes anything; it tells Black what he must do if he is to have any chance of avoiding defeat. This may seem a deliberately unsympathetic and indeed peripheral example. If it is either, we are entitled to it nevertheless. But it is open to question how peripheral it is. *Any* necessary statement which enjoins a rule of inference, $A \vdash B$, can be seen as prescribing action: in any situation where (you have grounds for believing) A, tailor your conduct, both linguistic and non-linguistic, so as to have it square with B.

The point is made by Wittgenstein himself[1] and does not seem

[1] See, for example, *RFM* I. 17; III. 15–16.

particularly consequential. All it appears to require is that the normative content of a particular necessary statement may ramify through a
wide field of human activity, and require a fairly painstaking description if we wish to give a detailed account of it. Indeed, it would be
fatuous to expect the 'rules of the language-game' to confine their
influence to linguistic activity in the narrowest sense, to saying and
writing. For the use of language simply cannot be peeled away, in the
manner which such a restriction would demand, from the wider sur-
roundings of human activity in which it is embedded. If $A \vdash B$ is
prescriptive both of linguistic and non-linguistic conduct, that is no
more than a reflection of the fact that our acceptance, or rejection, of A
and B will very likely have practical 'extra-mural' implications.

At this stage, however, Wittgenstein's idea that a normative rôle
distinguishes necessary statements from contingent ones begins to look
problematic. For if accepting a statement as necessary places certain
constraints on one's subsequent linguistic and non-linguistic behaviour,
so it appears does accepting any contingent statement as true. To present
any contingent statement as true is to require one's audience to modify
their subsequent behaviour in all the ways consequent on its acceptance.
In particular, constraints are imposed on subsequent linguistic performance: to present 'Jones is bald' as a correct assertion prescribes that if
Jones' appearance is to be described correctly, then no description must
be allowed inconsistent with his being bald. It seems very doubtful,
therefore, whether we are able easily to understand the idea of a normative statement in such a way as to have the contrast between normative
and descriptive correspond to the intuitive boundaries of the distinction
between necessity and contingent truth. Certainly, to accept a statement
as necessary enjoins certain definite courses of linguistic and non-
linguistic behaviour. But, in the absence of some further refinement, it
appears that in that sense accepting any putative statement of *fact* is
normative too; we have to organise our beliefs to accommodate it, to form
certain expectations, to allow or withhold certain forms of description in
certain relevant situations—in general, to do everything which accepting
the statement requires. And there is so far apparent no particular *kind* of
thing required of us if a statement is accepted as necessary, no specific
type of linguistic or non-linguistic action, to distinguish the prescriptions
involved from those consequent on accepting any contingent statement
as true.

What we are looking for is interpretation of a contrast between
normative and descriptive statements; they are supposed to emerge as
two quite different sorts of things, despite the surface grammatical
similarities. But, on the face of it, neither in being prescriptive, nor in
the types of conduct which they prescribe, are statements which are
accepted as necessary distinguished from statements which are simply
accepted. The clear respects in which necessary statements are normative are not distinctive of necessary statements.

These considerations, however, if sound, are not yet fatal to the capacity of the idea of normativeness to meet Wittgenstein's double purpose. There is no need for a conventionalist to deny that contingent statements have any prescriptive content, nor to pretend that the kind of prescription enjoined by accepting a contingent statement is generically different from what is prescribed by accepting a statement as necessary. What he has to be able to deny is, rather, precisely that necessary statements have any *descriptive* rôle; their function must be *purely* prescriptive. To substantiate this suggestion would suffice both to characterise what in use is distinctive of a statement's being necessary, and to undermine the temptation to think of necessity as a kind of 'superlative' species of truth.

But can the suggestion be substantiated? Suppose we take the idea of necessary statements as 'rules of grammar' as literally as possible: we have to try to think of them wholly as object-language analogues of what metalinguistically we should express as imperatives, explicitly stipulating circumstances under which certain explicitly mentioned descriptions are appropriate, or are to be withheld. Since it seems right to deny that such an imperative describes anything, or may have truth significantly predicated of it, it seems an appropriate course for our enquiry to take at this point to ask whether this putative equivalence can be defended by attention to our actual employment of necessary statements. We have to enquire whether such a metalinguistic imperative could indeed do duty for any particular necessary statement, whether the semantic content of an (apparent) assertion of any necessary statement would be equally satisfactorily captured by announcing, or endorsing, such a metalinguistic directive.

It is not straightforward to see how exactly to construct the examples which will enable us to give the suggestion a fair hearing. What would be the metalinguistic directive corresponding, for example, to $7 \times 7 = 49$? As a first shot, something, presumably, like this:

> You are to accept 'ϕ is constituted of 7 groups of 7 objects each' as a correct description of a totality if and only if you are prepared to accept 'ϕ is constituted of 49 objects' as a correct description of it.

There are, obviously, various problems with this. To begin with, it is unclear that it captures the full *generality* of $7 \times 7 = 49$; at best, a very liberal reading is going to be required of the notions of 'totality' and 'objects' if the directive is to apply to every situation in which we might want to apply such an arithmetical equality. We are also bound to be uneasy about the use made of quotation; as it stands, the directive could be understood and complied with by someone who did not understand any arithmetic at all. And, thirdly, it is arguable that the terms of our question are in any case too narrow: that it will not meet the conventionalist's purposes just to show that the issue, or endorsement, of an appropriate directive can always do duty for the apparent assertion of a necessary statement—he has, further, to give some explanation of the rôle of *unasserted* occurrences of necessary statements in compound contexts, and in particular of their capacity to feature significantly as the antecedents of conditionals.

One might have more enthusiasm for the task of sorting out these difficulties were it not for the fact that an over-riding question-mark shadows this whole strategy for substantiating the idea that necessary statements lack descriptive content; a problem which it is not clear that a solution to those difficulties would go any way towards meeting. The problem is that whatever there is to be said *for* the idea that an appropriate directive might always do duty for any particular necessary statement seems to apply equally to all contingent statements of an appropriate generality. Suppose we are satisfied of the soundness of some contingent generalisation, $(x)[Fx \leftrightarrow Gx]$; then (to parody the first approximation above) we shall presumably be content to endorse the metalinguistic directive:

> You are to accept 'x is F' as a correct description of an individual if and only if you are prepared to accept 'x is G' as a correct description of it.

Conversely, it is plausible that if we are prepared to issue, or endorse such a directive then we will be prepared to (quasi-) assert, or accept, the object-language analogue $(x)[Fx \leftrightarrow Gx]$.

Because our 'first shot' towards construing necessary statements as directives is suspect in several ways, the point is only suggestive. But what it suggests is that our willingness, if indeed we proved willing, to substitute the issue, or endorsement, of some sort of directive for any apparent assertion of a necessary statement would be insufficient to make good the idea that the latter class lack descriptive content, do not play a genuinely fact-stating rôle. At any rate, if a similar substitution would be acceptable in the case of some contingent statements as well, then the price the conventionalist will have to pay, if the point is to have its desired significance in the case of necessary statements, will be to forego the idea that such *contingent* statements have a descriptive function. And even if he were willing to pay that price,[1] it would remain that this particular mode of interpretation of the idea of *pure* normativeness would not result in an adequate characterisation of what necessity is, of what is distinctive in the use of a necessary statement. For the resulting characterisation will, it appears, put at least some generalisations in what is, intuitively, the wrong place. (It is, in fact, unclear whether *any* contingent statements will be put in the right place.)

We are brought back sharply, then, to a point raised in Chapter VIII, § 5. Wittgenstein stresses from time to time that we should not be

[1] As Wittgenstein, independently, seems to be in *PB* xxii, especially 228 (cf. *PG*, App. 6).

misled by the assertoric form of mathematical statements. But what would *show* that it would not instead by misguided to prefer their formulation as directives of some sort? What is needed is exactly a demonstration that the idea of truth is inappropriate to them in just the way in which it is inappropriate for any directive. But no way is so far apparent for showing that save by demonstrating that their function is exactly that of directives. And the obvious strategy, just sketched, for substantiating that idea looks to be extremely unpromising. A large part of the trouble, at bottom, is that assertion is itself *imperatival*; to present a statement as true is to enjoin certain definite forms of linguistic and non-linguistic conduct. It seems ominously likely that this fact will frustrate any attempt by the conventionalist to make his case along the sort of lines illustrated.

There are, to be sure, strands in Wittgenstein's thought which were meant to supply an independent purchase on why necessity should not be thought of as a kind of truth. Such was the point of the attempted parallels with aspects of measuring. There was to have been the same incongruity in thinking of necessary statements as true as there is in thinking of a tape measure as true, or of a unit of length as accurate. But the analogies with measuring proved unsatisfying in various ways. And the interpretation of the conception of necessary statements as 'antecedent to truth' to which we were eventually led by considering them is, despite its suggestive title, of no help in the present context; for, as we have noted, the thesis that when we accept a statement as necessary, we owe no responsibility to the investigation-independent truth-values of the contingent statements among which it will enable us to construct inferential liaisons, is quite consistent with holding that whatever notion of verification *survives* a rejection of investigation-independence will be applicable to necessary statements also.[1]

To summarise, then, our reflections up to this point. That those statements which we accept as necessary are in some way regulative of our linguistic and non-linguistic practices seems intuitively right. But the same seems prima facie to hold for those statements which we accept as contingently true; and the *kinds* of regulative influence which both sorts of statements exert, the sorts of thing which accepting them requires of us, seem to match. Moreover, no strategy is so far apparent for demonstrating, by reflection on (potential) aspects of the use of necessary statements, that this regulative influence, in contrast with the situation in the case of contingent statements, somehow exhausts their function; that is, that they have no descriptive rôle.

We shall continue with these matters in the next chapter.

---

[1] As is familiar, in the *Tractatus* the concept of literal sense, and hence of truth, was restricted to statements effecting a division, non-empty on both sides, among logically possible states of affairs. The conception that necessary statements are improperly regarded as a kind of truth was thus a feature of Wittgenstein's philosophy throughout his life. But which doctrines of the *Tractatus* survived to explain this continuity?

XXI

Normativeness

Sources

RFM: I. 164; II. 28, 39; III. 35–6; V. 3–5, 46
BGM: VI. 22–3
  OC: 87–8; 94–9; 121–4; 136–8; 210–12; 308–9; 318–21;
      401; 493–7; 628
LFM: lectures IV; XXXI, p. 290

1.   At the conclusion of the last chapter we were struggling to see how conventionalism might make good the thesis that the normativeness of necessary statements undercuts their capacity for truth. There may seem to be a fairly obvious explanation of the difficulties which this thesis encounters—albeit an explanation unwelcome to the conventionalist. The suggestion would be precisely that we cannot begin to understand the normativeness of necessary statements *except* by reference to the idea of their truth. For in order to give an exact and complete account of what any particular necessary statement prescribes, we cannot do better than to incorporate everything involved in the injunction: explain away, or otherwise satisfactorily discount, anything which prima facie *conflicts* with that statement. Our treatment, for example, of '7 × 7 = 49' as necessary does indeed involve our acceptance of certain criteria for applying the concepts of miscalculation, miscount, chemical change, illusion, etc.; but this involvement can be explained only by reference to our understanding of what it is for results prima facie to go against the arithmetical law, to seem to be other than as it requires. If we ask, for precisely which concepts does '7 × 7 = 49' enter into the rules of application?—what is the principle determining which concepts fall within the range of the above 'etc.'?—the only acceptable general answer is: any concept essentially involved in any of the kinds of judgments which would suffice to discount a prima facie falsification of '7 × 7 = 49'. It is only in terms of a conception of what would (per impossibile) *falsify* a necessary statement that we can explain the link between the various concepts into whose criteria of application it enters qua necessary statement, the various concepts of whose application it is regulative.

At first sight, that might seem an incoherent suggestion. For is it not a feature of necessary statements in general that we cannot *imagine* their being false? So how, it might be wondered, can we be supposed to have any conception of the kind of circumstances which would prima facie falsify a particular necessary statement, the kind of circumstances of which, according to the suggestion, the statement's normative rôle is to prescribe certain forms of redescription? But this question rests on a muddled notion of imaginability. When we feel we have a clear understanding of, for example, some so far unresolved mathematical conjecture, it seems quite unexceptionable to suppose that we have a conception both of what it would be like for the conjecture to prove to be correct and of what it would be like for it to prove to be incorrect. In the nature of the situation there is, to be sure, an inexactness in each conception; the inexactness consequent on our not knowing exactly how to effect either construction. But an inexact concept is still a concept. Suppose, though, that we actually get a proof of the conjecture. Must the effect of the proof inevitably be that we altogether lose hold on our former conception of what it would be like to disprove the conjecture, that the inexact concept collapses into no concept at all? It seems not. What will likely happen is that the concept will survive but will be downgraded, redescribed. It may indeed become, or may be declared to be, unimaginable that there could be a disproof of the conjecture; but the old concept will survive as a concept of what it would be like to *seem* to have a disproof. What will no longer be imaginable is that a seeming-disproof should prove on sufficiently rigorous examination actually to be a disproof. But the original suggestion requires only that we have a concept of what it would be like for a situation prima facia to conflict with a particular necessary statement; so all appears to be in order. The muddle was to conflate its being imaginable that so-and-so with its being imaginable that things *seem* so-and-so.

The same sort of point can, of course, be made to defend the intelligibility of prima facie disconfirmation of a necessary statement not by an apparent disproof but by apparently discordant results when the statement is applied in contingent contexts. No doubt it is unintelligible that there could actually be 7 exclusive groups of apples on the table, each consisting of 7 apples, and yet there not be 49 apples there; but that is not to say that it is unimaginable that it could somehow seem as if that were the situation.

Still, there may be an inclination to press the objection. For it is plausible to suppose that the distinction, between the imaginability of a concrete situation which went contrary to a particular arithmetical rule and the imaginability of one which merely seemed to do so, has the required content only if the rule in question is relatively *complex*. If we are concerned with something of the degree of simplicity of, for example, '2 + 2 = 4'—any example, in fact, involving only numbers small enough not to require that we count in order to determine their application—then it just does not seem imaginable that things might even *seem* to go contrary to such an example. Even Escher,[1] surely, would be baffled to make it look as if there were just 2 exclusive groups of 2 apples each on a table and yet not 4 apples in all. (His trick is always to supply the viewer with a *route* of attention through his pictures, such that at any particular point on the route some points of the picture are unattended to.) But if not even a prima facie falsification is imaginable of arithmetical equalities of this degree of simplicity, then it cannot be correct to suppose that their normative content has to be explicated by appeal to the idea. Grasping the rôle of such statements *cannot*, indeed involve grasp of any such idea. There is therefore no need to suppose that, in order to explain such statements' normativeness, the conventionalist will have implicitly to bring in the notion of their truth—since he will have to invoke their conditions of prima facie falsity—in a manner which surrenders his desired analogy with rules (to which the idea of even prima facie *falsity* is categorically inappropriate). And *if* he can supply some other account, there is no advance reason to doubt that it will be capable of extension to more complex cases too. (The alternative, of rejecting the idea that simple arithmetical equalities are normative at all, is, of course, not open to anything resembling Wittgenstein's conventionalism.)

To summarise: the thrust of the original objection was that since the normative content of any particular necessary statement has to be explained by reference to the idea of a prima facie falsification of it, and since the notion of prima facie falsification has no application to the kinds of rule—rules of Chess, for example—to which we do indeed feel it to be incongruous to apply the notion of truth, the normativeness of necessary statements, properly understood, in no way vies with but rather supports their candidacy for description as 'true'. And the counter was that that cannot be the way to explain the purported normative content of a wide class of necessary statements, for which the idea of a prima facie falsification is simply not intelligible.

This counter leaves us completely in the dark about what the conventionalist's next move ought to be, about just what account he is to give of the normativeness of necessary statements if it is to have the consequences which he desires. But it is in any case open to doubt whether the counter is acceptable. Arguably, a prima facie falsification of a very simple arithmetical equality seems unimaginable only because we concentrate on trying to imagine a situation in which everything is clear to view, in which all the apples, or whatever, involved are simultaneously and continuously observable by a single observer. If we aim a little lower, the situation changes. For example, a conjurer seems to deal exactly two cards into one box, and exactly two more cards into another; but when we are invited to inspect the contents of the boxes, we can find only three cards in all; and there seem to be no hidden

---

[1] I am thinking of the section, 'Impossible pictures', in 31.

compartments. 'Well then, it must have been an illusion that he really dealt two cards into each; he must have palmed a card, or something . . .'! (It wouldn't be nearly as good a trick if he worked with numbers too large to be taken in at a glance.) 'Prima facie falsity' need not mean: seeming-falsity for an optimally placed single observer.

Take another example: our old reliable, 'Nothing is simultaneously red and green all over'. Here it might seem that however non-optimal the conditions of observation, and however subtle the conjurer's art, there is no way in which one could have even the illusion of presentation with a counterexample. And from the point of view of a single observer, that is surely right. But suppose of a pair of observers, each competent by ordinary criteria in the use of colour words, that one reports of a particular object at a particular time that it was red and the other reports of the same object at the same time that it was green. 'Well then, they were looking at different sides of it, or one of them made a mistake about the time, or there is after all some misunderstanding of these concepts, or one of them suffered some queer illusion . . .' With very 'small' arithmetical equalities, prima facie falsifying circumstances are envisageable by inbuilding an appropriate lowering of the standards of observation, so to speak. And here we get the same effect by distributing the observation through more than one observer; and that that is the effect is vindicated by the reflection that, since nothing can simultaneously be red and green all over, we cannot let the observers' reports stand, but must postulate that the situation can be explained away along one of the sorts of lines indicated.

There seems no reason to doubt that one, or a combination, of these strategies can be applied to any example where it is our initial inclination to say that even a seeming-falsification is simply unimaginable. The suggestion, therefore, that the normative content of necessary statements is without exception to be explained by reference to the notion of circumstances constituting a prima facie falsification of them is still in the field. And the suggestion is inimical to the conventionalist's idea that necessary statements are not, properly viewed, truth-bearers not just in the fashion noted, that it undermines their assimilation to the sorts of explicit rule of which it it already our inclination to say that they cannot be true, but in another: the normativeness of necessary statements, so interpreted, is exactly that which we should expect of any statement of whose truth we were absolutely *certain*. So conventionalism, it appears, must find some riposte to the suggestion.

2.   A possible line of reply might be as follows. Consider any genuine rule—of a game, say—indicatively formulated; for example, 'The players move alternately'. Nothing in the statement of such a rule distinguishes it from a descriptive hypothesis, such as an observer of our practise while playing the game might arrive at, purporting to

codify a certain behavioural regularity. So what makes it one or the other? The answer seems irresistible that the difference resides in our—the players'—intentions. For the statement to have the status of rule requires that we possess a communal intention that in the course of the game we bring about no situation which would falsify that statement conceived as a descriptive hypothesis (and the very best kind of evidence that an observer could have for thinking that the statement was regulative of our practice, and not merely descriptive of it, would be evidence of our having a disposition, should we ever do anything which did conflict with the hypothesis, to remedy matters, to bring our behaviour back into line with it). As a directive, the content of the statement is that no situation should be brought about of which a correct description would be inconsistent with the same statement, construed as itself a description.

This is no more than an application of the truism that the content of any rule, or imperative, must be explicable by reference to a statement, or class of statements, which will be true if and only if it is complied with. The proposed line of reply, then, would be this: the objection misinterpreted as conditions of prima facie falsification of necessary statements what are actually conditions of prima facie falsification of the *genuine* statements whose truth-conditions are linked in the manner just noted with the content of the directives which it is actually the rôle of necessary 'statements' to articulate. Necessary statements, when used as such, do not possess, any more than any other directive, conditions of prima facie falsification. But there is a notion of prima facie falsification for certain associated genuine statements, statements which are falsified just in case the directive issued by the corresponding necessary statement is not complied with.

Let us develop that a little. The content of any directive is that no situation should be brought about, by action or inaction, whose correct description will be inconsistent with the truth of a certain corresponding statement: that statement which is true just in case the directive is complied with. Necessary statements, however, on Wittgenstein's view, are first and foremost directives of *description*, rules of linguistic practice. So their content pertains to the allowability of certain forms of description: it is that no situation should be admitted as correctly described which, if so, would constitute a falsification of a certain appropriately corresponding statement. The idea that we have a conception of conditions of prima facie falsification for necessary statements, and indeed the whole conception that necessity is a kind of truth, is nourished by a failure to note, or an equivocal shift within, the double rôle of the sentences in question: viz., that they are capable of functioning both as directives and as expressive of those statements whose continued acceptability will depend on the former directives being compiled with. The latter are ordinary, genuine statements; as such, we have a conception of their conditions of falsification. The content of the associated directives, however, is exactly that nothing is to be admitted as a genuine realisation of such conditions. Thus the idea of the objection, that we have conceptions of prima facie falsification-conditions for necessary statements, is, when properly understood, in no way in tension with their assimilation to rules.

If this suggestion can be made good, then we have the means to resolve what might otherwise appear to be a discord in Wittgenstein's own exposition. On the one hand we find assertions like these:

> The effect of proof is, I believe, that we plunge into the new rule (III. 36)

> The mathematical proposition has the dignity of a rule (I. 169)

and

> The proposition proved by means of the proof serves as a rule. (II. 28)

On the other, there are passages like these:

> What is proved by a mathematical proof is set up as an internal relation and withdrawn from doubt (V. 5)

and

> *What* is unshakeably certain about what is proved? To accept a proposition as unshakeably certain—I want to say—means to use it as a grammatical rule: this removes uncertainty from it. (II. 39)

If it is to be sense to talk of our withdrawing a proposition from doubt, or uncertainty, it seems it must be envisageable that situations might arise in which a doubt about it would be prima facie legitimate; otherwise the decision to 'withdraw it from doubt' will not have any practical consequences. So the apparent discord is exactly that complained of by the objection: how can such a situation be coherently envisaged for anything properly described as a rule, a pure directive? Well, now we have a suggested answer.

3.   The answer, however, is so far unclear just at the crucial point. The relation between the original example—the game-rule—and the associated contingent hypothesis is for our purpose perspicuous enough. The hypothesis describes the (continuing) state of affairs which unwavering compliance with the rule will effect. So it is indeed true just in case the rule is complied with. But *what* contingent statements are supposed to correspond in that way to the rules which necessary statements, according to the suggestion, function to articulate? If we take the analogy with the game-rule straightforwardly, it would appear that they ought to be hypotheses of the continuing existence of a state of affairs constituted by communal behaviour

running in accordance with the associated rule. That statements to such an effect can be formulated is not in question. But that it is any aspect of the use of mathematical sentences to effect such statements seems completely wrong; and is, besides, implicitly rejected by Wittgenstein's own denial that such sentences state behaviourial regularities.[1]

Clearly, we need a more sympathetic interpretation of the analogy. But, it appears, it must not be so sympathetic as to absolve the sought-for contingent statements from meeting these two constraints: (i) their truth must depend on our continued behaviour in accordance with the associated rules; and (ii) they must be statements whose assertion the very sentences used to express the associated rules could also be used to effect. The first constraint must be met if means are to be supplied for giving an account of the specific normative content of any particular necessary statement; and the second constraint must be met if the sort of explanation proposed of the apparent possession by necessary statements of conditions of prima facie falsification is to be possible.

Are there, then, any suitable contingent statements in sight? As an initial shot for the case of elementary arithmetical equalities, which are always in the forefront of Wittgenstein's thinking, we might suggest: those empirical generalisations, which Mill thought arithmetical statements actually express, about what happens when counting, aggregation, etc., are carried out correctly by operational, that is non-arithmetical criteria, and it is correct to say by ordinary observational criteria that nothing has been lost or added, etc. That there is *a* use of arithmetical sentences to express such generalisations is, arguably, attested to by the fact if a child called a simple arithmetical equality into question, we should not consider it inappropriate to answer him by manoeuvres with apples and pears. But it is unclear that such general-isations meet the first constraint. In a fit of communal insanity we might make wholesale alterations in our arithmetic overnight; but that would effect no change in the 'very general facts of daily experience' (I. 118) which underlie our arithmetic. Besides, we have to do no less than supply a method for locating an appropriate corresponding contingency for *any* necessary statement; and it is unclear how to generalise this suggestion.

Still, the example is suggestive in one way. A prima facie counter-example to a Mill generalisation is exactly the sort of situation of which it is the normative force of our arithmetic to bid us search for some warranted redescription which dissolves the tension. Now, it *is* a con-tingent matter whether or not that search will succeed in any particular case. Suppose, then, that the conventionalist avails himself—as he is perfectly entitled to do—of whatever specific notion of conditions of prima facie falsification the original objection would use in developing its point with respect to any particular necessary statement. And let us suppose that we have an adequately firm conception of the

[1] Most explicitly, *RFM* II. 65.

areas—illusion, mistake, incompetent observers, etc.—in which we are
to search in our efforts to explain away a prima facie falsifying situation.
Then, in terms of these two conceptions, a generalisation can be arrived
at for any particular necessary statement, P, in terms of the following
rubric:

> if ever a prima facie falsifying situation arises for P, then an expla-
> nation can be found in one of the appropriate categories which will
> dissolve the appearance of conflict.

Because of its open existential character, such a generalisation will not,
to be sure, be strictly falsifiable. But, provided the question of the
availability of an appropriate explanation is tied down to practical
criteria, a generalisation of this kind, like any open existential, is
capable of constituting a defeasible, contingent statement.

Now, do such generalisations meet the two constraints? To take the
second first: it there a use of any particular necessary statement to assert
such a generalisation? If the question is taken to mean, do we in general
have a practice of using sentences expressive of necessary statements to
do *nothing* but assert such a generalisation, it appears that the answer
should probably be negative. However, it is, plausibly, part of the
content of an endorsement of a statement as necessary that one believes
such a generalisation to be acceptable; for a doubt about it would be
exactly a doubt about whether the necessity of the original statement
ought to be endorsed. So, arguably, it is an *aspect* of the use of necessary
statements to affirm such generalisations.

What of the first constraint? Does the continuing acceptability of
these generalisations depend on our continuing behaviour in accordance with the rules which, according to the conventionalist picture, it
is the essential function of necessary statements to express? Obviously,
it depends on what exactly it is that these rules enjoin; whereas the
whole point of meeting constraint (i) is to be in a position to *supply* an
account of the normative content of necessary statements. So in order to
answer the question, we have to rely on our intuitive preconceptions
about what it is that accepting a statement as necessary requires of us.
Suppose, then, we ask: if we were to *abandon* the necessity of a particular statement, and so absolve ourselves from those requirements, would
that involve that the generalisation in question was no longer acceptable? If we are content with the line taken in the preceding paragraph
with respect to constraint (ii), then it would appear that the answer is
affirmative in one case at least. The case is when we abandon the
necessity of a particular statement by way of enthroning an alternative,
incompatible statement, as necessary. For thereby we will implicitly
affirm an alternative contingent generalisation. Not, so far as I can see,
that it would be straightforwardly *inconsistent* to continue to accept both
generalisations. But if, in addition to considering ourselves in a position

to defend the belief that whenever a situation prima facie conflicts with a necessary statement, P, a conflict-dissolving explanation can be achieved, we also believe that whenever a situation prima facie conflicts with an erstwhile, but now discarded, necessary statement, Q, incompatible with P, a conflict-dissolving explanation can be achieved, then we commit ourselves to the belief that whenever a situation prima facie *conforms* to P, a *conformity*-dissolving explanation can be made out. So *however* the facts seem to be in relation to P, we are constrained to believe that a further investigation, if taken sufficiently far, will lead us to discard that impression. One suspects some profound incoherence in the idea that we might have reason to believe that such was the situation. At any rate, we would not wear it; so the point goes through that in the case in question we should no longer regard the contingent generalisation corresponding to the discarded necessary statement as acceptable.

But there is another possibility: we might reject the necessity of a statement without enthroning anything in its place. And in that case no advance reason is apparent for supposing that we could not reasonably continue to believe that the corresponding generalisation was true. So to cease to do whatever it was that accepting the statement as necessary required us to do need not, it appears, affect the acceptability of the contingent generalisation in question. The proposal thus looks incapable of meeting the first constraint.

This, however, is too hasty. After all, if we abandon a particular game-rule, that does not require that our subsequent behaviour will *actually* refute the corresponding descriptive hypothesis; nor does it require that we could not continue to have reason to believe in that hypothesis. ('Old habits die hard.') All that need be involved is that conduct become permissible which would formerly have been malpractice. We must beware of interpreting the first constraint in an over-exacting way: it is not a condition upon the capacity of a conventionalist's corresponding contingency to serve in an explication of the normative content of a particular necessary statement that its continuing acceptability should require our continued acceptance of the statement in question *as* necessary. In fact, the way is open to him to press his case as follows. On an intuitive level, what is essentially involved in discarding the necessity of a statement is that we admit as a *possibility* something which was formerly denied that status. This is what corresponds, in the case of a game-rule, to admitting conduct as permissible which would formerly have been malpractice. But what is this possibility? It is exactly that a situation in prima facie conflict with a necessary statement may under certain circumstances be correctly described as *actually* being in conflict with it; that is to say, no explanation serving to dissolve the appearance of conflict can satisfactorily be made out. It is allowing that this possibility obtains in a particular situation which is to be compared, in the case of the game, with actually availing ourselves of

the new option to break the old rule. If we do the latter, the descriptive
hypothesis corresponding to the old rule is falsified; and if we do the
former, that is precisely to discard the contingent generalisation which
corresponds to the erstwhile necessary statement in the terms of the
proposed rubric.

It appears, then, that—albeit with a few creaks and shudders—the
sort of analogy which Wittgenstein wanted to commend can be spelled
out somewhat as follows:

(i)   (*a*) The affirmation of an indicatively expressed game-rule
      may be taken either as an endorsement of the rule or as an
      assertion of a corresponding descriptive hypothesis, true just
      in case the rule is complied with.
      (*b*) The affirmation of a statement as necessary serves both to
      endorse a linguistic rule and to affirm a contingent generalisa-
      tion, to be explained by reference to the conditions of 'prima
      facie falsification' of the necessary statement along the lines
      described.

(ii) (*a*) The content of the game-rule is that no situation be brought
      about which would falsify the corresponding descriptive
      hypothesis. To dignify a description of practice to the level of a
      rule of practice is (at least) to form a communal intention to do
      nothing which will falsify that description.
      (*b*) The normative content of a necessary statement is that no
      situation be admitted as correctly described which, if so, will
      constitute a counter-example to the corresponding contingent
      generalisation. To dignify a statement to the rank of necessity
      is to form a communal intention to admit as correctly described
      no situation which, if so, will refute that generalisation. Any
      situation which would—while falling short of *conclusive* falsifi-
      cation—previously have constituted sufficient practical
      grounds for rejecting that generalisation, is now denied that
      status; there are to *be* no such sufficient practical grounds.
      (The idea, adumbrated in previous chapters, that the rôle of
      necessary statements is to regulate the transition between
      appearance and reality is thus incorporated at this point.)

(iii) The possession by necessary statements of conditions of prima
      facie falsification in no way vies with their construal as rules,
      understood as stated. What a realisation of such conditions
      prima facie threatens is actually the acceptability of the corres-
      ponding contingent generalisation. For such a situation will
      incorporate an initial appearance, or justified presumption,
      that: the speakers, or observers, are competent; that the con-
      ditions of observation were normal; that no relevant change
      took place in the objects observed . . . and so on.

That this is in essentials the picture of necessity whose justice Wittgenstein thought he saw, and which he wished to commend, seems to me highly likely. But if so, it has a striking consequence. According to the conception which this picture is meant to oppose, the necessity of a large class of necessary statements is (modulo, perhaps, the rule-following considerations) a matter for cognitive apprehension, a species of truth about which we can legitimately aspire to something approaching an indefeasible certainty. For someone who cleaves to this conception, therefore, the symptoms of necessity in *use*—what, in the kind of application which we make of it, will show that we regard a statement as necessary—will typically be exactly the *same* as they will be on Wittgenstein's account. Recognition of a necessity will be exactly a recognition that if a situation prima facie conflicts with the statement in question, then it is in the *essential nature* of the case that a conflict-dissolving explanation can in principle satisfactorily be made out. So on both the recognitional view and on Wittgenstein's view the key symptom of our treating a statement as necessary will be that ordinary practical standards of defeasibility will be suspended in relation to a generalisation to the effect that prima facie conflicts can always be satisfactorily resolved. It therefore emerges that the response to be given by Wittgenstein's account to the first 'sub-question' (§3, Chapter XX) must be negative: Wittgenstein has to concede that when the normativeness of necessary statements is properly understood, the key facts of linguistic usage which serve to indicate that a statement has that kind of normative rôle are in no way in conflict with the recognitional conception of necessity which he wants to oppose; on the contrary, they are exactly what we should expect if that conception were right. So *all* we so far have is a picture. There is no proving its superiority over its rival by an appeal to what is distinctive of necessary statements in use. If it is superior, the grounds for saying so must be sought elsewhere. They must be sought, in fact, by an exploration of the first line of enquiry which we distinguished in §2 of the last chapter: we have to determine whether *deciding* always has an essential part to play in our admission of new necessary statements, unparalleled in our acceptance of new contingent ones. That is the question to which we shall turn in Chapter XXIII.

4.   As noted, in Wittgenstein's thought about these matters the idea of the normativeness of necessary statements was intended both to undercut our inclination to regard them as a kind of truth *and* to contribute towards an explanation of what necessity essentially is. It remains to consider the standing in relation to the latter objective of the ideas with which we have just been concerned. Note that it follows from the immediately proceding reflections that any shortcomings here will be shared by the orthodox, recognitional conception. Both views will regard our acceptance of the necessity of a statement as signposted by an

abrogation of the ordinary practical standards of defeasibility for a generalisation postulating the availability of a satisfactory, conflict-dissolving explanation for any situation in which things are prima facie other than as the statement in question requires. Any doubt whether Wittgenstein's picture points to a satisfactory account of what in their linguistic practice reveals speakers' acceptance of the necessity of a particular statement is thereby a doubt whether necessity, on the opposing recognitional conception, *has* any distinctive linguistic man-ifestation; a doubt, that is, whether grasp of the necessity of any particular statement, transcending grasp of its mere acceptability, is any essential part of a language-master's competence.

There are two principal doubts. First, the recognitional conception, though it has a long history of association with the notion that a near-indefeasible certainty always enshrouds our apprehension of a necessity, is not strictly committed to it. Indeed the plain fact is that in mathematics, or logic, of any degree of sophistication there will be large numbers of accepted theorems about the validity of each of whose proofs we possess a high degree of certainty—but *not* a certainty so strong that it will shout down any claim to have found error in the proofs; nor, therefore, strong enough to lead us to regard as deceptive any appearance that circumstances conflict with the theorem. Witt-genstein's play with normativeness misrepresents our use of accepted necessary statements in this category. In his keenness to contest the putatively wholly cognitive warrant for our certainty about the truth of necessary statements, he overlooks—at least in his published writ-ings—that the recognitional conception had in any case no business pretending to such certainty in very many cases; that the domain of statements of which we are maximally certain does not include but merely partially intersects with the domain of statements whose neces-sity we conceive ourselves to have recognised. It is only of a restricted class of accepted necessary statements that Wittgenstein's proposal holds out any prospect of an account of the distinctive use. To be sure, if this particular restricted class did not exist—if about no statement did we feel the kind of certainty which characterises our response to the simplest necessary statements—it is doubtful whether we should ever have evolved the conception of *necessary* truth. But neither Wittgen-stein, nor any proponent of the recognitional conception, seems to have any clear account to offer of what in our use of necessary statements outside this class distinctively manifests our acceptance of their necessity.

The second doubt concerns the success of Wittgenstein's account even for statements in the restricted class described. For what explana-tion can it offer of the distinction between necessity and *dogma*, or supremely deeply held conviction? The measure of the depth, or stub-bornness, of a conviction is exactly the length to which the adherent is prepared to go to explain away prima facie discordant data, the extent to which ordinary practical standards of defeasibility for the hypothesis that there is an explanation to accommodate the data with the belief are waived.

Intuitively, the required distinction would pivot on whether the adherent was willing to allow that there *could* be a situation in which no accommodating explanation was in principle available, in which the challenge which it posed to his belief could not be met. The Flat Earth Society, for example, presumably allow that their guiding conviction could run foul of the facts; they merely hold that it does not run foul of the facts, or appearances of facts, which motivate the theory of their opponents. But a moment's reflection shows that this is not good enough. The question immediately resurges in the form of a problem of interpretation. Corresponding to any corpus of well-entrenched belief is an *epistemic* notion of possibility, applying to just those circumstances which a holder of those beliefs is in a position coherently to envisage. So if, seeking to distinguish whether a man holds a particular statement to be necessary or is merely supremely deeply convinced of it, we describe a prima facie falsifying situation and ask of him whether such could really be the way of things, a negative answer is not yet decisive; we have in addition to be able to exclude the purely epistemic interpretation of his denial. But if we could do that, we should have already to be in possession of an account, tied to practical criteria, of the distinction between epistemic impossibility and that notion of impossibility which is the obverse of necessity. We should have to be in a position to explain what, if a man professed to regard some hypothesised situation as impossible, would *show* whether he was giving expression to his acceptance of a relevant statement as necessary or merely to his deep conviction of its truth. So the suggestion presupposes what it purports to provide.

We are brought back, then, to the problem which we confronted at the end of the last chapter: intuitively, the kind of normative content possessed by necessary statements, the kind of thing enjoined of one if a statement is accepted as necessary, did not seem to be any different to what is required if a statement is simply accepted as true. Wittgenstein's picture promised to solve the problem; for while a belief in a contingent statement will indeed involve a belief that situations prima facie conflicting with it will be capable of being explained away, it will not in general involve disclaiming the ordinary standards of defeasibility for the latter belief. But the stronger the belief, the greater the inclination to do so; and the difficulty remains of giving practical content to the idea that necessity alone is the ultimate destination of travel in that direction.

It is not clear, then, that normativeness, interpreted as proposed, is a feature either of all or of only the statements which we accept as necessary. Indeed it is unclear that it can serve as an operational criterion of necessity at all. For suppose that ordinary standards of defeasibility *are* brought to bear on the hypothesis postulating the existence of a conflict-dissolving explanation of data prima facie in conflict with a particular statement. Nevertheless that can hardly be *sufficient* to entitle us to conclude that the statement was not regarded as necessary. For if someone is prepared to reject such a hypothesis in a particular situation, it is intuitively a possibility that he may be doing so precisely by way of effecting a shift in the concepts involved in the statement in question; that is, precisely by way of de-throning it from the status of necessity. To be sure, in our own case there is a great temptation to say that we know perfectly well whether that is what we are doing in a particular case; that it will always be clear to us whether, if we reject such a hypothesis, we do so by way of modifying the concepts involved in the associated statement or revising our beliefs about the world. But what criteria is an interpreter to use to determine which we are doing? What will show that one rather than the other account is correct? Again, it looks likely to be futile to refer the question back to our explicit avowal on the point; for the question will re-arise as one of interpretation of what we say. And unless this problem of interpretation can be solved, and—more generally—unless some aspect of their application can be located to differentiate statements which we regard as necessary from statements of *every* other sort, *we* have no grounds for confidence that the idea that there is a communal understanding of the notion of necessity is any more than a fiction.

A natural suggestion is that we should seek to locate the distinction not in some differential aspect of the use of necessary statements but in the sorts of consideration which lead us to accept them. But, again, a moment's reflection seems to make it clear that the suggestion is worthless. For unless we have some antecedent account of the very distinction to be explained, how can we set about determining whether there *is* any fundamental contrast between the respective types of methods of assessment on the basis of which we put a statement on one side of the distinction or the other? That might seem a spurious circularity. Surely there just are such things as *proofs*, of a sort to which no contingent statement is susceptible. But consider what such a proof will actually be. If it consists in a derivation of the statement in question from other statements, the problem recurs in the demand for an explanation of what it is for those statements to be regarded as necessary; and the answer cannot appeal to the idea of proof of this sort, on pain of infinite regress. (It makes no difference, of course, if the proof takes the form of establishing an assumptionless sequent in a system of natural deduction; the problem is then to explain what it is for the rules of the system to be other than the reflection of *contingent* generalities.) If, on the other hand, the proof is of the kind which we should give of a simple arithmetical equality to a child, by manipulations of an abacus for example, nothing in the method used serves to distinguish the proved statement from something corresponding to it as Mill-generalisations correspond to arithmetical equalities. And if the method of 'proof' is simply to reflect on the statement in question, nothing has yet been done to distinguish the status of the accepted statement from that of any statement which, its negation striking us as enormously unlikely or difficult to imagine realised, we just proceed to accept as true. So it seems that the distinction, if there is one, must be drawn in terms of the use which we go on to give to the proved statement. And to draw it in such terms is just what looks to be problematic.

Considerations of this kind seem to drive us towards the scepticism about the distinction between necessity and contingency advocated par excellence by Quine. It is without doubt the most major shortcoming in Wittgenstein's thought about the status of logic and mathematics that he did not seem clearly to see, or at any rate face squarely, these problems which attend the attempt to substantiate the distinction of which he was so anxious to reject, and improve on, the received account. He allows, to be sure, that the contrast between 'rules of description' and 'descriptive propositions' should not be regarded as absolutely sharp; that it 'shades off in all directions' (v. 5). But

> . . . that in turn is not to say that the contrast is not of the greatest importance.

Well, even granting some blurring of the edges, what in our use of a particular proposition will show that it ought definitely to be placed on one side of the contrast? And what, in different circumstances, would have showed that it ought definitely to have been placed on the other?

Even if every necessary statement which we accept was regarded as indefeasibly certain, the conception that necessity consists in normative use would need backing up with an operational account of the distinction between necessity and dogma, or supremely deeply held conviction; and an operational account of the distinction between conceptual revision and revision in beliefs. Wittgenstein does not tackle these issues in *RFM*.[1]

5.   I do not know if either distinction can be satisfactorily explained. But before we plunge into a Quinean scepticism, we ought at this point to consider the possible bearing on the matter of a train of thought which I tried to develop in Chapter XVII: a quasi-transcendental argument to the effect that the practice of any form of predictive theorising *requires* that there is such a thing as necessity.

The argument was that the connection between a predictive theory and the specific expectations which it licenses cannot always be hypothetical. Let Θ be a statement of such a theory, A a description of

---

[1] In *On Certainty*, indeed, as the source passages make plain, he gravitates towards the view that the first distinction is empty.

the initial conditions of an experiment and B a description of its possible outcome. Then the claim is that statements of the sort:

$$\text{If } \Theta, \text{ then if } A, \text{ then } B,$$

effecting a connection of the theory with a specific expectation, cannot be universally regarded merely as well-founded hypotheses. The reason for saying so was that our confidence in such a statement cannot be based on the experience of a regular association between the truth, or reasonable believability, of its antecedent and that of its consequent in any case where the latter formulates a (sort of) prediction which we have never made before. Nor, for the same reason, can it be conceived as based on the derivability of that statement from certain ulterior principles for whose acceptability there is that kind of reason; for, in the sort of case in question, there can be no ground of that kind for accepting the conditional serving to articulate the soundness of the derivation. Nevertheless such statements must be accepted, at least in the sense that we *move* in accordance with them —whether or not we ever contemplate an explicit formulation —or we will not move at all. And if, further, we have the resources to formulate such conditionals, the reasonableness of our assent to them cannot coherently be supposed to require any sort of antecedent inductive corroboration.

This argument, however, falls short of a demonstration that among such conditionals which we accept there must be some which function normatively in the manner which the Wittgensteinian proposes to regard as characteristic of necessity; a demonstration, that is, that our acceptance of some such conditionals will go to the point of a radical suspension of ordinary standards of defeasibility for hypotheses postulating, for any situation prima facie in conflict with them, the availability of a satisfactory conflict-dissolving explanation. Can such a demonstration be given?

The first task is to characterise the notion of prima facie falsification for such statements. Compare the case with what was said on 'Nothing can simultaneously be red and green all over'. In order to explain a notion of prima facie falsification for that example, we had recourse to the idea of distributing the evidence through more than one observer. We envisaged, for example, receipt of a pair of reports, from people whom we know to be competent both as observers and speakers, of an object which they have had the opportunity to observe in good light, etc., that at a particular time it was, respectively, wholly red and wholly green. The situation, in general, is to be that we would be completely, and wholly reasonably, confident about accepting each report in isolation, were it not that we had the other. The normativeness of the statement, however, comes out in the fact that under no circumstances are we prepared to do so. Rather, we advance the hypothesis of the existence of a conflict-dissolving explanation: a hypothesis to the effect that there are facts to be known such that, if we knew of them, we could legitimately discount one, or the other, of the reports, even if it had been the only one which we had received.

Having an (implicitly) disjunctive consequent, the hypothesis is itself prima facie threatened in the described situation; for the reasons which we have to accept each report in isolation are reasons for denying that if we knew sufficiently many facts, we should have reason to discount that report. To pool the reasons would thus be to be in a position to deny the hypothesis. Merely to advance the hypothesis is thus *already* to insulate it from one defeasibility-criterion which we should apply in other cases: cases where, presented with independent reason for denying each of a pair of conditionals with shared antecedent, we should count that as reason for denying the conditional consisting of the disjunction of their consequents on the same antecedent.

In terms of this model it is straightforward to construe prima facie falsification conditions for the sort of conditional, articulating the predictive consequences of a theory, in which we are interested. Suppose that Θ is a complete statement of what we know to be a powerfully supported, widely predictive, trouble-free theory. But we have a report from an able researcher than an experiment rigorously conducted under the conditions described by A did not fulfil the prediction B. Suppose, in fact, that our background information is in every respect such that but for the report, we should have every reason to continue to accept the theory; and but for our confidence in the theory, we should have every reason to accept the report. If the conditional, if Θ, then if A, then B, has a normative status for us, then, rather than calling it into question by holding it open that it might be reasonable to accept both the theory and the report, we shall in this situation propose the hypothesis of the availability of a conflict-dissolving explanation: we shall hypothesise that if we knew sufficiently many further facts, then either we should be in a position rightly to consider our reasons for believing the theory to have been discredited—even if we had never had the researcher's report—or we should be in a position rightly to consider that our reasons for accepting the report had been discredited—even if we had never had any reason to believe the theory. Again, just by advancing this hypothesis, we waive a standard of defeasibility which we should apply in certain other cases. (The hypothesis, for example, so corresponding to 'If you go out in this weather, you will catch a cold'.) And, on Wittgenstein's account, our regarding the conditional in question as necessary will come out in our tending to stick to the hypothesis, however unpromisingly a subsequent investigation may go. (Perhaps every attempt to duplicate the recalcitrant findings is unsuccessful; but neither does any reason emerge for thinking that all did not go exactly as the researcher and his assistants reported, or that they did not properly observe certain relevant controls, etc.)

Suppose now that we are practitioners of such a theory, working in a language in which we can express an axiomatisation of the theory, an indefinite stock of observationally testable statements, and which otherwise has all the expressive resources of English. And suppose that we are prepared to treat *no* hypothesis of the sort illustrated, corresponding to a conditional statement, in the manner which according to the proposed account signposts acceptance of the conditional as necessary. That is, for each such hypothesis, we are prepared to grant that circumstances can be envisaged in which there would be prima facie reasons to accept the antecedent(s) of the conditional, and to deny the consequent, and in which there would be no reason to think that if we knew sufficiently further many facts, we should be in a position to discredit either set of prima facie reasons. Evidently, that is just to hold that circumstances are envisageable in which it would be reasonable to deny the conditional. In particular, for each conditional of the sort described, if Θ, then if A, then B, it is to hold that circumstances are envisageable in which it would be reasonable to assert each of Θ, A, and the negation of B.

The question therefore arises, what is our attitude to the suggestion that *all* such circumstances might obtain simultaneously, that a situation might arise in which, if we knew sufficiently many facts, we should be in a position to assert the antecedents and to deny the consequents of *every* conditional numbering Θ among its antecedents whose formulation the language allows? It is immediately clear that to allow that this is a possibility is to concede that we have not assigned a coherent pattern of use to the conditional construction, to concede that circumstances may arise in which we are paradigmatically justified both in asserting and in denying a particular conditional, and in which the warrant for doing each will survive the most painstaking investigation. A simple example suffices. Consider the pair:

$$(a) \quad (\Theta, A) \to B, \text{ and}$$
$$(b) \quad \{\Theta, {}^sA, \Theta \to (A \to B)\} \to B.$$

In the circumstances whose possibility we are granting, we should be in a position to assert the antecedents and to deny the consequent of each of these. So how can we avoid the conclusion that in such circumstances the former should be both asserted and denied?

If we are to be in a position, then, to believe that the use which we make of the conditional construction is coherent, we have to reject the suggestion. Clearly, though, it will not suffice simply to *hypothesise* its incorrectness. If that is all we do, we are bound to allow that a situation could arise in which it would be reasonable to reject *that* hypothesis; and that just is to grant that the original suggestion puts a genuine possibility. So our belief that we have assigned a coherent use to the conditional construction requires the belief, if that is the right way to put it, that the 'hypothesis' in question is indefeasible, that nothing can count as possession of information in which it will be reasonable to reject it.

We will have to hold, therefore, that in any situation in which certain conditionals numbering Θ among their antecedents are reasonably deniable—that is, in which it is reasonable to assert everything in the antecedent set and to deny the consequent—other such conditionals will not be. And we must hold this to be indefeasible. We hold it as indefeasible, that is to say, that there are disjoint sets, X and Y, of conditionals of the kind in question such that if each member of X is reasonably deniable in a particular situation, not every member of Y is so. But for every member of X to be reasonably deniable amounts, in the kind of case with which we are concerned, to each of their antecedents being reasonably assertable and each of their consequents being reasonably deniable. So corresponding to each pair of such sets, X and Y, standing in this relation, there is a conditional of the very sort in question: a conditional numbering Θ among its antecedents. To hold it indefeasible that there are sets of conditional statements standing in the described relation, is therefore to hold as indefeasible that conditional which so corresponds to any particular such pair of sets. If X and Y are, for example, $\{(a)\}$ and $\{(b)\}$ as above, the conditional in question is:

$$\{\ \Theta, A, -B\ \} \ \rightarrow\ --\{\{\ \Theta, A, \Theta \rightarrow (A \rightarrow B)\ \}\rightarrow B\ \}$$

And the example illustrates the simplest strategy for finding an appropriate Y for any particular set of conditionals X: viz. add to X any conditional which results from appending to the antecedent clauses of any member, x, of X, the conditional, x, itself.

The argument is, then, that in order to be in a position coherently to entertain the belief that we have assigned a coherent pattern of use to conditionals articulating the consequences of a particular theory, we have to assign to *some* such conditionals the normative rôle which the Wittgensteinian regards as constitutive of necessity. But we only have a theory at all, that is, we only have reason to form expectations in accordance with its consequences, so long as we are in a position coherently to entertain that belief. That some statements be treated as necessary is therefore a precondition of the whole activity of expectation-yielding theorising. ('The deep need for the convention')

It would be absolutely fair to protest that, granting the soundness of these reflections, all that emerges is a requirement of a communally *dogmatic* treatment of certain statements; that the problems of dissociating necessity from dogma, and of supplying practical grounds for identifying any particular statement as having the sort of rôle described—when any *sentence* may be rejected—remain untouched. But we will be less inclined to allow the second problem to ramify into doubts about the *reality* of Wittgenstein's kind of normativeness if we are already persuaded that there must be such a thing; and, whether or not reflections of the kind we have just been concerned with may be extended to apply to other sorts of intuitively necessary statement, it seems certain that they will not extend to any particular belief about the shape of the earth, the superiority of one form of political organisation over others, or the supremacy of a particular race.

XXII

Arithmetic and Imagination

Sources

RFM: ΙΙ. 42; ΙΙΙ. 1–7; v. 42–4
PI: ι. 395–7
Z: 263 sq.
LFM: lecture xv, pp. 145–7

1.   In the course of the preceding chapter, we noted that, in order to make out a notion of prima facie falsification for certain of the most fundamental-seeming necessary statements, it appeared that we had to 'distribute' the evidence through more than one observer, or —as with the conjuring-trick example—to qualify the conditions of observation in various relevant ways. This is the substance to the intuitive feeling that it is simply *unimaginable* that things could be other than as required by, for example, the very simplest arithmetical equalities; and it is this apparent unimaginability which is surely the single most potent source of the dissatisfaction which conventionalist theories of necessity have tended to inspire.[1] Such theories, there is a temptation to think, must at best be guilty of omitting some essential ingredient; for if all necessity is merely the immediate product, in a sense yet to be made out, of explicit stipulation, how is it that the opposites of so many necessary statements are, in the relevant sense, unimaginable—that none of us can conceive, save when matters are in some way not fully open to his view, how things might even appear to be so?

It has, on the other hand, been fashionable for some time to regard appeals to what human beings can, or cannot imagine, as the thinnest ice on which to attempt to rest any substantial point about necessity; so considerations about unimaginability have not in general played the large rôle in the literature on necessity that they have undoubtedly played in the secret thoughts of many anti-conventionalists. A conspicuous exception, however, is Edward Craig's treatment of arithmetic in *The Problem of Necessary Truth*.[2] Nowhere, to the best of my knowledge, is this particular kind of reservation about conventionalism

---

[1] Perhaps, therefore, I had no business envisaging, in §1 of Chapter XXI, its invocation—albeit in a muddled form—against an *objection* to conventionalism.
[2] 15.

more carefully developed. It is with the character and force of the resulting objection that this chapter is concerned.

2.   Whether or not a group of objects has been counted correctly on a particular occasion is something which may be settled by reference either to arithmetical or to purely perceptual criteria. There is such a thing as noticing someone making a particular mistake in the course of counting; and there is such a thing as judging that he must have miscounted somewhere along the line just because of the arithmetically discordant character of the results which he has achieved. What account, Craig asks, is the conventionalist to give of the relation between these two sorts of criterion? In particular, is it a possibility that our arithmetical criteria for miscounting should ever require us to issue verdicts which are incapable of independent, perceptual corroboration?

In effect, the question puts to the conventionalist Dummett's dilemma discussed in Chapter V. If it is answered negatively, the conventionalist appears to be conceding that our arithmetic correctly codifies a domain of independent, in principle perceptually accessible fact. For if, whenever counting results are not as our arithmetic requires them to be, there is an in principle independently noticeable mistake,[1] then whenever there is *no* mistake, results will be as our arithmetic requires them to be. Therefore, whenever there is no such mistake, they will not be as any 'alternative' arithmetical rules require; and the verdict, that there must have been a miscount, which the use of an alternative arithmetic would enjoin in such circumstances, will be incorrect. On the other hand, if the question is answered affirmatively, Dummett argued (in effect) that intolerable consequences would follow about the interpretation of the quantifier in 'Some mistake must have occurred'.

I have attempted to explain the 'antecedence to truth' doctrine as a means whereby the conventionalist could coherently answer the question negatively. Craig, however, takes it that the answer must be affirmative; and his argument is then that this answer is decisively objectionable for reasons distinct from those which Dummett suggested (which Craig does not explicitly consider). It follows—provided we are content that the conventionalist should respond to the dilemma in the way I suggest—that Craig was barking up the wrong tree. The fact is, nevertheless, that to grant that Craig's argument establishes what it purports to establish would require so severe a qualification of the sense in which arithmetic might legitimately be regarded as conventional as to preempt any meaningful arithmetical conventionalism altogether. To make that claim good will be the task of this and the next section.

---

[1] In order to shorten the exposition, I ignore here and often in what follows the other kinds of 'saving hypothesis' which might be advanced.

If the conventionalist were to opt for the 'possibility' that our arithmetic should require us to assert that someone had miscounted on a particular occasion without it being in principle possible to confirm that assertion in a result-independent, purely perceptual way, then, Craig points out, he would be committed to the contingency of propositions like:

> *M:* There was the best possible perceptual evidence that someone counted a group of boys correctly and made the answer 5, that someone counted a group of girls correctly and made the answer 7, that someone counted the whole group of children correctly and made the answer 13, and that no child joined the groups, or simply appeared in then, during the counting operations.

To have the 'best possible perceptual evidence' for a proposition involves, for Craig,

> have carried out an exhaustive investigation and still to have found only positive and no negative evidence. (p. 13)

Craig anticipates the complaint that no such investigation could in fact be exhaustive, that there is in principle no end to the tests which we can carry out to check on any perceptually based statement. But he is prepared to grant, for the sake of argument, that *M* could be regarded as non-contingent for that reason. His essential thesis is, however, that any proposition standing as *M* stands to one of our arithmetical equalities is impossible for a *further* reason, which his conventionalist is in no position to acknowledge: such a proposition would also express an impossibility

> because one would demonstrably be able to find contrary evidence in a finite time, or because there is some particular and crucial piece of evidence which one will never be able to get. (p. 13)

For Craig's conventionalist, if *M* is impossible, it is so only because the 'best possible' perceptual evidence, i.e. evidence based on exhaustive tests, can never be achieved for *any* proposition. But what, he holds, is a possibility is that the body of perceptual evidence should remain of the general character described by *M*; that is, that it should continue to support all *M*'s constituent propositions, however much we add to it; and this not because there is some crucial piece of evidence which we are unable to get. This is what Craig denies.

Let us defer consideration of Craig's grounds for this denial, and consider instead the consequences if he is right.

Dummett made use of the fiction of people who count as we do but have no such thing as calculation. Such people are to be thought of as having the same non-arithmetical criteria as we have for when counting is done correctly; they will have the same concept of what it is to try to count properly, what kind of safeguards one has to observe. Indeed, everything is to be built into the fiction which is needed to ensure that there would be no situation in which, leaving our arithmetical criteria out of account, we should suspect, or be confident of, the occurrence of a miscount or some other peculiarity, but they would not; or vice versa. Intercommunal agreement is always to be possible, at least to the extent that *we* agree among ourselves, about whether any particular single process of counting seems to have been carried through correctly, whether the situation is one in which a miscount is particularly likely, and whether anything other than a miscount seems to have occurred of the sort which we regard as liable to throw our calculations out.

Evidently, people of this sort need not as yet have assigned any use to one kind of judgment which our arithmetic empowers us to make: a judgment to the effect that, despite appearances to the contrary, a miscount, or *something* peculiar, must have taken place in such-and-such a situation. There will thus be an uncontroversial sense in which their adoption of an arithmetic may be regarded as modifying their use of language in this area. Not that they need have had *no* criteria before for saying that things were other than as they seemed; but an arithmetic will at least supply them with a kind of criterion which they have not used before. Now, the essence of the antecedence doctrine is that, whatever arithmetic this community may be persuaded to adopt, while their choice may indeed be an *impractical* one, there will be no sense in which it can be regarded as right or wrong. Rather, they will precisely be *determining* a conception of when the transition from appearance to reality in counting-situations may be straightforwardly negotiated. Different arithmetics will regard different sorts of results as discordant; so the range of circumstances in which they will require the postulation of conflict-dissolving explanations will differ correspondingly. But there is intelligible no 'Olympian' standpoint from which we can enquire which of two conflicting arithmetics requires the proposal of some such hypothesis *only* when some such hypothesis is actually true.

The ultimate argument for this attitude, it will be recalled, resides in the rule-following considerations. The idea of ratification-independent predeterminacy in truth-value is rejected for hypotheses of the relevant sort. Such a stance is quite beyond the bearing of Craig's argument. It will also be recalled that a more specific, subsidiary play was made with the *indefeasibility* of the saving hypotheses which, if they adopted an arithmetic alternative to ours, the community in question would be forced (repeatedly) to advance. Suppose, to pursue the Dummett-Craig example, that the two communities, accepting that $7 + 5 = 12$ and $7 + 5 = 13$ respectively, collaborate to count the children. The results of the three stages are: 7 girls, 5 boys, and 12 children. No one has been able to see any error in the counting; the children have kept perfectly still; there has been no suggestion of an extra child slipping in, or of one leaving, etc. In this situation *we* are quite content, on the basis of both perceptual and arithmetical criteria, to allow that all three counts were carried out correctly. *They,* on the other hand, while agreeing that there is so far nothing apparent which by the perceptual criteria ought to be regarded as calling the correctness of the counts into question, regard the results as anomalous. They insist that either some mistake did occur, or that a child dropped out, or *something* . . . One way or another, things must somehow have been other than as they seemed to us all. Then the point about indefeasibility is simply that there is no proving them wrong. On the contrary, in any particular situation it *may* turn out that they are right; it cannot be ruled out that they will succeed in finding a mistake, or some other peculiarity—which *we* will then have to regard as having been compounded by something else along the same lines, in order to explain how the 'right' result was nevertheless obtained. So there is no convicting an 'alternative' arithmetic of conflict with contingent fact in any particular situation.

Again, Craig's arguments seem to do nothing to call this point into question. To allow that M is impossible, or, in general, that it is impossible that, no matter how painstaking our investigation, all the perceptual evidence accumulated should continue to comply with what we regard as a spurious arithmetical equality, is quite consistent with granting that, whenever the use of such an equality requires advance of a hypothesis of the relevant conflict-dissolving sort, that hypothesis will be strictly indefeasible—and may even be perceptually corroborated. So both strands of the antecedence doctrine—its deep source in the rule-following considerations, and its specific suggestion about how the use of alternative principles of inference could avoid anything demonstrable as conflict with contingent facts, or as utilisation of deviant non-inferential criteria for the application of certain relevant concepts—seem to go unchallenged by Craig's essential thesis.

Nevertheless, that thesis, if correct, would appear to introduce an important asymmetry. We have continually represented the conventionalist as holding that our principles of inference possess an at most *practical* superiority over alternatives. If Craig is right, however, it becomes unclear whether this can be the whole truth of the matter. For the fact is that, if M, and anything differing from it only in that some numeral other than '12' is substituted for the occurrence of '13', is, as Craig thinks, impossible, M^*, differing only in that the occurrence of '13' is replaced exactly by one of '12', clearly *does* express a possibility. It *is* a possibility that counting-experiments should seem to go as our arithmetic requires them to do; and, moreover, that they should continue to seem to do so, no matter how painstakingly we investigate them. It is a possibility which is realised all the time.

The consequent asymmetry can be brought out like this. Suppose we put the two arithmetically-divergent communities into competition. The competition is to concern only situations in which one community is content to accept the results of a series of countings, while the other is constrained by its arithmetic to advance the hypothesis of a miscount, or of some other peculiarity. The acquiescent community then challenges the other to supply independent perceptual corroboration of the hypothesis. The challenged community wins just in case it is able to supply such independent corroboration; the challenging community wins just in case it is able to refute the hypothesis advanced; and in any other situation, the challenge is drawn. Then the effect of the antecedence doctrine is that no challenging community can ever win! The best it can do is get a draw. But, if M and any relevant variant of it other than M^* are impossible, it follows that when we, the normal community, are challenged, we can *always* win; for if it is impossible that things should seem and continue to seem to be contrary to our arithmetic, however painstakingly or ingeniously we investigate them, then a sufficiently painstaking investigation will always bring up *something* which will independently confirm one among the range of hypotheses which we might have chosen to advance. A win is there for the taking, provided we were fortunate enought to choose the right hypothesis. Our opponents, however, are not so happily placed. When they are challenged, they *may*, to be sure, be able to locate independent corroboration for their hypothesis; but the possibility of M^* requires that it is also a possibility that such corroboration cannot be found. For if it is a possibility that things should seem and continue to seem, no matter how painstakingly investigated, as a proposition of *our* arithmetic requires them to be, that just is the possibility that none of the hypotheses to which our opponents might have recourse in order to explain what for them are recalcitrant counting results be capable of independent perceptual corroboration. So whereas *we*, when challenged, can always get a win provided we make a fortunate choice, it is a possibility that the best our opponents can do, when challenged, is to get a draw. And in fact it is practically certain that this is almost *always* the best that they will be able to do.

Of course, it would not be material to the conventionalist if the ground of this asymmetry were purely *inductive*. If the reason for saying that we are always capable of independently corroborating our saving hypotheses were just that that is how it always turns out when we try hard enough, the point would amount to no more than one formulation of the sense in which the rules of our arithmetic enjoy an undoubted practical superiority over any particular alternative to them. But that is not the situation. For M and its ilk are argued by Craig not to be merely false, but to be *impossible*. If he is right, the asymmetry is therefore not inductive in origin, but metaphysical; and our arithmetical rules are *essentially* superior to any conceivable alternative.

3.　　　If such a thesis were correct, then it would appear, as I intimated, that the clarion of conventionalism had better be very much muted. It would remain that an alternative arithmetic could not be decisively convicted of misdescription of contingent facts in any particular situation. But neither would it be capable of corroboration by the most rigorous examination of contingent fact. The impression that things were as a non-standard arithmetic required them to be would be capable of being sustained only if our investigation were not the most exacting possible, or if certain relevant facts were de facto inaccessible to us.

In order to see more sharply the arguable significance of the asymmetry, consider an example in which it would be absent. Suppose we were to use some sort of sophisticated machinery to set up a rule of conversion between metres and yards by a comparison between standard exemplars of those lengths, regarding the apparatus—for whatever reason—as likely to produce a much more accurate assessment of the relationship than we could achieve simply by experimental measurings and statistical generalisation. Clearly, the apparatus would have some leeway; that is, there would be a small range of verdicts any one of which it could return without arousing our suspicion that it might have malfunctioned, this range being determined by our cruder, directly experimental findings. Moreover whatever rule the apparatus gave us, provided it came within this range, would enter into our criteria for mismeasurement in certain circumstances; if something were measured simultaneously in both yards and metres, it would now be counted as sufficient for at least one of the readings not having been completely accurately taken if the results did not stand in the ratio postulated by the rule. And it may be that, before we had the rule, we should have been content about the correctness of both sets of readings.

So to treat the rule would simply be a consequence of our conception of the superiority of the apparatus which produced it. But the claim, made on the basis of the rule, that a mismeasurement occurred, is still to be understood as the claim that something occurred which we could in principle identify—if our eyes were sufficiently sharp, our hands sufficiently steady, etc. The apparatus could not coherently be conceived as superior if the inaccuracies for whose occurrence the issued rule would serve as a criterion had to be regarded as inaccuracies in a quite different *sense* from the sort of thing which we were antecedently empowered to recognise.

Let R_1 and R_2 be any pair of rules falling within the described range. And suppose R_1 is issued by the apparatus. We should, nevertheless, have been perfectly ready to accept R_2 had it been issued. Now, it is clear that there is here no problem about supposing that the proposition could be true which corresponds to R_1 and R_2 as M corresponds to $7 + 5 = 12$ and $7 + 5 = 13$ respectively. That is, so long as the notion of 'best possible perceptual evidence' is tied to ordinary, practical criteria distinct from the rule, there can be the best possible perceptual evidence that someone correctly measured a bar of iron in yards and got the answer y; while simultaneously someone correctly measured the same bar in metres and got the answer z; that y and z should stand in the ratio prescribed by R_2, rather than by R_1; and that there was no fluctuation in the length of the bar while they were measuring. And a corresponding possibility obtains, mutatis mutandis, if R_2 is the adopted rule.

Each possibility is relatively unproblematic for two reasons. First the kind of errors which, when there is the best possible perceptual evidence that the measurements have been performed accurately, someone may nevertheless have made, are in that case exactly going to be errors which in practice may be too *subtle* for us to discern. By contrast, there is no sense in the idea that, while we might have the best possible perceptual evidence that no *miscount* had occurred in a particular situation, one might nevertheless have occurred but in a way too subtle for our coarse senses to discriminate; a miscount cannot, as it were, be too *slight* to be accessible to our perceptions. In other words the sort of saving hypotheses whose proposal in certain circumstances will be consequent upon our acceptance of a metric rule within the range in question will not be answerable to independent perceptual corroboration in the same way as those consequent on our acceptance of an arithmetical rule.

Secondly, and relatedly, what is at stake between R_1 and R_2 really is plausibly viewed as a choice between explicit conventions. Since, because of the oracle-like status of the machine, we shall regard whatever rule is issued—provided it falls within the described range—as improving on the assessment which we can make directly, there need be no constraint on the rule to square even with statistical generalisations about the results which we get by hand, so to speak. Any discrepancy may rather be treated as an indication of how accurately we have tended to measure. But once a metric rule is treated in that way, what is at stake between it and a sufficiently close rival is relegated to perceptually undetectable status. It follows that there can be no *proof* of the superiority of such a rule over a near rival by any means involving only attention to its ordinary application. And that, surely, *is* for the proposal of such a rule to be the proposal of an explicit convention.

To avoid misunderstanding, I am not suggesting that *our* rules of metric conversion are so used as to dominate statistical methods in the kind of way indicated. The point is only that they might be. The purpose of the example is rather, first, to clarify the conception of arithmetical necessity to which Craig believes the conventionalist is committed, and, second, to bring out the apparent implications of the untenability of that conception. Craig's arithmetical conventionalist emerges as one who, in effect, attempts to maintain that what is at stake between our own and alternative arithmetical rules is in all relevant respects comparable to what is at stake between rules of conversion related as are R_1 and R_2. But, if the foregoing is right, any such attempted comparison will immediately fall foul of the fact—and its consequences—that the superiority over alternatives of at any rate some of the most simple arithmetical equalities which we accept is susceptible to a kind of proof—for example, by drawing objects on a blackboard —which involves direct appeal to their typical applications.

All that follows, it might be thought, is that Craig has lumbered the conventionalist with an unplayable position. Indeed, we questioned at the start whether Craig ought to have robed his conventionalist in the belief that it was possible that our present arithmetic should impel us to saving hypotheses which were incapable of independent perceptual corroboration. But we are now in a position to see that Craig's argument puts a serious challenge to *any* form of conventionalism—even one which, contrary to his initial supposition, rejects that belief.

That may still seem a mystifying suggestion; for is not such a conventionalist in a position to grant precisely what Craig argues, viz. the impossibility of M and all its family save M^*? But the point is not difficult to see. Suppose the machine were to issue a wildly eccentric rule of conversion, radically disparate from our experimental findings; but that, such is our faith in the machine, we accept the rule anyway. Clearly, the effect of our accepting such a rule will be to set up a striking inverse correlation: the more assiduously we attend to the ordinary practical safeguards which we impose in order to try to ensure that reliable measurements are made, the less likely we are to get results which the rule will allow us to regard as reasonably accurate. Conversely, when we get results of the sort which the rule would require, the situation will be at best that there is room for serious doubt whether ordinary practical safeguards were properly observed. A tension would thus be introduced into the whole conception of accurate measurement; our concept of what it was to make every effort to ensure that one measured accurately would become inversely correlated with the likelihood of doing so. It is doubtful whether the ordinary conception of measurement as a means for determining a property of the object measured could survive in such a situation. For there would be no *dependable* way of getting the sort of result which the rule would require: if careful measurement would not do so, haphazard measurement, as now, could be expected to produce only haphazard results. The length of an object is what we find if we measure it accurately. But what would it be to measure accurately in this situation? What would be the ordinary, result-independent conception of a well-conducted measurement? How could it be explained? Some such notion has to be explicable if there is to be anything for the rule of conversion to regulate. The idea of an accurate measurement cannot be fixed *solely* by rules of metric conversion, or we shall have no operational criterion for saying of a single measurement that it was sufficiently accurately taken, that it may be relied on. Therefore, if—as seems certain—the effect of the eccentric rule will be to undermine all such operational criteria, there is a serious doubt about its capacity to play any part in anything communicable as a coherent scheme of metric concepts.

Now, if Craig is right, there can be no alternative arithmetical rule corresponding to $7 + 5 = 12$ in the manner in which R_1 and R_2 corresponded; any alternative rule will be in the situation of the sort of wildly eccentric metric rule just considered. For the proposition which corresponds to it as M corresponds to $7 + 5 = 13$ will, like M, express an impossibility; the impression that a series of counts have gone as such a rule requires will survive, in favourable circumstances when all relevant facts are accessible to us, only so long as we do not impose all the practical safeguards on correct counting which we habitually employ. So there will be a similar inverse correlation. As soon as we really set about taking these safeguards—insisting that the counting is done in a slow, deliberate manner; insisting that all the objects remain open to view; making sure that nothing is slipped in or out, etc—the impression that the situation conforms to the rule is going to dissolve. Only by waiving these safeguards will it be possible to get a situation which seems to go as the rule requires. And, as with haphazard measurement, waiving the safeguard will not lead to results which *consistently* conform to the non-standard rule. There is therefore an exactly comparable doubt about the resulting status of the ordinary notion of a reliable count of a group of objects. If the harder we try to count groups of objects correctly, and to ensure that nothing occurs which might disrupt the outcome, the less likely we are actually to do so, how is it to be explained under what circumstances we may take a *single* count of the objects in some group as reliable? Again, unless some such notion is explicable, it seems that there can be no practice for arithmetical rules to regulate. So, comparably, there is a doubt about whether an alternative arithmetic could form part of a coherent, communicable system of numerical concepts.

Of course, if this doubt is correct, it will arise even if propositions standing to deviant arithmetical rules as M stands to $5 + 7 = 13$ are merely always *false* rather than impossible, are merely always in fact defeasible if sufficient care is taken. But in that case it would be open to the conventionalist to accept the doubt as an elaboration of what he has conceded all along, the undoubted practical superiority of our arithmetic. The doubt would be exactly whether, *as things are,* an alternative arithmetic would be usable. But if we are confronted not with falsity but with a discernible impossibility, the conclusion has to be that any alternative arithmetic is *essentially* unusable. Our arithmetic would not merely be essentially 'game-theoretically superior' to any alternative, in the manner noted; more, there simply could not be any practical alternative to it. In that case, what interesting, residual sense would there be in which it could be regarded as conventional?

It ought now to be clear why Craig's argument confronts even a conventionalist who, availing himself of the 'antecedence to truth' points, refused to acknowledge that he was committed to the possibility of *M,* with a point which he has to answer. Such a conventionalist will hold that it is open to us, *whatever* arithmetic we accept, to assert it to be impossible that things should obdurately seem, however carefully investigated, to be contrary to some rule of that arithmetic. It will be open to us to do this because the claim that if a sufficiently painstaking investigation is made, sooner or later perceptually accessible facts will come to light which dissolve the appearance of conflict with the rule in question, is a strictly indefeasible claim. Well, strictly indefeasible it may be, at least in the sense that there is no defeating it finally on any particular occasion when it is made. But there is no point in stressing that indefeasibility, unless conventionalist capital can be made out of it; that is, unless it, in association with the rule-following considerations, is going to be sufficient to justify the view that the conception of what is in principle capable of independent, perceptual corroboration is 'up for grabs', that it is an idea which it is open for alternative arithmetics to determine in alternative ways. The whole point of Craig's argument, however, is that that appearance of latitude is illusory; there are no arithmetical determinations of that conception which are both alternative to our own and capable of issuing in a coherent conceptual framework; nor is the absence of alternatives a mere absence *in practice,* contingent on features of the world in which we have to live. Rather, no alternative determination could in principle be manageable; for any alternative *must* essentially play havoc, it appears, with the ordinary practical conception of a reliable single count.

There is, I suppose, some room for debate whether a circumstantial restriction might not be imposed on that conception which would enable users of an alternative arithmetic to avoid severing altogether its connection with ordinary practical safeguards. The crudest version, for example, would be if they were to hold that a single count could be accepted as reliable provided *both* that ordinary practical safeguards had been observed *and* that the count was not made in the context of a series of counts whose collective results might prima facie violate an accepted arithmetical equality. The corresponding ploy with measuring would be: a single measurement may be regarded as reliable provided both that ordinary practical safeguards are observed and that it is not taken in the context of other measurements in conjunction with which it might prima facie violate an accepted rule of conversion. But it would be sheer desperation for the conventionalist to rest his case on the belief that a satisfactory such restriction would always be available. In any case, we do not have to chase down this particular rabbit; for if, following Craig, we really can apprehend it as a metaphysical impossibility that a particular situation might continue to seem to conform to any particular deviant arithmetical rule, provided our investigation is painstaking enough and no relevant facts are de facto inaccessible to us, then any such situation must ultimately come to seem to conform to *our* arithmetic if it seems to conform to any arithmetic at all. It follows—leaving the second, dismal possibility out of account—that the sort of saving hypothesis which the practitioners of an alternative arithmetic will be constrained to advance, while incapable of being *convicted* of falsity—and, indeed, possibly true—in any particular case, are bound *in general* to be incapable of independent perceptual corroboration also. This is just to reiterate one half of the asymmetry which we noted in the context of the imaginary competition. But for a hypothesis of this character to be incapable, no matter how painstaking the investigation, of independent perceptual corroboration *is* for it to be false. So we can infer that the users of an alternative arithmetic will be constrained to propose saving hypotheses which, taken by and large, are actually incorrect, even if there is no conclusively demonstrating the point in any particular case.

If that is right, then the antecedence doctrine is in dire trouble. Indeed, it threatens to collapse completely. Once the intended point of the stress on indefeasibility is lost, the force of the doctrine must rely on its appeal to the rule-following considerations. But it is no longer clear that that appeal can have its intended effect. It is all very well to regard each successively postulated saving hypothesis of the arithmetically deviant community as further determinative of their concept of the capacities of the purely perceptual criteria for miscounting, etc. We built it in to the fiction, however, that, so far as is possible, there is to be intercommunal agreement about everything *save* arithmetic. So if, as he intends, Craig's arguments make no special appeal to the arithmetic which *we* actually use, their force ought to be appreciable by the deviant community also. But then, assuming Craig argues soundly, that is just for *them* to be in a position to recognise that their arithmetic is by and large steering them false, that they will not in general be able to find the sort of independent perceptual corroboration for their saving hypotheses whose availability in principle is recognised by both communities as necessary if those hypotheses are to be acceptable.

Craig's argument, then, though he directs it against a needlessly clumsily formulated version of conventionalism, actually cuts right to the heart of the essential conventionalist thesis of the unanswerability of our 'choice' of necessary statements to contingent fact. It remains to see whether the argument is sound.

4.   The crucial moves occur in sections v and vi of Craig's paper. He begins by considering the analogue of M for $0 + 0 = 1$.

> What we have to imagine is an observer to whom it (i) seems just as if no boys were present, and (ii) seems just as if no girls were present, and then (iii) seems just as if one child were present, whilst (iv) seeming to him just as if no child appeared in the region in question during the relevant period. (p. 17)

But it seems impossible to imagine this, for

> in describing an observer of whom (i), (ii), and (iii) were true, we have
> exactly described an observer of whom (iv) was false . . . if one were to
> describe such a series of perceptual experiences as would make (i), (ii),
> and (iii) true, one would have described a perceptual experience such that
> to anyone who had it, it would seem as if a child had appeared, quite
> irrespective of any arithmetical beliefs he might hold. (p. 17)

It follows that no one can have the 'best possible perceptual evidence'
for the truth of each of (i)–(iv). For the successive experiences counting
as evidence for (i)–(iii) collectively constitute an experience which
counts against (iv); so the observer does not have the best possible
perceptual evidence for (iv).

The same, Craig contends, goes for $1 + 1 = 3$. It seems impossible to
comprehend—

> how it could seem to an observer just as if one boy were present, just as if
> one girl were present, and then just as if three children were present,
> without at some stage seeming to him as if a child appeared. (p. 19)

And

> the same would be true of any instance where the number of objects
> involved was small enough for the observer to take them in 'at one go', that
> is without having to scan them successively, leaving others temporarily
> unattended to. (p. 19)

Next Craig considers 'distributing' the evidence over two observers.
There seems no difficulty in the idea that throughout the relevant
period it might seem to the first observer just as if there was one boy
present, one girl present, and no others; and to the second that there
were three children present. If each reports accordingly, then, assuming their general competence and sincerity, we have the sort of situation
for $1 + 1 = 2$ which I have been describing as 'prima facie falsification'. But obviously this falls short of having, in Craig's sense, the best
possible perceptual evidence for $1 + 1 = 3$; for neither observer has any
special reason to think that the other has done his part competently. But
as soon as we allow the observers to confer and attempt to persuade each
other of the correctness of their respective reports, then, Craig is easily
able to show, the case in effect reduces to that of the single observer
already considered. That is, it seems impossible to envisage how it
could continue to appear to either observer both that he had answered
his question correctly, and that his colleague had done the same with
the other.

Craig's notion of best possible perceptual evidence, then, has these
two features: in order for it to be possible that there should be the best
possible perceptual evidence for a particular proposition, it has to be

possible both that it should seem to a single observer in all respects just as if the proposition were true and that the observer should be especially favourably placed. In the sort of example with which we are concerned, the latter constraint is to be interpreted to mean that the situation should be completely surveyable to the observer: it has to be the case that any change, or movement, in the objects involved will be immediately apparent to him. It is worth noticing that in order for a situation to be completely surveyable in this way, it is not necessary that the number of objects involved be so small as to be capable of being 'taken in at a glance', without counting. Thus Craig is able to extend his argument to *M* itself:

> Let the boys carry cards with numbers on them from 1 to 5; and let the girls have similar cards showing on the upper half the numbers from 1 to 7, on the lower half the numbers from 6 to 12. So the observer finds himself facing a pattern like this:
>
> $$1\ 2\ 3\ 4\ 5\ 1\ 2\ 3\ 4\ \ 5\ \ 6\ \ 7$$
> $$6\ 7\ 8\ 9\ 10\ 11\ 12$$
>
> What we have to imagine is that while it seems to our observer just as if the top line goes from 1 to 5 and then from 1 to 7, with no numerals missed out, it also seems to him just as if the line consisting of the boys' cards and the lower line of the girls' cards runs from 1 to 13, likewise with no omissions, and without any card having two numbers on its lower line. No such series of experiences can be imagined. (p. 21)

Two points arise. First, not every arithmetical equality which we accept is susceptible to this sort of completely surveyable application. So Craig's argument will not extend to every such equality; in some cases, the best possible perceptual evidence will not be 'the best possible perceptual evidence'. That, of course, is of no consequence in a context in which our concern is with the tenability of an absolutely *general* conventionalism about necessity. Secondly, however, there is at least a prima facie substantial question whether Craig is within his rights to interpret the notion of best possible perceptual evidence in such a way as to require that all the evidence be available to a single observer. Craig acknowledges the question:

> Have we not agreed that, provided we divide the observation of the counting experiment between a number of people, each one can easily be imagined to have the best possible perceptual evidence for events which, taken together, amount to something that goes against arithmetic? And if this is agreed, is the fact that the evidence has to be spread among different observers of any importance? (p. 21)

Rather than attempt to defend the legitimacy of the single-observer condition on the notion of best possible perceptual evidence, Craig proceeds to attack directly the coherence of the supposition that each of several observers could have the best possible perceptual evidence for a proposition which, taken in association with those attested by the others' observations, would complete a counter-arithmetical conjunction like *M*. The coherence of the supposition requires both that it seem to each observer that he is correctly reporting his part of the situation, and that it *continue* to seem that way to him, no matter how rigorous an investigation is conducted into the matter. Craig contends that the impression of coherence will founder, on examination, upon the horns of the following dilemma: *either* investigation will sooner or later bring something to light which calls into question the supposition that one in particular of the observers has the best possible perceptual evidence for his part of the conjunction, *or* ground will disappear for the belief that all the observers are perceiving the same spatio-temporal region. And, of course, if the latter happens, the situation is no longer apparently counter-arithmetical.

Generalised, Craig's strategy of argument is as follows. If, first, only a pair of observers are involved, then it must be possible on investigation to elicit a proposition, to the effect that an object of a particular sort is in a particular place, which seems to one of them true and to the other false. If this is how it goes on seeming to them, no matter what tests are carried out, then the supposition that they are perceiving the same region is called into question. (In fact, it would be more natural to doubt that both understood the proposition; but that conclusion has in this context the same value as Craig's, namely that the situation is no longer apparently counter-arithmetical.) The alternative is that one or the other, after various appropriate tests, begins to doubt his original report; again, then, there is no longer the best possible perceptual evidence for each constituent in the counter-arithmetical conjunction; for it has ceased to seem to one of the observers as though his part of it is correct. If, on the other hand, more than two observers are involved, then, in certain examples, it is possible that their reports are pairwise consistent. Nevertheless, even here, Craig contends, it must be possible to elicit a proposition of the same kind as before about which at least two of the observers will disagree, although at least one of them may have offered no explicit opinion on it in his report. The argument then proceeds as for the first case.

To sample the flavour of the argument and the kind of cogency it possesses, consider an example of the second sort. A, surveying the whole of a table, reports that it seems to him that there are exactly three glasses on it. B and C, each in a position to survey only opposite halves of the table, each report that it seems to him that there is exactly one glass on his half. Clearly, the reports are pairwise consistent. But it seems impossible to explain how it could seem to A as though there were three glasses in all without it seeming to him that there is more than one glass on one half of the table or the other. So he must be prepared to make a judgment which either B or C will dispute. The way to elicit the crucial proposition is then as follows. Suppose it seems to A that there are three glasses on B's half of the table and none on C's. Then A and C are in dispute whether there is a glass on C's half. Suppose, alternatively, that it seems to A that there are two glasses on B's half of the table and one on C's. Then let A for example place his left hand (as it seems to him) on one of the glasses on B's half and his right hand on the other; and let him ask B whether it seems to him that he, A, has a glass under his left hand and a glass under his right hand. If the answer to either question is 'no', then there is the sought-for proposition. Alternatively, if the answer to both questions is 'yes', then it seems impossible to understand how it can continue to seem to B just as if there were only one glass on his half of the table without its also seeming to him that A's hands are positioned differently to the way they seem to A. The proposition in question will then concern the location of A's hands.[1]

It is plausible enough to suppose that such a treatment can be extended to any example in which a counter-arithmetical conjunction is formed by a set of observer's reports, provided both that each observer had only to attend to a completely surveyable field, and that the fact that their reports collectively constitute something counter-arithmetical is itself surveyably appreciable. But, again, there would be no point in disputing this generalisation, or in stressing that, anyway, it will not be possible to extend the approach to *all* the arithmetical equalities which we accept. It suffices, in order to pose a challenge to conventionalism, that Craig's contention of the impossibility of any proposition standing to an arithmetical equality which we accept as M stands to $7 + 5 = 12$ can at least be demonstrated in a single case; and so, it appears, it can.

5.   But how much did the demonstration really establish? The situation was merely that it seemed impossible to imagine how due investigation could ultimately avoid locating a disagreement of the sort described. What, for example, if someone were to ask for *proof* that it could not seem to an observer that there were three glasses on a table without its also seeming to him that there was more than one glass on one half of it? Could we do better then throw the challenge back at him, inviting him, say, to draw a situation which seemed otherwise? Craig concedes that his argument has a kind of inconclusiveness. The argument is—

> of a kind that can of its nature never be stated fully. Its conclusion is that a certain situation cannot coherently be imagined, and the method has been to inspect some of the most immediately plausible attempts to imagine it. Necessarily one cannot in this way cover with certainty all the attempts that might be made, and it remains possible that some important approach has been overlooked. But while I cannot be said to have proved that the situation described in our original proposition (M) cannot be

[1] The way I have developed the example does not coincide exactly with Craig's treatment of it; but it is absolutely in keeping with his general approach—see pp. 23–7 of his paper.

imagined, nonetheless what I have written so far strongly inclines me to think that it cannot, and I believe that the techniques used in the last few pages would prove adequate to deal with any putative counter-example that may be proposed. (p. 27)

So it is, just barely, available to a sufficiently steely conventionalist to insist that no strict demonstration has been given of the impossibility of one of the relevant class of propositions.

That, though, would surely be quite the wrong response. For conventionalism, if it really is a correct account of the nature of necessity, must be capable of being recognised as such; so it cannot rely on an article of faith that something is possible which Craig's arguments strongly suggest is not. Rather it must supply an interpretation of those arguments which, as it were, brings them into the fold.

The most immediate tactic would be to attempt to discern in those arguments some implicit appeal to the arithmetic which we actually have. Then the apparent impossibility of propositions like M could be seen, in a conventionalistically acceptable way, merely as a reflection of that determination of the capacities of perceptual criteria for miscounting, etc., which, by our adopting the arithmetic which we have, has been effected. On the face of it, to be sure, there is no such appeal. Every arithmetical judgment made by an observer in the examples is described as having been arrived at by purely perceptual criteria. Still it might, I suppose, be contended that the appeal comes precisely at the point at which we are invited to regard a certain situation as *unimaginable*; for, the suggestion would be, there are no facts about what can or cannot be coherently imagined save as are determined by the body of statements which we accept as necessary. (Wittgenstein seems to come close to such a view; see, for example, III. 4–6). But this, it seems to me, is to ignore the specific character of the challenge which Craig's arguments pose. If we grant that we are so much the prisoner of the arithmetic which we were taught as children that we are quite unable to appraise any situation of the relevant kind without implicit appeal to it, then it is of course to be expected that we shall declare it to be unimaginable that a situation could persistently seem to be counter-arithmetical. For to seem that way *persistently*, no matter how thoroughly investigated, is to be so. So we are bound to say that that appearance must, sooner or later, break down; that it is unimaginable that it should not break down. What Craig's arguments appear to show, however, is that in some cases at least we can specify in advance exactly *how* the impression will break down. And that cannot be explained just by appeal to the idea that we are tacitly applying a rule whose consequence it is that the impression must break down somehow. So the feeling remains that there is more to the situation than this particular manoeuvre can account for; that, whether or not a verdict of the unimaginability of a counter-arithmetical situation can ultimately be substantiated only by appeal to an accepted arithmetical rule, Craig's pattern of argument will give a special content to the unimaginability of at least some such situations which could not issue just from such an appeal.

We shall get a better idea about how the conventionalist should, I believe, respond to these considerations by enquiring just what the source is of our confidence in the arithmetic which we accept. The answer, whatever pragmatically-inclined philosophers may suggest, has little to do with its general success in application; though we are, of course, totally confident that it will continue to enjoy such success. Rather, asked to justify the special certainty which we place in our arithmetic, we should naturally call attention to its susceptibility to *proof*. But what sort of proof exactly? Certainly, neither the possibility of a proof in axiomatic set theory, nor of a slog via the recursion for '+' in first-order number theory, lies at the source of our confidence in the necessity of $7 + 5 = 12$; the special unassailability which that statement strikes us as enjoying has nothing at all to do with its susceptibility to proofs of those sorts—for it struck us that way long before we had any inkling of such techniques. The correct explanation, for all arithmetical equalities of this degree of simplicity, has to do instead with exactly what Craig skilfully elaborates: our apparent inability to conjure to our satisfaction a situation which would counterexemplify them. The kind of reasoning which his argument illustrates, though he conceives of it as strongly suggesting a metaphysical conclusion about the objective potentialities of our perceptual criteria for the application of specific number concepts, is actually nothing other than the most fundamental sort of proof-strategy which we have for statements of this sort: a strategy grounded directly in the kind of application which we make of them. Any convincing proof of such statements *must* seem to show something about the potentialities of the perceptual criteria for counting correctly, etc.; if no such proof could be given, their status really would be comparable to that of any (consequences of) explicit conventions. But it is not. The intuitively natural response to a query about one of them is not, for example: 'That is simply what we *mean* by "5", "+", . . .', but: 'How could things possibly be *otherwise*?'

In brief: the invitation to describe how a proposition standing to any accepted arithmetical equality as M stands to $7 + 5 = 12$ might be true is, in effect, the invitation to depict a state of affairs which would counterexemplify that equality. For the truth-conditions of ascriptions of number, in contrast with those of ascriptions of length, have sufficient *observationality* to entail the following: if it is a possibility that, however thorough and painstaking our investigation, things obdurately seem in all respects as if a particular conjunction of ascriptions of number is true, then that is exactly the possibility that the conjunction *be* true. To take the invitation up successfully, therefore, would be to achieve a demonstration of the possible falsity of the arithmetical equality. So the necessity of $7 + 5 = 12$ and the impossibility of M, and of any relevant variant of it save M^*, stand or fall together. The central argument of Craig's paper thus consists of nothing other than a rehearsal of the most fundamental type of ground which we have for accepting elementary arithmetical equalities as necessary.

The kind of response which the conventionalist must make is therefore clear: he must dispute that proof of this kind is properly viewed as leading us to any sort of *discovery*. But that proof is ever properly so conceived is something which he has to be prepared to dispute in general; conventionalism is indefensible in any case unless the claim, that our admission of the necessity of a statement is never correctly viewed wholly as a matter of recognition, can be substantiated. So we are brought back to our outstanding question, whether some element of decision can always be discerned in our admission of a statement as necessary.

To interpret Craig's argument, then, as leading us towards the discovery of certain impossibilities of which the conventionalist can give no satisfactory account is tacitly to beg the question. For if the argument is rather to be interpreted as the conventionalist would propose to interpret any proof, then the intended significance of the 'discovered' impossibilities is lost. Of course, even if the argument can be so interpreted, that circumstance will make it no clearer how the users of an alternative arithmetic could avoid repeatedly postulating saving hypotheses which were incapable of independent substantiation; or what sort of notion of a reliable single count they would be in a position to employ. But that these matters are unclear will be viewed by the conventionalist as no more than reflection of our inability to make clear to ourselves how a situation might persistently seem to be counter-arithmetical, no matter how rigorously investigated. Nothing here will constitute the *discovery* that such people, in the essential nature of things, will be led to misdescription of contingent fact. We can, to be sure, in effect *prove* that they will; the proof just consists in the most fundamental kind of demonstration which we are able to give of the soundness of our arithmetic. But, for the conventionalist, this proof, as indeed does all proof, lacks the sort of metaphysical potency which would empower it to make possible our recognition that our arithmetic is the only one which, for people who count as we do, can avoid wholesale misdescription of the world. So the antecedence doctrine survives: it remains the case that there is no *recognising* that the use of any alternative arithmetic will be bound to require false contingent beliefs; no recognising, therefore, that our arithmetic uniquely codifies the truth-value liaisons between contingent judgments of number. There is, indeed, no recognising such things; that is the whole point. We are no less entitled to assert them, but it is an error to see that entitlement as more than a reflection of the manner in which *we* have determined our concepts of the capacities of the perceptual, pre-arithmetical criteria for the acceptability of contingent numerical judgments. If, as Craig makes plausible, we are unable to conceive of how any alternative determination might be viable, then that is how things are with us; it is a further, tendentious step to inflate our imaginative limitations into a metaphysical discovery. Craig's kind of argument, properly interpreted, is to be seen rather as leading us to a 'grammatical' decision.

6.   Everything comes to depend, then, on the defensibility of a generalised conventionalist conception of proof. But it is worth noticing that that conception has a special plausibility for the sort of reasoning with which we have been concerned. The reasoning takes the form of challenging ourselves, if such-and-such is to be thought of as a possibility, to describe how the possibility might be realised in a particular situation; and then we find that all our best efforts to meet the challenge seem to break down—something emerges each time to prevent us regarding what was described as having the appropriate status. But it is one thing to find, quasi-experimentally, that this is how things always seem to go, and that we can form no conception of how they might go otherwise; quite another to elevate to the status of necessity the claim that the putative possibility is spurious. Could we not instead keep an open mind? If we do, is there clearly some *fact* about whose acknowledgment we are being unduly cautious? At any rate, our efforts to respond to the challenge give us no guarantee that we shall not sometime in the future want to describe some situation in terms of the form of words whose truthful application to any situation we are now on the verge of proscribing. If that came about, would we all then be *wrong*?—or would we have had to have changed the meanings of certain of the expressions in that description? What does either account really amount to?

It is unclear that the recognitional conception has any satisfactory answers to these questions. And it is this unclarity which gives a certain plausibility to the conventionalist's opposing picture: the effect of our investigation is that we lose patience with the alleged possibility; if caution was on option, we throw caution to the winds and determine that nothing is to count as a realisation of the description in question. We plunge into a new rule.

Whether such a conception can be sharply distinguished from its cognitivist counterpart, and whether it is superior, are questions with which we shall be concerned in the next chapter. Let me conclude here with an observation on a different point. The conventionalist has to face the question: if necessity is not a matter of recognition, what account should be offered of why there are any necessary statements? The best answer we have so far been able to come up with is that based on the 'transcendental' argument of the preceding chapter. If that argument is sound, then, provided a language is sufficiently rich, it is bound to furnish means for the construction of sentences to which an absolutely normative rôle will be granted. But, of course, as far as the production of *discoveries* is concerned, philosophical reasoning of that sort is no better placed than logic or mathematics. So, from a conventionalist point of view, the effect of that reasoning, if it is accepted, is to be described not as leading to any discovery but rather as further determining our conception of the character which a sufficiently rich language must have. And once it is viewed like that, it may seem questionable whether it can provide the kind of intellectual satisfaction which the original question was seeking. Besides, the argument is only of a 'thin end of the wedge' type; the range of statements which we accept as necessary far exceeds the class to which reasoning of that sort seems applicable. So what answer should be given?

Craig complains that arithmetical conventionalism—as he understands it—pays insufficient heed to the 'phenomenology of perception'.

> The consideration of the experiences of perceiving groups of objects is crucial to any explanation of the status of arithmetic in our thinking. Conventionalism tries to get by without it, and that, ultimately, is why it fails. (p. 28)

I believe that Craig is fundamentally right in the contention that considerations of that character are indeed crucial, and that the appropriate answer for the conventionalist to give to the question must therefore involve due stress of them. But once we see that such considerations merely embody the most fundamental sort of proof which we can give of elementary arithmetical equalities, this answer, generalised, simply amounts to referring the question back to whatever we conceive as our most fundamental grounds for accepting the necessity of any particular statement. So the answer the conventionalist will give in any particular case is exactly the same as that of his opponent; he will invite us to scrutinise whatever grounds we have for accepting the necessity of the statement in question. And if it is thought that such an answer can no longer have the relevant force, once a cognitivist conception of proof is rejected, then there is no answer. The best that can be said is: that is what we do.

XXIII

Deciding

Sources

The 'geometrical' cogency of proof:
RFM: II. 38, 59–60; v. 36, 51
BGM: VII. 74
LFM: lecture VII, p. 70.

Proof and certainty:
RFM: I. 67, 154; II. 78 (p. 101, 'sleep-walking' passage),
      86–7
 OC: 26, 39, 43–51, 77, 212, 217, 643–5, 650–8
LFM: lectures VII, p. 72; xv, p. 148

1.   We come finally to the question, then, whether there is invariably an element of decision in our judgments of necessity. Three things ride on the warrantability of an affirmative answer. The tenability of the central conventionalist thesis, that necessity is no species of truth, and the possibility of handling along the lines sketched at the end of the preceding chapter the sort of phenomenological objection raised by Craig, are two of them. The third is the status of Wittgenstein's conception that proofs are essentially instruments of concept-determination.

The central difficulty confronting sympathetic interpretation of the concept-modification thesis has all along been that, generally speaking, when a proof is accepted, nothing occurs which requires us to modify the pattern of explanation which we should antecedently have given of those concepts to which successive stages in the proof make appeal; the concepts, that is, possession of which, as we should ordinarily think, enables us to *recognise* the validity of the proof. This is not invariably so; but it is typically so. Where a proof is intuitively of the most compelling sort, this is exactly because it seems to us to contain in essentials no novelties; each constituent step will strike us as being of a previously acknowledged pattern. It is true that the rule-following considerations supply a sense in which, consistently with this, each constituent judgment in our overall judgment that the proof is valid may nevertheless be regarded as novel; but that notion of novelty fails to discriminate between acceptance of necessity and acceptance of contingent truth, and ought therefore to provide no motivation for a *special* thesis about proof.

There is obviously a major obstacle to any conventionalist conception of proof here. It was because of this that I proposed (in Chapter VI) that Wittgenstein's ideas about proof and concept-modification stood in need of re-expression: Wittgenstein should have said not that proof *changes* the 'grammar' of our concepts but that there is no objective sense in which accepting any particular proof leaves all the concepts involved just as they were; this repudiation of objectivity being, admittedly, just what the rule-following considerations require for absolutely every judgment which we make. However, if deciding could be made out to play the part which the conventionalist needs it to play, then it would be possible to do better than this. It would be possible to reinstate Wittgenstein's thesis in something like the terms in which he habitually expressed it. The modification of concepts involved in our accepting a proof would be exactly comparable to that which would be occasioned if we were simply to adopt as an implicitly definitional postulate that statement which the proof proves. But the trouble is that if our acceptance of a new proof really does occasion no modification in the kinds of explanation which we should consider appropriate for the various concepts to which the proof makes appeal, that looks near enough sufficient to *constitute* the circumstance that no such element of decision is involved.

It is clear, at any rate, that if there is a defensible conventionalist account of proof which somehow steers round this obstacle, it will have to meet the following pair of constraints. First: the element of decision to whose universal rôle in our acceptance of proofs it will call attention must be such that to take a decision of this kind is not to do anything which will require re-explanation of the concepts in play in the proof; the decision involved must be one which people can be expected to take, or assent to, on receipt of a normal explanation of those concepts. Secondly: the element of decision called attention to must be unparalleled by anything which we do when we accept contingent statements. Otherwise, merely to call attention to such an element of decision is so far to do nothing which warrants the conclusion that our acceptance of necessity is non-cognitive in such a way as to render inappropriate the ordinary view that necessary statements are a sub-class of true statements, as much susceptible to recognition as are true contingent ones.

2.   It might appear to be an essential third constraint on a defensible conventionalist account that it provide some means for avoiding the trap into which radical conventionalism stumbled: to attempt to conceive of our acceptance of each new necessary statement as the product of a quite unconstrained decision is to sign away the capacity of necessary statements to function in just the normative manner which the

conventionalist thinks is distinctive of them. Applying any convention is irreducibly a matter of inference; we have to be able to agree about what its requirements are in a particular situation, to agree about what follows from it. But the statement that such-and-such is a consequence of a particular convention will itself be, where acceptable, necessary; so if the acceptability of every necessary statement requires an arbitrary decision, the capacity of the original convention to inform our practice is lost. There is, for example, no substance to our agreement about the rules of a game if further, arbitrary decisions are needed at every turn in order to determine what moves are, or are not, consistent with them.

Actually, it is worth while taking a little more trouble over this difficulty; for, as presented, it might be thought to slide over a relevant distinction. There is no question but that a convention, if it is to inform our practice, must be able to do so without repeated arbitrary decisions about what its requirements are at any particular stage. But that, it might be thought, does not immediately entail that, in order for a convention to be able to play that rôle, its subscribers must be able to agree explicitly about what its consequences are. In order for a convention to function normatively, is it not enough that its users have a communal conception of what circumstances do, or do not, constitute following it, and a communal intention to do nothing of the latter sort? That is less than saying that they have to be able to agree in *articulate* judgments about whether or not a particular 'move' is consistent with it.

What would it be, though, for it to be a matter of arbitrary decision how to *describe* the requirements of a particular convention in a particular situation, without its being comparably arbitrary just what kind of action was required? If the communities' agreement *in* the kind of action required is to be in no way miraculous—in the sense in which our mutual agreement about every nth place of a series of free choices would be miraculous—but is genuinely informed by the convention, how could it nevertheless be arbitrary whether they should *describe* a particular move as conforming with the convention? Of course, it is possible that the community may lack the linguistic resources to make articulate judgments about what their conventions do, or do not, require of them; but that supplies no sense in which a particular, explicit verdict can be conceived as arbitrarily given—it merely means that they are in no position to give a verdict at all. So provided that the community does possess linguistic resources adequate for articulating statements concerning the consequences of certain conventions, the acceptability of a particular such statement can no more be a matter of arbitrary decision than is the actual character of the behaviour which conformity to a particular convention requires.

In any case, there is, I believe, something suspicious about the idea that there could be people whose behaviour in a particular region was regulated by certain conventions, although they lacked the linguistic resources to formulate explicit judgments about the requirements of those conventions. Consider, for example, a conditional articulating the requirements of the rules of Chess for a particular position; say, 'If Black is to comply with the rules here, he must move the Queen to KB_1 or resign.' Now—to take the most radical kind of case—it does seem superficially conceivable that creatures might play Chess, cats for example, who have no language at all; at least in the sense that, for an observer, the most natural explanation of their behaviour would be that they were playing Chess. But this appearance, arguably, is only superficial. Lacking any language, the animals would not be able to negotiate the admissibility of a move if disagreement arose among them; nothing which they could do would properly be describable as an appeal to the rules. But in claiming that they were playing Chess, the observer would be saying more than that the rules of Chess, indicatively stated, seemed to provide satisfactory predictive hypotheses for this aspect of the animals' behaviour. He would, to repeat, implicitly be saying something about their *intentions*; for the rules of Chess to function as rules, it is necessary that players intend to keep by them. But what is it to have such an intention? What is required, centrally, is a readiness to accept departure from the rules as a ground of legitimate criticism of one's behaviour; and it is by no means clear how entirely non-verbal behaviour could provide evidence of such an attitude. What is needed, if the hypothesis that the cats are playing Chess is to be substantiated, is the capacity on their part to do something which amounts to an invocation of the rules; and, plausibly, only if they are equipped to make articulate judgments, like the conditional cited, about the requirements of the rules can anything properly described as such an invocation take place. There is thus a large question whether to describe such creatures as playing Chess would not be wholly an underdetermined, anthropomorphic account, the intentions whose attribution it would presuppose being capable only of linguistic manifestation. Indeed, even if we fantasise a measure of linguistic proficiency for the cats, no relevant difference will be effected unless it extends not merely to the capacity to articulate the rules of the game, but to describe situations which arise in the course of playing it, and explicitly to assess them for conformity to, or breach of, the rules; otherwise, the question whether they are playing Chess, rather then merely behaving in a way predictable by reference to the rules of Chess construed as descriptive hypotheses, will still lack determinacy.

In general, if the capacity to make articulate judgments about the requirements and consequences of rules is missing, it is unclear how there can be a sufficient basis in the behaviour of prima facie participants in an activity for supposing that its rules are genuinely *normative* for them at all. Such articulate judgments, however, are not in general contingent statements. We noted, to be sure, when we discussed this matter in Chapter XIX, that conditionals like that above about Chess are strictly contingent because of the demonstrative reference, 'here', to a particular position. But that could be parsed out in favour of an explicit citation of the crucial features of the position. The resulting sentence will then be a pure theorem of Chess, and it will be in effect a demonstration of this theorem which we shall supply if Black chooses to contest the matter. In any case, no such reservation is appropriate about the statement that the theorem in question *is* a consequence of the rules of Chess and such an explicit citation; and agreement about what this statement affirms is equally essential if an appeal to the rules of Chess is to serve to assess the position. So there seems to be no alternative to a full endorsement of the original objection to radical conventionalism: if a necessary statement is to function normatively for us, we must be capable of non-miraculous agreement about what accepting the statement as necessary requires of us in particular situations; and this agreement must be capable of expressing itself in our assent to statements, themselves necessary, articulating the consequences of the statement in question.

Now, it unquestionably is a constraint on a satisfactory conventionalist account of necessity that it should avoid this fatal objection to the 'radical' version. So the required element of decision must be located and characterised in such a way as to leave uncompromised the ordinary, practical likelihood that we shall agree about the validity, or otherwise, of new chains of argument. Plausibly, however, an account which meets the first constraint proposed above will meet this constraint also. For if the account succeeds in explaining how it is that, when we accept a proof, the decision involved need nevertheless occasion no re-explanation of those concepts in terms of which the proof is assessed, that can only be because the account is able to make room for the idea that it is exactly our ordinary understanding of those concepts which guides us towards accepting or rejecting the particular proof; and once room for that idea is provided, the danger that the decision involved will have to be construed as arbitrary, and the objectionable consequences of such a construal, will be averted.

3. When the recognitional conception is taken to involve that a *special* certainty is enjoyed, at least in a wide class of fundamental cases, by our apprehension of necessity, it is easy enough to call it into question; rather harder to see how it might actually be demonstrated to be wrong. One way of calling it into question was illustrated towards the end of Chapter XVII: all we have the right to be certain about, it would be urged, when presented with a calculation on paper, for example, is that everything *appears* to be in order; all we have is an ordinary empirical certainty that an apparently correct computation of, say, the product of a pair of numbers yields such-and-such a result. But is there not room for a *sceptical* doubt whether we might not have failed to detect some error? Slips and confusions do on occasion occur. What, then, is

the warrant for the absolute certainty, which we profess at least when calculations are sufficiently simple, that if the operation in question is correctly performed, that *must* be the result? Might we not, on waking as it were, find it absolutely incredible that we should have been so deceived?

Of course, this scepticism, as much as many other varieties, is subject to the reproach that nothing would count as allaying the doubt which it professes to raise. But its rôle in this train of thought is only dialectical—there is no suggestion that it cannot be rational to accept any calculation before an adequate rebuttal is devised. The suggestion is merely that if all we have is ordinary, empirical certainty about the features of the actual construction on paper with which we are presented, then the certainty which we tend to go on to profess about what *must* happen when that particular construction is carried out correctly is disproportionate. So there must be some other element involved in that certainty; and the conventionalist suggestion is precisely that this other element is a (well-motivated) decision, a decision to count no situation as correctly described which, if so, would constitute a counterexample to the connection between the correctness of the computation and its outcome about which we are presently certain.

But, of course, these considerations only commend a picture. They are powerless against someone who is prepared to insist that he simply *discerns* that that connection obtains. That, if the certainty professed is not to be disproportionate, there must be some additional aspect to the situation, beyond empirical certainty about the features of the particular construction, need not be contested; the recognitional conception will maintain—traditionally has maintained—that the additional aspect, the source of our absolute certainty, is just what our cognitive sensitivity to internal relations furnishes. We just are able to recognise that no other outcome is possible, provided the calculation in question is carried through correctly. And whether the operations of this cognitive capacity which we have are in turn susceptible to a kind of scepticism is, though doubtful, irrelevant.

A defender of the recognitional conception need not, indeed, go so far. It is open to him to disavow the absolute, or near-absolute, certainty traditionally claimed: to agree with Wittgenstein that

> . . . one cannot contrast *mathematical* certainty with the relative uncertainty of empirical propositions. For the mathematical proposition has been obtained by a series of actions that are in no way different from the actions of the rest of our lives, and are in the same degree liable to forgetfulness, oversight and illusion. (*OC* 651)

That, however, the claim to have recognised that the cat is on the mat is defeasible in the kind of circumstances on which a sceptic may dwell in no way compromises the cat's situation as a possible object of cognition. Why should it be any different with our knowledge that $121 \times 74 =$

8954? In order to be entitled to the view that what must happen when the product of 121 and 74 is correctly calculated is something which we are capable simply of recognising, it is unnecessary to claim that such recognition has a certainty surpassing that possessed by our recognition of empirical fact.

How should we proceed here? Clearly, we need a sharper formulation of what it is for the acceptability of a statement to be purely cognitive in the way in which the conventionalist believes our acceptance of necessity is not. A natural move would be to look to Dummett's contrast between genuine assertions and *quasi-assertions*: a declarative sentence, it will be recalled, is merely a quasi-assertion if it is not associated with determinate conditions of truth and falsity, but is such that assent to it commits the assentor to some definite course of linguistic or non-linguistic action. So characterised, there is nothing in the nature of a *quasi-assertion* to prevent its being associated with certain quite definite conditions under which assent to it is, or is not, *appropriate*; what is precluded is that such conditions of appropriate assent actually be conditions of truth also. On the face of it, though, the distinction between genuine- and mere quasi-assertions is of no direct help to us. To be sure, the question seems to be: is an endorsement of the necessity of a statement rightly regarded merely as a quasi-assertion? But no method is apparent for eliciting from Dummett's account a way of effectively resolving the issue in controversial cases, like ethical statements or those which presently concern us. We have some idea what has to *be* decided; but we do not know how to decide.

Still, let us persevere. To begin with, some modification is needed to Dummett's account if the distinction in question is to stand up from the point of view of a generalised anti-realism. From that standpoint, the reference to 'determinate conditions of truth and falsity' can no longer be generally apt; for we do not want to legislate all statements out of the class of genuine assertions for which the notions of truth and falsity would be problematic from an anti-realist point of view. The obvious modification is to replace that reference by: 'determinate conditions of warranted acceptance and rejection'. But how, in that case, are *genuine* assertions to be distinguished from those quasi-assertions for which we do have a notion of conditions under which assent to them is, broadly speaking, appropriate? There is a danger here that we shall get too easy a settlement in favour of the genuinely assertoric character of ethical and mathematical discourse. To avoid it, we need to impose some further refinement upon the loose, intuitive notion of the conditions under which assent to a particular declarative sentence may be appropriate. A natural proposal would be along these lines:

A declarative sentence expresses a genuine assertion if it is associated with communally acknowledged conditions of acceptability in such a way that a sincere unwillingness to assent[1] to it when such conditions are realised, and the agent is in a position to recognise as much, convicts him either of a misapprehension about the nature of the circumstances presented to him or of a misunderstanding of the sentence.

The point of the proposal is to capture the intuitive idea that, when we are concerned with genuinely fact-stating discourse, the rightness or wrongness of assent to a particular statement in certain circumstances is in no sense *optional*, or up to the individual. It is a reservation on this score which underlies the feeling we have that, for example, whether or not a particular situation is funny does not, as it were, reside in the situation—that there is no 'fact of the matter'; for if one person finds a situation humourous and another does not, neither need be subject either to a misunderstanding of what it is for something to be funny or to a misapprehension about certain relevant features of the situation. Whether the same holds for ethical, or aesthetic, judgments, is something which here we shall do well to resist considering.

This proposal looks to be disastrous for conventionalism. For the ordinary conception of valid proofs as cogent seems precisely to involve that error, or a lack of understanding (misunderstanding), are the only admissible categories of explanation for someone's inability to agree with us about a valid proof. How, then, can the judgment that it is valid be a mere quasi-assertion, and how can deciding have any essential part to play in our acceptance of that judgment? Of course, there are cases where it is natural to describe someone as deciding to accept a statement which we should ordinarily regard as, if correct, necessarily so. A Chess player may in a sense *decide* that his best strategy in a particular position is to move the Queen to KB1. But if deciding is meant to contrast with recognising in that kind of example, then it has a part to play precisely because, for whatever reason, an absolutely rigorous proof (analysis) is unavailable, and a view of some sort needs to be formed. If the player had an exhaustive analysis of the position, deciding would not come into it, it seems.

Any trained person faced with a, by ordinary criteria, sound and not too complicated proof will surely find it enormously natural to describe himself as *driven* along, willy-nilly, step by step. If the proof really is rigorous, no alternative will be apparent at any stage of it than to grant the soundness of the immediately preceding step. And if we cannot get someone, willing in principle, to agree with us about it, there seem to be no categories available in terms of which his behaviour could be accounted for save misapprehension of what is actually being presented to him, or misunderstanding of some of the key concepts. The strategy

[1] In order to avoid misunderstanding the point of this proposal, the reader should reflect that one's assent to a statement need receive no public expression; so the sort of social considerations which might generate a sincere reluctance to publicly express one's assent are not in the picture.

for remedying such a situation is therefore clear, even if we have no guarantee of its success in any particular case: we must patiently work through the detail of the proof with him and, if necessary, patiently re-explain whatever he has misunderstood or failed to grasp. Naturally, it can happen that somebody may persistently fail to respond to such efforts; some intellectual shortcoming, or even perceptual abnormality, may stop him grasping the proof. But that the corrective strategy, corresponding to our two categories of explanation of his inability, may fail does nothing to suggest some *third* category, beyond misunderstanding or misapprehension, in terms of which we can explain his performance. The situation seems, in that respect, to be exactly comparable to that of any decidably true contingent statement. There is, therefore, so far apparent no plausible sense in which the endorsement of new necessary statements can in general be thought of as conventional.

4.   If the necessity of a wide class of statements is genuinely recognitional, can we do other than invoke a special intellectual faculty enabling us to effect that recognition? That may sound an objectionably mysterious proposal; and it may seem, besides, that the rule-following considerations have already disposed of any such fiction—for do they not dispose of the purely conceptual *facts* sensitivity to which would be the essential manifestation of such a faculty? The proper answer is that they no more do so than they dispose of those facts accessible via sensory perception. To repeat: what, if accepted, those reflections force us to jettison is one central aspect, viz. the belief in *investigation-independence*, of the way we ordinarily think about the objectivity of what we are capable of discovering. But a purified notion of what it is for the correctness of a statement to be susceptible to discovery will survive, tied to what distinguishes the class of genuine assertions within the wider class of declarative sentences. In terms of the account of that distinction proposed above, for example, a discovery will have been made precisely when considerations in relation to a certain statement have been brought to light of such a kind that only a man's failure to understand the statement or his ignorance, or misapprehension, of those considerations can explain his sincere unwillingness to assent to the statement's correctness.

If necessity is amenable to discovery in such a sense, then postulation of a special, necessity-sensitive faculty can be avoided only if we can suppose that our ability to recognise necessity can be laid at the door of other less mysterious-seeming faculties which we undoubtedly possess. Our perceptual faculties, however, will not serve. Indispensable as they are in enabling our recognition of proofs, it is their province only to inform us of the sensible properties of the particular construction with which we happen to be confronted; and recognition that a construction conforms at all stages to a certain antecedent specification and has such-and-such an outcome falls short of recognition of necessity just in the way that recognition of an *experiment* that it conforms to a certain set of initial conditions and has such-and-such an outcome falls short of recognition of necessity. A recognitional conception of necessity which holds that recognition to be purely perceptual will thus be incapable of explaining the difference between a proof and an experiment, and so of explaining our alleged recognition of necessity at all.

It is justified to regard a putative faculty as mysterious only if there is some essential unclarity about what its possession enables one to do, or if it is unclear that the capacity in question could not be bestowed by other faculties of a previously acknowledged sort. So if it is right to regard necessity as recognitional, and if—for the reason just advanced, or others—we decide that this ability of recognition cannot be subsumed under our purely perceptual abilities, no alternative is apparent but to picture those aspects of our linguistic practice in which recognition of necessity is involved as manifestations of a faculty of the relevant special kind. I ought to stress at this point, however, that no more than a *schematic* use will be made of the idea of such a faculty in what follows; we shall employ the term merely to designate whatever capacity it is which enables us to recognise necessity—if recognise it we do.

We can now make the following proposal: what the conventionalist essentially needs to be able to demonstrate is that the hypothesis of such a faculty is at bottom *empty*; that nothing which we are actually able to do need be regarded as a manifestation of it. Of course, that just means that he has to be able to demonstrate that nothing which we are actually able to do need be regarded as *recognition* of necessity. The proposal provides no short cut around the prima facie case which has been made for regarding necessity as recognitional, as a species of discoverable truth. Some sort of fault will still need to be found with that case. The point of the proposal is merely to refine our idea of where the onus of proof lies: the conventionalist does *not* need to demonstrate that the position of a 'hard-line' cognitivist is actually incorrect—that necessity cannot be a matter of recognition; it suffices if he can demonstrate that there is no *need* to hold that position. This is just to apply a principle acknowledged in other areas. However striking, for example, the results of a series of experiments in telepathic communication, we feel that no evidence has been provided for supposing that at least some human beings do possess special powers of that sort unless we are confident that no more mundane explanation can be provided of the experimental findings.

The conventionalist, then, must demonstrate that there is a more 'mundane' way of explaining our capacity to agree about new necessary statements than to suppose that we are able to recognise that that is their status. And if he can do so, then he will of course have solved the decisive difficulty for radical conventionalism. His first task, however,

is somehow to undermine the suggestion, issuing from the above proposal about how assertions proper are to be distinguished within the
class of declarative sentences, that there is no alternative to regarding
necessary statements as coming within the former category.

5.   In part II of *RFM*, Wittgenstein talks of the cogency of proof as
depending on its 'geometrical' cogency. He has in mind the case of a
totally formalised proof, consisting of an ordered array of signs, whose
validity is a correspondingly formal notion concerning solely the
character of the transformations among the signs. But the essential
point extends to any proof consisting of a series of distinct, characteris-
able steps. It is that the success of any such proof depends upon its
establishing a *conditional*, which may or may not be its conclusion—it is
so in the case of an arithmetical calculation—to the effect that if such-
and-such transformations (here we give a description of successive
steps) are carried out on such-and-such an initial basis (here we cite
the axioms, or assumptions, or objects on which the reasoning, or
operation, goes to work), then such-and-such will be the result. The
conditional, to take a trivial example, so corresponding to the natural
deduction inference from '*A*' to '*A* V *B*' could be: if a single step of
Disjunction Introduction is applied to '*A*' as premiss, the result will be a
disjunctive statement of which '*A*' is the initial disjunct. In general, our
judgment that something is a valid proof is simultaneously a judgment
about the necessity of such a corresponding, descriptive conditional—a
conditional articulating how the proof gets its conclusion. We have no
proof if we merely have goodish grounds for saying that the conditional
is true; rather the construction has to give us the best possible grounds
for saying that the particular (type of) conclusion is what we *must* get if
the antecedent clauses of the conditional are genuinely implemented.
This is not to claim, of course, that we always explicitly consider such a
conditional, or assess the proof with it in mind. The point is that a *doubt*
about the necessity of such a conditional would be a doubt about
whether the proof had perhaps achieved its outcome *fortuitously*; which
is just to doubt whether it really is a proof. (There are several passages
getting at this idea between II. 20 and II. 40.)
    Let us concentrate on the case where we are concerned with a proof
consisting in a linear array of sentences in a formal symbolism; the
antecedent clauses of the relevant corresponding conditional will then
describe the assumptions on which we start and how successive lines are
achieved. Surely, now, there *is* a way in which someone can withhold
assent from such a proof without showing that he is mistaken about, or
has failed to apprehend, certain aspects of its 'physiognomy', or has
failed to follow or misunderstands it. In order intelligibly *not* to accept
such a proof, a person does not have to do anything which will indicate a
lack of understanding or misunderstanding, or a mistake. A quite

different possibility will be his adoption of what seems superficially an excessively, perhaps neurotically, cautious attitude to the thing. We get, for example, this sort of response:

> I can find no fault with the construction; it seems to me that all the steps are sound in just the ways that your descriptions of how they are achieved require, and that we appear to have wound up with a proof of just what you set out to prove. Repeated checks have served only to confirm these impressions; and I accept that further repetitions would amost certainly turn out the same way. However, you are asking me to affirm that whenever exactly the specified sequence of transformations is correctly followed through on exactly the specified basis, we are *bound* to achieve this (sort of) result—that no other (kind of) outcome is *possible* provided the blueprint is correctly implemented. And that very strong claim, I feel, I am not entitled to make.

A natural immediate response is that this attitude is not really a third kind of possibility, but is rather a special case of misapprehension. Surely, someone who took such a line would have *had* to have failed to grasp certain essential aspects of the proof; he would have had to have failed completely to follow it—or he would simply *see* that such-and-such is the result which we are bound to get. But suppose he is able to convince us, while still professing this attitude, that he knows perfectly well how to carry through the proof for himself; what then does his having failed fully to follow it consist in? It had better be something other than his simply taking the attitude in question; for it is the idea that following a proof involves recognition of an essential connection between basis, process and outcome which is under examination.

In fact, there seems no obvious incoherence in the idea that someone could be a perfectly able mathematician and yet consistently take this line about proofs, accepting their status as proofs and their conclusions just for the sake of doing mathematics, as it were. And it would so far be quite tendentious to suppose that he must be the victim of some sort of misunderstanding or misapprehension. His working criteria for the correctness of proofs could be the same as everybody else's; he could be willing to accept on the basis of any particular proof the same conclusion as everybody else; and, most important, he would be willing to assent to the corresponding descriptive conditional in any particular case. But he would refuse to regard the capacity of a proof apparently to survive arbitrarily close scrutiny as a sufficient ground for saying that such a conditional expresses an *essential* truth; his acceptance of proofs would always be provisional in the way our acceptance of any hypothesis is provisional—proofs would never impress him as possessing the special cogency with which they impress other men.

But, someone may want to enquire, what are this Cautious Man's reservations *about* exactly? Is it that he thinks an error might have escaped everybody's attention? And if so, how is his doubt even intelligible when we are concerned with the very simplest sort of deductive transition which we allow? For, faced with such a step, it seems quite unclear how to attach sense to the idea that we might somehow discover some sort of mistake in it.

Certainly, if this attitude is coherent at all, it must be possible to sustain it with respect to the very simplest ingredient steps of a proof. Suppose we are confronting, for example, a Conjunction Elimination step: the move is from '$(P \lor Q) \,\&\, (P \lor R)$' to '$P \lor Q$'. Then grasp of the validity of the step, we would ordinarily suppose, issues in and depends upon a willingness to accept the necessity of the conditional:

> If '$(P \lor Q) \,\&\, (P \lor R)$' is taken as a premiss, and a single Conjunction Elimination step is correctly made, then the result will be '$P \lor Q$' or it will be '$P \lor R$'.

So the Cautious attitude must, it appears, involve some sort of reservation about this conditional; but what sort of reservation can be made *intelligible?*

Suppose our man responds like this:

> It is not that I have any doubt that if I carry out a Conjunction Elimination on that premiss, and am satisfied that I have done so correctly, the result which I have will always square with the conditional. Indeed, I grant that I cannot imagine what it would be like for it to *seem* to me that I had correctly carried through such a step on that premiss while it simultaneously seemed to me that I had some other outcome than one of the two prescribed. Still, I stick to it that my confidence that that is how it would always be, and my inability to imagine that it might even seem otherwise, fall short of an entitlement to conclude that it is in the *essential* character of correct implementation of a Conjunction Elimination step on that particular premiss to issue in one of the two described outcomes, that the conditional is, as you seem to want to insist, *necessarily true*.

This may still seem not to be a fully intelligible attitude. But is the main reason for that not simply that everything has now been conceded on which our conception of the essential soundness of the step is based? It is granted to be certain, as a practical hypothesis, that the relevant conditional will be borne out by any amount of future experimentation; and it is granted that an imaginative inability is associated with the example—our man lacks the faintest idea how to construct even the *illusion* of a sound Conjunction Elimination step on that premiss with neither of the prescribed outcomes, still less how to give sense to the idea that such an impression might be no illusion. So the situation merely poses the following dilemma for the recognitional conception: either the Cautious attitude grants everything on which our so-styled recognition of the essential soundness of the particular step is based; or it does not. If it does, then in order to defend the recognitional conception, it has to be shown that the Cautious attitude is not genuinely coherent; it has to be shown that to grant so much is already to recognise the necessity of the conditional, and that no more conservative standpoint about its correctness is then left open. It is unclear how such a demonstration might proceed. If, on the other hand, we deny that a well-founded practical belief that any subsequent trial would go as the conditional describes, and a certain imaginative inability, do exhaust our epistemic situation when we confront the particular step and are persuaded of its soundness, what element in that situation has been left out of account?

On a phenomenological level, so to speak, it is, I think, unclear what more there is to being completely convinced by an articulated proof. To seem to oneself fully to understand a proof is both to believe that one can recognise as such a conditional corresponding to it in the manner described and that one's best efforts to reproduce the proof will always square with that conditional; and it is essential to the impression of full understanding that there should be a special unimaginability about the idea that the conditional might actually be counter-exemplified. Perhaps we do not, in this sense, fully understand all the proofs whose validity we are prepared to grant; but such is certainly the situation with sufficiently simple cases, and it seems to be the whole of the situation.

In summary, then, the suggestion is that it is possible for someone sincerely not to accept the validity of what we regard as a valid proof without his suffering from any independently identifiable misunderstanding of any of the concepts involved or from any independently identifiable misapprehension of its structure. What he does is grant the apparent correctness of all the steps; grant that it is indeed imaginatively obscure how he might come to revise that assessment; grant that there is every reason to believe that whenever the proof is reproduced in a satisfying way, it will lead to the same outcome; *but* dispute that there is anything in all that which justifies him in claiming to have apprehended any essential connection between basis, process and outcome—in claiming, indeed, of any statement in the vicinity that it 'cannot but' be true. So the proof is not accepted *as* a proof in what, for the recognitional conception, is the intuitively relevant sense; rather, it is accepted as a well-conducted experiment with a distinctive phenomenology which makes it seem that repetitions of the experiment will go the same way. Someone who accepts a proof only in this spirit need do nothing to betray a specific misunderstanding or misapprehension—unless we beg the question, regarding it as sufficient to convict him of one or the other that he refuses to regard the proof as bringing to light an essential conceptual connection.

The Cautious attitude can be adopted with regard to any particular statement (save, naturally, an explicit convention)[1] which we regard as necessary, whether or not we do so as a result of proof. The Cautious Man will grant the acceptability of the statement, and grant that whatever prompts us to accept the statement as necessary—including appeals to imaginability and the like—weigh with him too; but the upshot is only that he accepts the statement as well founded. There is, to be sure, some unclarity therefore about what it is that he is unwilling to do. For if he is, or can be, prepared to *assert* the statement in question, and regards himself as justified in doing so, then is that not tantamount to his having apprehended a necessity, whether or not he realises it? The truth is that the unclarity here actually works against the recognitional conception. An adherent of the recognitional conception ought to hold that there is a perfectly respectable statement which the Cautious Man is not prepared to make: the statement, namely, that the statement in question—the descriptive conditional, or whatever—is necessary. The question is therefore: what bearing upon the *use* of that conditional, or whatever, does an apprehension that it is necessary have? Whether or not Wittgenstein's notion of normativeness has any part to play here, the dialectical point is that *if* the recognitional conception can come up with a satisfactory answer, then the contrast between the kind of use to which it will have succeeded in pointing—what I shall dub *Nec-use*—and the kind of use distinctive of well founded hypotheses will supply the answer to the question, what is it, substantially, that the Cautious Man will not do. And his unwillingness to Nec-use the statement in question will stymie any attempt to argue that the Cautious Man's willingness to assent to it is, willy nilly, expressive of an unacknowledged recognition of a necessary truth.

If, on the other hand, the recognitional conception fails to make out the notion of Nec-use, then it will have to acknowledge that recognition of necessity can be manifested only by one's use of modal operators. And in that case it ought also to be acknowledged that it is unreasonable to require a further account of what it is that the Cautious Man will not do over and above: accept that the statement in question is necessary; and acknowledged, too, that there can now be no better reason for thinking that the Cautious Man has really grasped a necessary truth, but without realising it, than for accepting his word for it that he has not.

6.   On the basis of these considerations, a challenge of the required sort can now be mounted against the hypothesis that we possess a necessity-sensitive cognitive faculty. Suppose that some unfortunates happen to be born without it. They lack whatever it is that enables us to recognise necessity, but are capable of achieving practical certainty

[1] In which case there is nothing to be 'cautious' about.

about proofs, that is, about the general truth of the corresponding descriptive conditionals, and suffer the same imaginative limitations which we regard as symptomatising our grasp of proof and necessity. Will their deficiency, like tone-deafness or colour-blindness, be bound to reveal itself in certain circumstances? It is not clear that it will. For all they have to do is to apply ordinary criteria for the correctness of proof, or the soundness of informal reasoning, and then just deliberately cross the divide —eschew their 'caution'—readily indulging in Nec-use of the statement of the validity of the inference, or at least—if Nec-use is chimerical—in orthodox expressions of the necessity of that statement. So long as they keep to this policy, there seems no reason why they should not mingle unnoticed among normal people.

If that is allowable, we immediately have a stronger result. Such people need never *know* that they suffer from any disability. For they will be trained to give expression to judgments of the validity of proof in exactly the same way as everyone else: by avowals of unassailability, necessary truth, internal connections, and so on; and trained, too, in whatever kind of use of the statements in question the acceptability of such judgments is communally treated as enjoining. It need never be apparent to them that others in the same community give expression by such avowals to the discernings of a special cognitive faculty which they lack, or that the kind of use of a statement which such avowals occasion expresses the operations of such a faculty. And if that is right, the whole hypothesis of the faculty is seen to be empty; for nothing distinguishes its operations from the moves of people who possess only ordinary empirical faculties, certain imaginative limitations, and who just follow the convention of never sticking at the cautious attitude when logical, or mathematical, reasoning seems to square well enough, step by step, with paradigms.

On this view, the decision involved when we accept new necessary statements not by way of adopting an explicit convention but as the result of, in the widest sense, proof would be one of *policy*. We have the policy to confer Nec-use upon statements—or to be ready to do whatever it is that the recognitional conception conceives as the distinctive expression of our recognition of necessity—in a communally grasped range of circumstances. And the policy is a convention, and so gives sense to the idea that necessity is itself globally conventional, just in the sense that in any particular case somebody can opt out of it and assume the mantle of 'caution' without doing anything to evince, by ordinary criteria, a misunderstanding of the concepts involved in the reasoning presented to him or a misapprehension of its character. When we accept any particular proof, a decision is taken, therefore, just at the point where we go beyond the Cautious attitude, moving from an acknowledgment of everything which that attitude grants in the particular case to acceptance of a necessity.

Can such a conception of the rôle of deciding meet the first constraint

which we imposed on any workable conventionalist account, the constraint of explaining why it is that our acceptance of a new proof
typically does nothing to occasion re-explanation of the concepts
involved? That it can follows—assuming the coherence of the Cautious
attitude—from a very simple reflection: the Cautious Man need be
guilty of no demonstrable misunderstanding of any of the concepts in
play in our assessment of a proof; he is defined, indeed, as agreeing with
us about the apparent soundness of all the constituent judgments in our
verdict that the proof is valid. So even after we decide to accept the
proof, he can still be in a position to give exactly the explanations of the
relevant concepts which we should now give. Accordingly, in moving
beyond the Cautious attitude, we have done nothing so to change those
concepts that the explanations of them to which we should formerly
have assented no longer capture our current understanding.

That is not to say, however, that no conceptual modification will
have occurred. For in accepting the validity of any proof, or chain of
reasoning, we shall have ratified the necessity of certain statements.
And the conceptual connections which we shall thereby have ratified
will be novel not merely in the sense which the rule-following considerations supply for every judgment which we make, but in the sense that
even an appropriately purified notion of discovery has no application to
them. Rather, such connections will actually be *constructed* every time,
in accordance with our general policy in logical and mathematical
contexts, we take the step of moving beyond the Cautious attitude; for,
to stress, it is only an acknowledgment of everything which the attitude
does in any particular case acknowledge which is required of us if we are
to be 'faithful to our understanding' of the concepts involved in a proof.
If, in particular, a viable notion of Nec-use can be elucidated, then the
modification of concepts occasioned by our acceptance of a proof will be
manifested in our readiness to Nec-use statements which someone who
adopted the Cautious attitude to this proof alone would not be prepared
to Nec-use. So a stronger version of the concept-modification thesis
now returns to the fray than the version which, towards the conclusion
of our discussion of the matter in Chapter VI, it came to seem that
Wittgenstein should have advanced.

Finally, the practical likelihood that we shall agree in our assessment
of new proofs, and new putatively necessary statements in general, is in
no way puzzling, or put in jeopardy, when the element of decision is
located in the manner sketched; it is just the practical likelihood that we
shall agree in our assessment of those facts which the Cautious attitude
will acknowledge—so, to that extent, a special case of the practical
likelihood that we shall in general agree in new judgments when we have
received a common training —and that we shall then transform that
assessment in the necessity-creating manner which we have been
trained, as a matter of communal policy, to follow. The decisive objection to radical conventionalism is, in fact, deflected at two points. First,

necessary statements are dignified to that status not arbitrarily, at individual whim, but in accordance with a communally grasped general policy. Secondly, a distinction has in any case now opened up between accepting a necessary statement in the Cautious spirit and accepting its necessity; and what is strictly requisite, in order for general conventions to inform a communal practice, is only that the participants concur in judgments concerning what are or are not consequences of those conventions—they do not need to concur about the *necessity* of any such judgment.

These considerations may help to explain why Wittgenstein felt he could consistently assert both passages like those with which I prefaced Chapter III, suggesting that the acceptability of a proof is always in some sense a matter for decision and that the effect of accepting it will be to change concepts, and passages like these:

> By the construction of a sign I *compel* the acceptance of a sign. (ii. 29) (My italics.)

> If the calculation has been done right, then this must be the result. Must *this always* be the result, in that case? Of course." (iii. 35)

> A proof leads me to say: this *must* be like this. (ii. 30)

> Proof must be a procedure of which I say: yes, this is how it has to be; this must come out if I proceed according to this rule. (ii. 23)

The essence of a good proof is that it produces this kind of conviction. But this convincingness in a particular case is acknowledged by a holder of the Cautious attitude when he grants that he can do nothing to make intelligible the idea that a particular step in the proof might turn out to be wrong, or even to simulate the illusion of its being wrong. It is a decision, none the less, to move from that acknowledgment to acceptance of a necessity, a decision to treat the descriptive conditional corresponding to the proof, or those corresponding to smaller stages of it, in any way other than as having been experimentally confirmed by the proof.

We have, in summary, a conventionalist prospectus along the following lines. A declarative sentence effects a genuine (fact-stating) assertion—something whose correctness is subject to discovery—when and only when, it was proposed, the linguistic community acknowledges conditions of appropriate assent to it of such a sort that someone's sincere unwillingness to assent to it when those conditions obtain and he is in a position to appreciate as much will indicate either a misunderstanding or some sort of misapprehension of the character of the presented circumstances. It is possible, however—if smitten with the Cautious attitude—sincerely to refuse to accept any particular proof *as a* proof (to refuse to acknowledge that any necessity has been demonstrated) without convicting oneself of either kind of failing. It follows that necessity is not purely recognitional, that attributions of necessity are not genuine assertions, and that there is an element of decision involved in accepting any proof as a proof. We take a step, that is to say, not required of us by correct understanding of the concepts involved and correct apprehension of the proof's mechanics. Not that there need be any sense of freedom, or consciousness of a decision; the step is one which we are *trained* to take as much as any other. But, as we ordinarily conceive of the matter, fully to accept a proof always involves accepting a necessity, irrespective of the character of its premises and conclusion, or whether, like a calculation, it rather goes to work on nonpropositional objects; fully to accept a proof is to accept that the necessity has been demonstrated of the associated descriptive conditional. And to do this is a step beyond what is required by understanding the concepts involved or correctly apprehending the character of the transformations made. Dummett wrote that we do not know what it would be like for someone who by ordinary criteria already understood the concepts employed to reject a proof;[1] but this claim is correct only if by 'reject' we mean: find fault with. For the fact is simply that it is open to someone to refuse to grant the necessity of the associated conditional while conceding that its correctness is *practically* (I do not mean 'almost') certain, and that he can do nothing to make intelligible the supposition that we might discover an error in one of the steps.

'Cautious' is not a wholly apt characterisation of this attitude. For there need not be a *doubt* involved such that, if further facts were brought to light, it would be allayed; rather, what the attitude essentially involves in any particular case is merely an unwillingness to participate in the institution of proof. The core of the proposal is that *our* participation on any particular occasion is nothing imposed on us by due acknowledgment of the facts or correct understanding of the concepts in play.

7.   We have not so far considered how these ideas stand in relation to our second constraint: we insisted that the element of decision to be located in our acceptance of new necessary statements should be such as to have no counterpart in contingent cases. Any doubt on this score must be a doubt about whether the original proposal adequately characterised the notion of a genuine assertion. For, of course, if contingent statements in general—or even too many of them—turned out not to be genuine assertions by that proposal, the foundations of the suggested conventionalism will collapse. So we have to ask: does the proposal correctly locate a sufficiently wide class of *paradigm* genuine assertions? For its capacity to inform our conception of controversial cases obviously depends on its getting things like 'the cat is on the mat' right.

One reservation is immediately apparent. Surely, there is a way a man may always sincerely withhold his assent to an assertion, whatever

[1] 23 (2, p. 498).

the presenting circumstances, without betraying any misapprehension of them or misunderstanding of a relevant concept. For *scepticism* is always an option. The defeasibility of any statement for which we have no conclusive notion of verification, and of any claim to have conclusively verified a statement for which we do have such a notion, makes it possible to raise the doubt, however favourable the evidence seems to be, whether subsequent evidence might not force us to withdraw the statement in question. So why should a man not be struck by that doubt and, while acknowledging everything that prompts us to assent to the assertion, nevertheless refuse to give his assent?

The parallel looks to be close. Once again, the attitude in question is prima facie one of caution; but, as before, that characterisation is not entirely apt—for the nature of the reservation is such that no further facts can be brought to light which would dispel it. Rather, such a person is on the particular occasion simply refusing to participate in our linguistic institutions. If such a stand can be coherently adopted, then it seems it ought to be granted that there is the indicated kind of element of 'decision' in our willingness to make any assertion whatever; and in that case we cannot appeal to the presence of such an element in the case of endorsements of necessity to undermine their status as genuine assertions.

A possible response would be to question whether the kind of scepticism adumbrated really is a coherent attitude. After all, if somebody is just blankly unwilling to assent to 'the cat is on the mat' in the face of the appropriate kind of circumstances, that *does* evince some sort of misunderstanding; in order to distinguish his position from misunderstanding or error, the sceptic is forced to make his refusal articulate—he must acknowledge the appearances, so to speak, but emphasise what we all acknowledge, that the appearances can be deceptive. But what does the possibility to which he then calls attention consist in, if not in that a situation may develop to which, if it is to have the significance which his present attitude requires, he will precisely *not* adopt a comparable attitude? So the distinguishability of his present attitude from one based on misunderstanding of the statement or error about the presented circumstances requires that he call attention to a situation in which he will not take this stance; and if not there, why here?

This is not good enough. For one thing, it was not suggested that the original Cautious attitude could coherently be adopted absolutely universally, that somebody could coherently accept merely as practically certain everything which we regard as necessary, but only that it is available for a particular case. But, more important, the response misrepresents the character of the sceptical reservation. Certainly, in order to distinguish his attitude from one involving misunderstanding or error, the sceptic must point to a certain possibility; but this possibility need only involve the development of further *appearances*. The inherent defeasibility of the sort of statement with which we are concerned is not to be identified with the possibility that we may come to be constrained to deny it; it is enough that our pattern of experience may always be envisaged as developing in such a way as to require that it no longer be asserted.

What the conventionalist needs to do is challenge the assimilation of the Cautious attitude to a kind of scepticism. Can such a challenge be mounted? Compare the descriptive conditional corresponding to a simple proof with a conditional describing the initial conditions of a standard text-book experiment, what is actually done in the course of the experiment, and its regular outcome. Three principal kinds of scepticism are possible about the latter: *inductive* scepticism will concede that the conditional has been corroborated by each of a multitude of trials, but will call into question our right to assert it as an unrestricted generalisation; scepticism about *memory* will call into question our right to assert that the conditional has indeed been corroborated by a multitude of trials; and *perceptual* scepticism, concentrating on a particular trial, will concede that the initial conditions, etc., *seemed* to be duly observed, etc., but will profess a doubt about whether we ought to affirm that they really were so. But a sceptic of any of these three sorts had better be unwilling to assent to the conditional if he wants to deserve his title; otherwise his position is no different to that of any reflective person who, while acknowledging that developments are conceivable in each of three appropriate areas which would force retraction of an assertion of the conditional, is nevertheless prepared in the circumstances to assent to it.

The Cautious Man, in apparent contrast, will be quite prepared to grant that it is practically certain that all is in order with the proof; and he will concede that the mists close in when he tries to envisage definite circumstances in which that certainty might be confounded. So he will be prepared to accept the corresponding conditional and to treat it as unassailably as his confidence in it warrants. What he disputes is rather that the proof so brings out aspects of the essential character of the concepts involved in the corresponding conditional that it becomes justified to claim that certainty about it cannot in principle rightly be confounded. Who is to say that we shall not somehow, sometime want to describe ourselves as having found an error in the proof; and how, unless we are tacitly *legislating* about the matter, can we now discount the possibility that we shall be right to do so? The Cautious Man, it can be argued, is therefore asceptical just in the sense that, in contrast with the attitude adopted by any of the above three varieties of sceptic towards assertion of the generalised conditional concerning the experiment, he does not call into question our warrant to accept the descriptive conditional corresponding to the proof. What he does call into question is that the proof supplies a *wholly cognitive* warrant for making any of the moves which the recognitional conception supposes to be vindicated by our apprehension of a necessity.

Let us say that an attitude towards a particular statement in particular circumstances is sceptical just in case it involves a doubt about our entitlement to assert the statement in those circumstances of a kind which would apply to any situation in which a statement of that type is made; and would thus, if sustained, effectively debar us from ever making that type of statement at all. And let us recharacterise the class of genuine assertions as statements communally associated with conditions of such a kind that one who is sincerely unwilling to assent to such a statement when, by ordinary criteria, those conditions obtain, can make himself intelligible to us only by betraying a misunderstanding or some sort of misapprehension, *or* by professing some sort of sceptical attitude. This is not an ad hoc proposal. For however we attempt to characterise what it is in our use of genuine assertions which distinguishes them within the wider class of declarative sentences, we have to acknowledge, I believe, that circumstances can never so constrain assent to an assertion as to eliminate all possibility of a sceptical response. So if it is indeed, as the original account implicitly presupposed, by reference to the 'room for manoeuvre' which the conditions for appropriateness of assent to a declarative sentence leave open that the question whether it is a genuine assertion is to be resolved, our account of the distinction will have to include some sort of scepticism clause—or contrive to dispel the impression that scepticism really is always a possibility. Either way, it will be by their provision for some sort of non-sceptical room for manoeuvre that mere quasi-assertions will display their true colours.

In these terms, the prima facie case for saying that the Cautious attitude is not sceptical is simply that if someone adopts this stance in a particular case, it is still open to him, unlike the true sceptic, to accept the, as others view them, necessary statement(s) at issue; for he will not dispute that a proof provides grounds for asserting what others consider it proves, but that it provides the *kind* of warrant which others take it to provide.

This is, plainly, a weak case. A proponent of the recognitional conception will surely reply that since what the Cautious attitude is charged to distinguish itself from is scepticism about necessity, it is ascriptions *of* necessity to particular statements which ought to be put to the Cautious Man for his assent or dissent. And such ascriptions, as he has been characterised, will *not* secure his assent. So there is so far no basis for thinking that the Cautious attitude to necessity is not, properly speaking, a sceptical one. (It is true that the prima facie case could not be rebutted so easily if the Cautious Man spoke only a language containing no modal vocabulary; but then, a sceptic about whether anything was green would be hard pressed to find distinctive expression in a language with no word meaning 'green'.)

8. How should we proceed at this point? We seem to have two

relatively clear, opposing theories: according to one, our intellects afford us access to a special category of truth, afford us a special—though not, it may be admitted, indefeasible—knowledge of how things *must* be; according to the other, we have no such knowledge but only an ordinary empirical knowledge of how things are, coupled with the policy of promoting the propositional objects of such knowledge to the status of necessity in certain circumstances. The conventionalist argued that if the recognitional view were correct, there could at any rate be no distinctive symptom of its correctness; that nothing which we do distinguishes us from people of whom *his* theory would be correct, and that the recognitional view is therefore an empty hypothesis. But the cognitivist countered that unless the conventionalist's so-styled 'policy' can be distinguished from our general policy, in making statements of all kinds, to override the philosophical sceptic, no basis for thinking of necessity as conventional has yet been provided —save in so far as the acceptability of *any* statement is conventional. How is it to be decided which, if either, theory has the truth of the matter?

One natural path for further enquiry to pursue would be to examine varieties of scepticism in detail. Only after it has been determined whether the suggestion above—with, perhaps, some further refinement—captures what is essential to any form of philosophical scepticism, or whether some quite different approach is needed, can it be clear, it might be supposed, whether the Cautious attitude can resist being so classified. I shall not attempt such an enquiry here.

Two observations are, however, in point at this juncture. First, the cognitivist cannot charge the Cautious attitude with scepticism *unless* he admits the success of the conventionalist's challenge against the special necessity-sensitive faculty. Only if it acknowledges everything which the cognitivist conceives to justify endorsement of the necessity of a particular statement can the Cautious attitude be so viewed; and in that case the idea of a special cognitive ability, transcending our faculties for empirical knowledge and our imaginative susceptibilities, will implicitly have been dropped. The cognitivist will implicitly have allowed that recognition of necessity is nothing beyond the amalgam of empirical knowledge and imaginative limitation which the Cautious attitude acknowledges. If this were to prove to be the final configuration in the game, the letter of the law would award the victory to the recognitional conception. But it would be a victory dearly gained. For whatever exactly it was thought to consist in, our ability to grasp the 'truths of reason' was not at the outset conceived by believers to be anything rightly viewed as so anthropocentric, or so *slight*. The classical notion would have necessity to be something absolute, objective, imposed, and certain; and the cognitivist who now scents victory would be hard pressed at this stage of our enquiry to make out his title to any of these features as they were classically intended.

Secondly, there is a difficulty with the conventionalist's challenge against the special necessity-sensitive faculty which we have so far glossed over, a difficulty which would have to be met before the dispute could take the turn just described. The challenge required that someone lacking the hypothetical necessity-sensitive faculty, and so in a position when confronted with any particular proof cognitively to apprehend only those facts which the Cautious attitude acknowledges, need suffer no handicap—would be able to do everything which a possessor of the faculty could do, and without relying on others' behaviour as a guide. It is not absolutely clear that this requirement can be met. The suggestion was that in all appropriate contexts a man could follow the policy of promoting the object of his practical certainty to Nec-use, or treating it in whatever way the cognitivist holds to signal apprehension of a necessity. But if such a policy is to work, our man will have to possess a non-parasitic grasp of just what the 'appropriate contexts' are. He will have to be able to recognise for himself the circumstances under which the community is 'in the market' for an internal relation. But just which circumstances are those? In order to make it clear that this policy will be viable, an account is required of the relevant range of circumstances which makes no appeal to others' actual willingness to accept that an internal relation is therein susceptible to demonstration. And to supply such an account looks to be exceedingly difficult. Obviously, it is no good to talk blithely in terms of 'logical' and 'mathematical' contexts. For one thing, we sometimes accept statements as necessary on the basis of considerations which are happily placed in neither category ('Nothing is simultaneously red and green all over'). For another, what *makes* a context logical, or mathematical? Any attempt at a general account looks likely to be encumbered by something which Wittgenstein himself was keen to stress: the essential '*motley*' of mathematical questions and techniques.[1]

Wittgenstein faces the question in v. 17:

The problem: find the number of ways in which we can trace the joins in this wall:

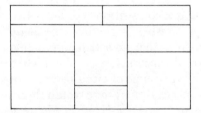

without omission or repetition, will be recognised by everyone as a *mathematical* problem. If the drawing were much bigger and more complicated, and could not be taken in at a glance, it could be supposed to change without our noticing; and then the problem of finding that

[1] *RFM* ii. 46, 48.

number (which perhaps changes according to some law) would no longer
be a mathematical one. But even if it does not change, the problem is, in
this case, still not mathematical—but even when the wall can be taken in at
a glance, that cannot be said to make the question mathematical, as when
we say: *this* question is now a question in embryology. Rather: *here* we
need a mathematical solution. (Like: here what we need is a *model*.)

Did we 'recognise' the problem as a mathematical one because
mathematics treats of making tracings from drawings?

Why, then, are we inclined to call this problem straight away a
'mathematical' one? Because we see at once that here the answer to a
mathematical question is practically all we need. Although the problem
could easily be seen as, for example, a psychological one.

Wittgenstein is obviously right in thinking that the surveyability of the
object to be considered is insufficient to make the problem mathematical; and right in dismissing the idea that what makes it mathematical is
that in approaching it we shall use techniques which mathematics
sometimes uses. But his answer is, for present purposes, unsatisfying.
All we *practically* need—that is, need for practical purposes—is *practical*
certainty; and why do we feel that a *mathematical* question is raised at
all?

The answer seems, irresistibly, to be that we are inclined to describe
ourselves as recognising that what is at stake is an *internal* property of
the particular figure. But that is just to say that we are here 'in the
market'; what we need to understand is why. Part of the answer, surely,
is that our conception of the *identity* of the figure is such that we are
convinced that whenever that very figure is reproduced, the solution to
the problem, when we are satisfied that it has been arrived at correctly,
will always be invariant. But how does that certainty differ from inductive certainty that whenever the initial conditions of an experiment are
properly implemented, and we are satisfied that it has been properly
conducted, the outcome will always be the same? Is it that we have a
convention that certain sorts of properties are to be counted as entering
into the identity of figures which exemplify them, so that the loss of
such a property is to be counted a case of change in identity? If so, to
which sorts of properties exactly does the convention apply? To introduce the idea of such a convention will not advance us here.[1]

Perhaps the most decisive single turn this dispute could take would
be if it were to transpire that no satisfactory general account could be
given of the range of circumstances in which the conventionalist's
'policy' would apply, the range of circumstances in which we are in the
market for an internal relation. There would then be no prospect of the
sort of case which the conventionalist needs to make, no prospect of
accounting for our agreement in new judgments of necessity save by
reference to the idea that necessity is recognised. That is a possibility
which this enquiry will leave open.

Set against both sides in this dispute is the view that necessity is a

[1] Or so it would seem. But the issue is dark.

radically *mistaken* notion, that there is no fully intelligible concept to be the subject matter of the dispute. The correctness of this view would require both that it is impossible to make out any satisfactory notion of Nec-use, and that it is unacceptable to suppose that acceptance of a necessity can be distinctively manifested only by the use of modal vocabulary. If such a view were as easily verified as its popularity would suggest, Wittgenstein's efforts in this area would be of interest only because of the depth of their roots in his philosophy of language. But the undoubted genuineness of the distinction between statements and rules, the obstacles which we have found in the path of any purely empiricist conception of scientific theorising, and the difficulty of decisively making good the two required theses just noted, prevent so easy a verification.

An improved formulation[1] of the philosophical problem of necessity would be: is the notion legitimate, and if so, is necessity recognised? Kreisel characterised *RFM* as 'a surprisingly insignificant product of a sparkling mind'; and Anderson was only slightly kinder when he wrote that the book was unlikely to enhance Wittgenstein's philosophical reputation. Recently Chihara has written in similar tones about the 1939 Lectures.[2] The review articles of these commentators are best shown the kind of charity which they did not show to Wittgenstein's personal notes and the contents of his lectures reported at third hand. *RFM*, and indeed Diamond's edition of the Lectures, both testify to a wealth of vigorous, profound and valuable thought on some of the most difficult questions in philosophy. The hard fact remains that Wittgenstein did not solve the twofold problem formulated above, but bequeathed both it and his contribution to its solution to a largely ungrateful posterity.

[1] Compared with that of Dummett; see above, § 1, Chapter XVIII.
[2] Kreisel 53, concluding remark; Ross Anderson 68, conluding remark; Chihara 13—an extraordinarily superficial piece of work.

References

Works alluded to in the text.
(For Wittgenstein's works, see the List of Abbreviations.)

1. Ayer, Sir Alfred — *Language, Truth and Logic* (Gollancz 1946).

2. Benacerraf, Paul, and Putnam, Hilary, eds. — *Philosophy of Mathematics, Selected Readings* (Prentice Hall 1964).

3. Bennett, J. F. — 'On being forced to a conclusion', *Proc. Aristot. Soc. supp. vol.* 1961.

4. Bernays, Paul — 'Comments on Ludwig Wittgenstein's *Remarks on the Foundations of Mathematics*', *Ratio* 1959; repr. in 2.

5. Bernays, Paul — 'Sur le Platonisme dans les mathématiques', *L'Enseignement Mathematique* 1935; tr. in 2.

6. Blackburn, Simon — 'Goodman's paradox', *Am. Phil. Quart. Monograph Series* 3, 1969.

7. Bostock, D. — *Logic and Arithmetic* (Oxford 1974).

8. Broad C. D. — 'Are there synthetic a priori truths?' *Proc. Aristot. Soc. supp. vol.* 1936.

9. Brouwer, L. E. J. — 'Consciousness, philosophy and mathematics', *Proc. Xth International Congress of Philosophy* (North Holland 1940); repr. in 12, and excerpted in 2.

10. Brouwer, L. E. J. — Thesis, 'On the foundations of mathematics', 1907; tr. in 12.

11. Brouwer, L. E. J. — 'The effect of Intuitionism on the classical algebra of logic', *Proc. Royal Irish Academy* Sect. A 57; repr. in 12.

12. Brouwer, L. E. J. — *Collected Works*, vol. 1, ed. A. Heyting (North Holland 1975).

13. Chihara, Charles S. — 'Wittgenstein's discussion of the Paradoxes in his 1939 Lectures on the Foundations of Mathematics', *Phil. Review* 1977.

14. Church, Alonzo — *Introduction to Mathematical Logic*, vol. 1 (Princeton 1956).

15. Craig, E. J. — 'The problem of necessary truth', *Meaning, Reference and Necessity;* ed. Blackburn (Cambridge 1975).

16. Davidson, Donald — 'Truth and meaning', *Synthèse* 1967.

17. Davidson, Donald   'The logical form oᴄ action sentences', *The The Logic of Decision and Action,* ed. Rescher (Pittsburgh 1967).

18. Davidson, Donald   'On saying that', *Words and Objections,* eds. Davidson and Hintikka (Reidel 1969).

19. Davidson, Donald   'Action and reaction', *Inquiry* 1970.

20. Davidson, Donald   'Semantics for natural languages', *Linguaggi nella Società e nella Tecnica* (Edizioni di comunità, Milan 1970).

21. Davidson, Donald   'In defence of Convention T', *Truth, Syntax and Modality,* ed. Leblanc (North Holland 1973).

22. Dummett, Michael   'Truth', *Proc. Aristot. Soc.* 1958; repr. in Pitcher, ed., *Truth* (Prentice Hall 1964).*

23. Dummett, Michael   'Wittgenstein's philosophy of mathematics', *Phil. Review* 1959; repr. in 2.*

24. Dummett, Michael   'The reality of the past', *Proc. Aristot. Soc.* 1968.*

25. Dummett, Michael   *Frege: Philosophy of Language* (Duckworth 1973).

26. Dummett, Michael   'The philosophical basis of intuitionistic logic', *Logic Colloquium 1973,* eds. Rose and Shepherdson (North Holland 1973).*

27. Dummett, Michael   'The justification of deduction', *Proc. of the British Academy* 1973.*

28. Dummett, Michael   'Wang's paradox', *Synthèse* 1975.*

29. Dummett, Michael   'What is a theory of meaning? (II)' in 32 below.

29a. Dummett, Michael   'What is a theory of meaning', in *Mind and Language,* ed. Guttenplan (Oxford 1975).

30. Dummett, Michael   *Elements of Intuitionism* (Oxford 1977).

31. Escher, M. C.   *The Graphic Work of M. C. Escher* (Pan Ballantine 1972).

32. Evans, Gareth, and McDowell, John, eds.   *Truth and Meaning: Essays in Semantics* (Oxford 1976).

33. Ewing, A. C.   'The linguistic theory of a priori propositions', *Proc. Aristot. Soc.* 1939.

37. Foster, J. A.   'Meaning and truth-theory' in 32.

35. Frege, Gottlob   *The Foundations of Arithmetic,* trans. Austin (Blackwell 1950).

* Reprinted in Michael Dummett, *Truth and Other Enigmas* (Duckworth 1978).

470 *References*

36. Frege, Gottlob — *Die Grundgesetze der Arithmetik,* excerpted and translated by M. Furth as *The Basic Laws of Arithmetic* (Berkeley 1964).

37. Geach, P. T. — 'Ascriptivism', *Phil. Review* 1960.

38. Gödel, K. — 'What is Cantor's continuum problem?' *Amer. Math. Monthly 1947;* repr. in 2.

39. Gödel, K. — *On Formally Undecidable Propositions of* Principia Mathematica *and Related Systems,* trans. Meltzer (Oliver and Boyd 1962).

40. Goodman, Nelson — *Fact, Fiction and Forecast* (Athlone 1954).

41. Grice, H. P., and Strawson, P. F. — 'In defense of a dogma', *Phil. Review* 1956.

42. Grice, H. P. — William James Lectures, Harvard University 1968 (unpublished).

43. Grice, H. P. — 'The causal theory of perception', *Proc. Aristot. Soc. supp. vol.* 1961.

44. Hacker, P. M. S., and Baker, G. P. — Critical notice of Wittgenstein's *Philosophical Grammar, Mind* 1976.

45. Hacker, P. M. S. — *Insight and Illusion* (Oxford 1972).

46. Heyting, A. — *Intuitionism: An Introduction* (North Holland 1971).

47. Heyting, A. — 'The intuitionistic foundations of mathematics', *Erkenntnis* 1931; tr. in 2.

48. Hume, D. — *Inquiry Concerning Human Understanding.*

49. Kant, I. — *Critique of Pure Reason.*

50. Kleene, S. C. — *Introduction to Metamathematics* (North Holland 1962).

51. Kielkopf, C. F. — *Strict Finitism* (Mouton 1970).

52. Kneale, W. — 'Are necessary truths true by convention?', *Proc. Aristot. Soc. supp. vol.* 1947.

53. Kreisel, Georg — 'Wittgenstein's *Remarks on the Foundations of Mathematics', Brit. Jour. for Phil. of Sci.* 1958.

54. Kreisel, Georg — 'Mathematical logic', *Lectures on Modern Mathematics,* ed. Saaty (John Wiley 1965).

55. Lewis, C. I., and Langford, C. H. — *Symbolic Logic* (New York: Dover 1959).

56. Locke, J. — *An Essay concerning Human Understanding.*

57. McDowell, J. H. — 'Truth conditions, bivalence and verificationism' in 32.

58. McDowell, J. H.   'On "The reality of the past",' *Action and Interpretation*, etc., eds. Hookway and Pettit (Cambridge 1978).

59. McDowell, J. H.   'On the sense and reference of a proper name', *Mind* 1977.

60. Mackie, J. L.   'Proof', *Proc. Aristot. Soc. supp. vol.* 1966.

61. Malcolm, N.   'Are necessary propositions really verbal?', *Mind* 1940.

62. Mill, J. S.   *A System of Logic* (Longman 1873).

63. Moore, G. E.   *Philosophical Papers* (Allen and Unwin 1959).

64. Moore, G. E.   *Lectures on Philosophy*, ed. C. Lewy (Allen and Unwin 1966).

65. Nietzsche, F.   *Twilight of the Idols;* tr. R. J. Hollingdale (Penguin 1968).

66. Quine, W. V. O.   *From a Logical Point of View* (Harvard 1953).

67. Quine, W. V. O.   *Word and Object* (Cambridge, Mass. 1960).

68. Ross Anderson, A.   'Mathematics and the "Language Game" ', *Rev. of Metaphysics* 1958; repr. in 2.

69. Russell, B., and Whitehead, A. N.   *Principia Mathematica* (3 vols, Cambridge 1910, 1912, 1913).

70. Russell, B.   'The limits of empiricism', *Proc. Aristot. Soc.* 1935.

71. Shwayder, D. S.   'Wittgenstein on mathematics', *Studies in the Philosophy of Wittgenstein*, ed. Winch (Routledge 1969).

72. Strawson, Sir Peter   'Propositions, concepts and logical truths', in his *Logico-Linguistic Papers* (Methuen 1971).

73. Strawson, Sir Peter   'Scruton and Wright on Anti-Realism Etc.', *Proc. Aristot. Soc.* 1976.

74. Stroud, Barry   'Wittgenstein and logical necessity', *Phil. Review* 1965.

75. Wang, Hao   'Eighty years of foundational studies', *Dialectica* 1958.

76. Wang, Hao   *A Survey of Mathematical Logic* (Peking: Science Press 1964).

77. Wright, Crispin   'Truth-conditions and Criteria', *Proc. Aristot. Soc. supp. vol.* 1976.

78. Wright, Crispin   'On the coherence of vague predicates', *Synthèse* 1975.

79. Wright, Crispin   'Language-mastery and the Sorites paradox', in 32.

80. Wright, Crispin   'Strawson on Anti-Realism', *Synthèse* 1979.

81. Yesenin-Volpin, A. S.   'Le programme ultra-intuitioniste des fondements des mathématiques', in *Infinitistic Methods* (Warsaw 1961).

82. Yesenin-Volpin, A. S.   'The ultra-intuitionistic criticism and the anti-traditional programme for the foundations of mathematics', in *Intuitionism and Proof Theory*, eds. Kino, Myhill, Vesley (Amsterdam 1970).

Bibliography

Works concerning Wittgenstein's later philosophy of mathematics other than those alluded to in the text. (Warning: the secondary literature, while still by no means extensive, is of *very* mixed quality).

Ambrose, Alice	'Finitism and the "limits of empiricism" ', *Mind* 1937.
Ambrose, Alice	'Proof and theorem proved', *Mind* 1959.
Ambrose, Alice	'Mathematical generality', *Ludwig Wittgenstein: The Man and his Philosophy,* ed. Ambrose and Lazerowitz (Allen and Unwin 1972).
Baker, G. P.	'Criteria: a new foundation for semantics', *ratio* 1974.
Benardete, José A.	*Infinity* (Oxford 1964).
Castañeda, H-N.	'On mathematical proofs and meaning', *Mind* 1961.
Chihara, Charles, S.	'Wittgenstein and logical compulsion', *Analysis* 1960–1.
Chihara, Charles S.	'Mathematical discovery and concept formation', *Phil. Rev.* 1963.
Cowan, J. L.	'Wittgenstein's philosophy of logic', *Phil. Rev.* 1961.
Diamond, Cora	'Riddles and Anselm's riddle', *Proc. Aristot. Soc. supp. vol.* 1977.
Engel. S. M.	'Wittgenstein's *Foundations* and its reception', *Am. Phil. Quart. 4,* 1967.
Fogelin, Robert J.	'Wittgenstein and Intuitionism', *Am. Phil. Quart. 5,* 1968.
Fogelin, Robert J.	*Wittgenstein* (Routledge 1976).
Goodstein, R. L.	'Mathematical systems', *Mind* 1939.
Goodstein, R. L.	'Wittgenstein's philosophy of mathematics', *Ludwig Wittgenstein: The Man and his Philosophy,* ed. Ambrose and Lazerowitz (Allen and Unwin 1972).
Harward, Donald W.	'Wittgenstein and the character of mathematical propositions', *International Logic Review* 1974.
Klenk, V. H.	*Wittgenstein's Philosophy of Mathematics* (Nijhoff 1976).

Lazerowitz, Morris   'Necessity and language', *Ludwig Wittgenstein: The Man and his Philosophy,* ed. Ambrose and Lazerowitz (Allen and Unwin 1972).

Levison, A. B.   'Wittgenstein and logical laws', *Phil. Quart.* 1964.

Levison, A. B.   'Wittgenstein and logical necessity', *Inquiry* 1964.

Nell, E. J.   'The hardness of the logical "must" ', *Analysis* 1960–1.

Pears, D. F.   *Wittgenstein* (Fontana 1971).

Peppinghaus, Benedikt   'Some aspects of Wittgenstein's philosophy of mathematics', *Proceedings of the Bertrand Russell Memorial Logic Conference—Denmark 1971* (Leeds 1973).

Rhees, Rush   'Questions on Logical Inference', *Understanding Wittgenstein,* ed. Vesey (Macmillan 1974).

Rhees, Rush   *Discussions of Wittgenstein,* ch. 9: 'On continuity: Wittgenstein's ideas, 1938' (Routledge 1970).

Richardson, John T. E.   *The Grammar of Justification: An Interpretation of Wittgenstein's Philosophy of Language* (Sussex University Press 1976).

Sloman, A.   'Explaining logical necessity', *Proc. Aristot. Soc.* 1968–9.

Steiner, Mark   *Mathematical Knowedge* (Cornell University Press 1975).

Waismann, Friedrich   *Introduction to Mathematical Thinking* (Ungar 1951).

Waismann, Friedrich   *The Principles of Linguistic Philosophy,* ed. Harré (Macmillan 1965).

White, R. I.   'Riddles and Anselm's Riddle', *Proc. Aristot. Soc. supp. vol.* 1977.

Wood, O. P.   'On being forced to a conclusion', *Proc. Aristot. Soc. supp. vol.* 1961.

Wrigley, Michael B.   'Wittgenstein's philosophy of mathematics', *Phil. Quart.* January 1977.

Index

431, 442, 450; and anti-realism, 221 ff; and the impossibility of an 'Olympian standpoint', 424–5; as attenuating the distinction between a choice-series and a rule-governed series, 169 ff; as backing up the 'antecedence' doctrine, 272 ff; as challenging Wittgenstein's avowed philosophical methodology, 288–9; as apparently undermining the applicability of the law of excluded middle more widely than intuitionist criticisms, 169; construed anti-realistically, 29–38; construed as inductive scepticism, 25–7; lead to no particular theory of necessity, 393–4; leave room for a purified notion of discovery, 450; not alone sufficient to do justice to Wittgenstein's conventionalist streak, 386, 393, 442–3; not sufficient to meet Craig's objections to conventionalism, 432; possible revisionary consequences of, 227–38, 251

rulers, soft, *see* alternative ways of measuring

Russell, B., 82, 132

S4 principle, 366 ff

scepticism, about induction, 25 ff; and 'Caution' 462–3; *see also* Cartesian doubt

semantics versus pragmatics, *see* force, theory of

sensation 'S', 381–3

set theory, 66, 247, 263–4; axiomatic, 438

Shwayder, D. S., 39, 42, 46, 134

sieve, *see* Eratosthenes

Sorites paradox, *see* paradoxes

speech acts, 286

Strawson, P. F., 123 n, 224 n, 360

strict finitism, 117 ff; *see* anti-realism, intuitionism, surveyability

strict finitist, concept of length, 82 ff; explanation of logical constants, 138; number theory, 136 ff; view of C-conditionals, 185; view of conversion rules, 96 ff

Stroud, Barry, 370, 379, 380

surveyability, 117–41 *passim*; an incoherent concept?, 137; as a manifestation of strict finitism, 128; as a revisionary element in Wittgenstein's thought, 118; as non-revisionary, 140; Wittgenstein's distinct definitions of, 118 ff; Wittgenstein's emphasis on a result of his repudiation of unacknowledged

internal connections, 129 ff

synonymy, and the explanation of analyticity, 359 ff; and the concept modification thesis, 45–8; Quine's critique of, 359 ff

T-theorems, 253, 258, 282; assertibility-conditional analogues of, 257

tautologies, recognition of, 344–5; truth-table explanation of, 251

telepathy, 451

tense-links, 393

theorems: Cantor's, 54, 117, 247; Fermat's last, 17, 50 ff, 127, 334; Fundamental Theorem of arithmetic, 54; Gödel's, 28, 153; prime number theorem, 138

theoreticity, of a predicate, 280

theory of force, *see* force

theory of meaning, systematic, 238, 252–9, 279–92; and changes in meaning, 290–1; and the status of object-language principles of inference, 252 ff; as a model of epistemic capacity, 281; as a theory of speaker's understanding, 280; as empirical, 291; as explaining logical validity, 238, 252–9, 279–93; as explanatory of language mastery, 283; as having a single central notion, 286; compatible with anti-realism, 255; gives no account of necessity, 254; homophonic, 291; modest, 281; the rule-following considerations' challenge to, 288–9; whether compatible with the indefinite variety of language-games, 285 ff

Theory of Types, 295

therapy, *see* concept modification, philosophy

transfinite numbers, *see* set theory

translation, 67–8; radical, 281, 390

Truth: analogy with winning a game, 246 'dethroning' of, 125, 224–5; its distributivity over disjunction, 251; purely in virtue of meaning, 343–63 *passim*; recursive definition of, 253–4; timelessness of, 173 ff, 182–4, 191–6; strong correspondence theory of, 9

truth-value, the concept jeopardised by Quine's critique of synonymy, 360

'ultra-physics', mathematics and logic, as, 61, 69

understanding, the light of, 354;